Moscow Lectures

Volume 8

Moscow Lectures is a new book series with textbooks arising from all mathematics institutions in Moscow.

More information about this series at https://link.springer.com/bookseries/15875

Victor V. Prasolov

Differential Geometry

NATIONAL RESEARCH
UNIVERSITY

Skolkovo Institute of Science and Technology

Victor V. Prasolov
Independent University of Moscow
Moscow, Russia

ISSN 2522-0314 ISSN 2522-0322 (electronic)
Moscow Lectures
ISBN 978-3-030-92251-1 ISBN 978-3-030-92249-8 (eBook)
https://doi.org/10.1007/978-3-030-92249-8

Mathematics Subject Classification: 53-XX

Translated from the Russian by Olga Sipacheva: Originally published as дифференциальная геометрия by MCCME Moscow, 2020

This Springer imprint is published by the registered company Springer Nature Switzerland AG.
The registered company address is: Gewerbestrasse 11, 6330 Cham, Switzerland

Preface

Differential geometry studies properties of curves, surfaces, and smooth manifolds by methods of mathematical analysis. Riemannian geometry is a section of differential geometry which studies smooth manifolds with an additional structure, Riemannian metric. The main part of this book is devoted to exactly Riemannian geometry. The exception is affine differential geometry, projective differential geometry, and connections more general than the Levi-Civitá connection, which originates from a Riemannian metric.

The book begins with the simplest object of differential geometry, curves in the plane. The most important characteristic of a curve at a given point is curvature. The first chapter considers both local properties of curvature (the Frenet–Serret formula and osculating circles) and global ones (the total curvature of a closed curve and the four-vertex theorem). The total oriented curvature of a closed curve is invariant with respect to a regular homotopy, and vice versa: if the total oriented curvatures of two curves are equal, then these curves are regularly homotopic (this is the Whitney–Graustein theorem). To each curve corresponds its evolute, that is, the locus of centers of all osculating circles of this curve, that is, the envelope of the family of normals. With respect to its evolute, the given curve is an involute. But the inverse operation of assigning an involute to a curve is ambiguous: every curve has a whole family of involutes (a curve orthogonal to a family of normals can be drawn through every point of a normal). We give two proofs of the isoperimetric inequality between the length of a closed curve without self-intersections and the area which it bounds. A part of the differential geometry of plane curves is not related to a Riemannian metric, that is, remains beyond the scope of Riemannian geometry. It includes enveloping families of curves, affine unimodular differential geometry, and projective differential geometry. The chapter about plane curves is concluded by elements of integral geometry: we derive a formula expressing the measure of a set of straight lines intersecting a given curve.

The second chapter studies curves in spaces, first in three-dimensional space and then in many-dimensional ones. Given a curve in three-dimensional space, we define curvature and torsion, derive the Frenet–Serret formula, and define osculating planes and spheres. We also define the total curvature of a closed curve and prove

Fenchel's theorem (that the total curvature of a closed curve is at least 2π) and the Fáry-Milnor theorem (that the total curvature of a knotted closed curve is at least 4π). At the end of the chapter, we consider curves in many-dimensional space, define quantities generalizing curvature and torsion for such curves, and derive the generalized Frenet–Serret formulas.

The third chapter is devoted to surfaces in three-dimensional space. On a surface in $\mathbb{R}^3$, the first quadratic form is introduced; this is the inner product of tangent vectors to the surface. With a curve on a surface, we associate a Darboux frame and use it to define the geodesic curvature, the normal curvature, and the geodesic torsion of a curve on a surface. A geodesic on a surface is a curve with zero geodesic curvature; any shortest curve on a surface joining two given points is geodesic. On a surface in $\mathbb{R}^3$, the second quadratic form is also defined. In a basis with respect to which the matrix of the first quadratic form is the identity matrix and the matrix of the second quadratic form is diagonal, the diagonal elements of the second quadratic form are the principal curvatures. The Gaussian curvature of a surface is the product of principal curvatures. The principal curvatures cannot be expressed only in terms of the first quadratic form, but the Gaussian curvature can. The Gaussian curvature of a surface can also be defined in a different way, in terms of differential forms on the surface. The integral of the Gaussian curvature over a polygon on the surface can be related to the integral of the geodesic curvature over the boundary of this polygon and the sum of exterior angles of the polygon (the Gauss–Bonnet formula). We define the parallel transport of a vector along a curve and the covariant differentiation of vector fields and introduce the Riemannian curvature tensor. Using geodesics, we define the exponential map of a tangent space to a surface. To study properties of geodesics, we derive the first and the second variation formula. Using Jacobi vector fields and conjugate points, we find out when the length of a geodesic is not globally minimal. We prove the theorem on the local isometry of surfaces of constant Gaussian curvature. At the end of the chapter, we introduce the Laplace–Beltrami operator, which is a generalization to surfaces of the Laplace operator in the plane.

In the fourth chapter we discuss two topics, hypersurfaces in many-dimensional space and connections on vector bundles. The study of connections on vector bundles is based on certain prerequisites concerning manifolds and vector bundles over manifolds. Thus, beginning in Chap. 4, the reader is supposed to have background knowledge of manifolds, tangent vectors, differential forms, vector bundles over manifolds, sections of bundles, and the inverse function theorem; all the necessary information can be found in the books [Pr2] and [Pr3]. For a hypersurface in Euclidean space, we define the Weingarten operator and use it to introduce the second, third, etc. quadratic forms. We define connections first on hypersurfaces and then on any manifolds and vector bundles over manifolds. In parallel, we introduce geodesics with respect to a given connection. We also define the curvature tensor and the torsion tensor of a given connection and introduce the curvature matrix of a connection.

The fifth chapter is concerned with the general theory of Riemannian manifolds. A Riemannian manifold is a manifold whose every tangent space is equipped with

An inner product (Riemannian metric). On a Riemannian manifold, there exists a unique torsion-free connection compatible with the Riemannian metric (the Levi-Cività connection). The Riemann tensor of this connection has several symmetries (satisfies several identities). Geodesics on Riemannian manifolds have some specific features in comparison with geodesics for arbitrary connections; in particular, any such geodesic is locally a shortest curve. A Riemannian manifold is said to be geodesically complete if all geodesics on this manifold can be extended without bound. According to the Hopf–Rinow theorem, geodesic completeness is equivalent to the completeness of the Riemannian manifold as a metric space. The Riemann tensor can be described by using the sectional curvatures corresponding to two-dimensional subspaces. For Riemannian submanifolds, as well as for hypersurfaces, we can introduce the second quadratic form and prove generalizations of Gauss' and Weingarten's formulas. An important class of Riemannian submanifolds is formed by totally geodesic submanifolds (a submanifold M is totally geodesic if each geodesic on M is also a geodesic on the ambient manifold). In the many-dimensional case, just as in the case of surfaces, we obtain the first and the second variation formulas and use them to introduce Jacobi fields and define conjugate points. The chapter is concluded by a discussion of the holonomy (transformations of the tangent space obtained by the parallel transport of vectors along closed curves) and an interpretation of curvature as infinitesimal holonomy.

The sixth chapter discusses the differential geometry of Lie groups, that is, manifolds endowed with a group structure consistent with the smooth structure. With each Lie group, its Lie algebra is associated, which is the tangent space at the identity element in which the multiplication of elements is defined as taking the commutator of the left-invariant vector fields corresponding to tangent vectors. The Lie algebra of a Lie group is mapped to this Lie group by the exponential map. For a Lie group and a Lie algebra, adjoint representations are defined; the adjoint representation of a Lie algebra is used to define the Killing form. On a Lie group, there exist various connections and metrics related to the group structure in various ways. On a compact Lie group, invariant integration can be defined. Some properties of Lie groups are possessed by more general spaces, namely, by homogeneous and symmetric ones.

The last (seventh) chapter is devoted to some of the applications of differential geometry: comparison theorems, relationship between curvature and topological properties of manifolds, and the Laplace operator on Riemannian manifolds.

Basic Notation

k is the curvature of a curve in the plane (p. 3) or space (p. 48)
e_1, e_2, e_3 is the Serret–Frenet frame, p. 47
$\varkappa$ is the torsion of a space curve, p. 48
E, F, and G are the coefficients of the first quadratic form, p. 66
g_{ij} are the coefficients of the first quadratic form, p. 67

g^{ij} are the entries of the matrix inverse to the matrix (g_{ij}), p. 68

$\varepsilon_1, \varepsilon_2, \varepsilon_3$ is the Darboux frame, p. 68

k_g is geodesic curvature, p. 69

k_n is normal curvature, p. 69

$\varkappa_g$ is geodesic torsion, p. 69

L, M, and N are the coefficients of the second quadratic form, p. 72

H is mean curvature, p. 73

K is Gaussian curvature, p. 73

Γ^k_{ij} are the Christoffel symbols, p. 85

$\mathbb{S}^2$ is the sphere, p. 87

$\nabla_W V$ is the covariant derivative of a vector field V in the direction of a vector field
   W, p. 93

R^l_{ijk} is the Riemannian curvature tensor, p. 98

$\exp_p(V)$ is the exponential map, p. 100

Moscow, Russia   Victor V. Prasolov

Contents

Chapter 1
Curves in the Plane

The simplest object of differential geometry is a curve in the plane. The definition of a curve varies between different areas of mathematics. In many cases, it is natural to represent a curve as the trajectory of a moving point. In doing so, one should distinguish between a parameterized curve $\gamma(t) = (x(t), y(t))$ and a nonparameterized curve, which is the image of a parameterized curve, i.e., a set in the plane. The functions $x(t)$ and $y(t)$ are not arbitrary. They are usually assumed to be smooth. But even under this assumption, a nonparameterized curve may have corners. This can be avoided by requiring the derivatives $x'(t)$ and $y'(t)$ not to vanish simultaneously. In that case, the parameterized curve is said to be *smooth*. A smooth nonparameterized curve is the image of a smooth parameterized curve.

Geometry deals with both closed and nonclosed (that is, joining two different points) parameterized curves. A smooth (not necessarily closed) parameterized curve in the plane is a map $\gamma : [a, b] \to \mathbb{R}^2$ such that $\gamma(t) = (x(t), y(t))$, where x and y are smooth functions, and $v(t) = \frac{d\gamma}{dt}(t) \neq 0$ for all $t \in [a, b]$ (we assume that the derivative have finite limits at $t = a$ and $t = b$). A smooth parameterized curve $\gamma : [a, b] \to \mathbb{R}^2$ is said to be *closed* if $\gamma(a) = \gamma(b)$ and $v(a) = v(b)$.

The *length* of a curve $\gamma : [a, b] \to \mathbb{R}^2$ can be defined as the limit of the lengths of polygonal chains with vertices on the curve. In more detail, we choose a partition $a = t_0 < t_1 < \cdots < t_n = b$ of the interval $[a, b]$, consider the polygonal chain $P_0 P_1 \ldots P_n$, where $P_i = (x(t_i), y(t_i))$, and find the limit of the lengths of these polygonal chains as the maximum of the numbers $\Delta t_i = t_i - t_{i-1}$ tends to zero.

It is easy to show that the length of a curve $\gamma : [a, b] \to \mathbb{R}^2$ is equal to

$$\int_a^b \sqrt{(x'(t))^2 + (y'(t))^2} dt.$$

© The Author(s), under exclusive license to Springer Nature Switzerland AG 2022
V. V. Prasolov, *Differential Geometry*, Moscow Lectures 8,
https://doi.org/10.1007/978-3-030-92249-8_1

Indeed, the length of a polygonal chain $P_0 P_1 \ldots P_n$ equals

$$\sum_{i=1}^{n} \sqrt{(x_i - x_{i-1})^2 + (y_i - y_{i-1})^2},$$

where $x_i = x(t_i)$ and $y_i = y(t_i)$. Using the mean value theorem, we can rewrite this sum in the form $\sum_{i=1}^{n} \Delta t_i \sqrt{(x'(\xi_i))^2 + (y'(\eta_i))^2}$, where the ξ_i and η_i are some numbers between t_{i-1} and t_i. The limit of this sum has the required form.

The *area* of the figure bounded by a closed curve $\gamma : [a, b] \to \mathbb{R}^2$ without self-intersections can be defined as the least upper bound for the areas of polygons contained in it and the greatest lower bound for the areas of polygons containing it. A figure for which these two numbers are equal is said to be *squarable*. The area of nonsquarable figures is not defined. Any figure bounded by a smooth closed curve is squarable.

The *oriented area* of a figure bounded by a parameterized closed curve without self-intersections equals the area of this figure in absolute value. When the curve is traversed counterclockwise, the oriented area is positive, and when it is traversed clockwise, the oriented area is negative.

A formula for the oriented area of a figure bounded by a parameterized (possibly self-intersecting) curve can be obtained from the following formula for the oriented area A of the triangle with vertices $(0, 0)$, (x_1, y_1), and (x_2, y_2):

$$A = \frac{1}{2}(x_1 y_2 - x_2 y_1).$$

By analogy with this formula, the oriented area of a polygon with consecutive vertices (x_i, y_i), $i = 0, 1, \ldots, n$, can be defined as

$$\frac{1}{2} \sum_{i=0}^{n} (x_i y_{i+1} - x_{i+1} y_i),$$

where $(x_{n+1}, y_{n+1}) = (x_0, y_0)$. The polygon may have self-intersections.

The formula for the oriented area of a figure bounded by a closed curve $\gamma : [a, b] \to \mathbb{R}^2$ can now be obtained by choosing a partition $a = t_0 < t_1 < \cdots < t_n = b$ of the interval $[a, b]$, considering a polygon $P_0 P_1 \ldots P_n$, where $P_i = (x(t_i), y(t_i))$, and passing to the limit of the oriented areas of such polygons as the maximum of the numbers $\Delta t_i = t_i - t_{i-1}$ tends to zero.

We set $x_i = x(t_i)$ and $y_i = y(t_i)$ and use the mean value theorem: $x_{i+1} = x_i + x'(\eta_i)\Delta t_i$, $y_{i+1} = y_i + y'(\xi_i)\Delta t_i$. As a result, we obtain the following formula for the oriented area of a polygon $P_0 P_1 \ldots P_n$:

$$\frac{1}{2} \sum_{i=0}^{n-1} (x(t_i) y'(\xi_i) - y(t_i) x'(\eta_i)) \Delta t_i,$$

where $\Delta t_i = t_{i+1} - t_i$ and $\xi_i, \eta_i \in [t_i, t_{i+1}]$. If the numbers Δt_i are positive and the maximum among them tends to zero, then we obtain the integral expression

$$\frac{1}{2} \int_a^b (xy' - yx')dt$$

for the oriented area of the figure bounded by the curve γ. Integrating by parts, it is easy to show that

$$- \int_a^b yx'dt = \int_a^b xy'dt.$$

Indeed,

$$\int_a^b yx'dt + \int_a^b xy'dt = (xy)|_a^b = 0,$$

because the curve is closed.

Thus, the oriented area A of a figure bounded by a smooth closed curve can be calculated by any of the following three equivalent formulas:

$$A = - \int_a^b yx'dt = \int_a^b xy'dt = \frac{1}{2} \int_a^b (xy' - yx')dt. \tag{1.1}$$

Problem 1.1 A closed curve γ bounds a convex figure. The endpoints of a chord of length $a + b$ move on the curve γ. A point M of this chord divides it in the ratio $a : b$. As the chord moves, M traces a closed curve γ'. Prove that the area of the figure bounded by the curves γ and γ' equals πab.

1.1 Curvature and the Frenet–Serret Formulas

Let $\gamma(t) = (x(t), y(t))$ be a smooth curve. It is often convenient to replace the parameter t by the *arc length parameter* $s = s(t) = \int_0^t \|v(\tau)\| d\tau$, where $v(\tau) = \frac{dy(\tau)}{d\tau}$. The arc length parameter is the length of the arc of γ enclosed between the points $\gamma(0)$ and $\gamma(t)$.

For the arc length parameter, we have $\frac{ds}{dt} = \|v(t)\|$. Therefore, $\frac{d\gamma}{ds} = \frac{d\gamma}{dt} \cdot \frac{dt}{ds} = \frac{v(t)}{\|v(t)\|}$, whence $\left\| \frac{d\gamma}{ds} \right\| = 1$.

The endpoint of the vector $v(s) = \frac{d\gamma(s)}{ds}$ moves along the unit circle; therefore, $\frac{dv}{ds} \perp v$. Let n be the unit vector in $\mathbb{R}^2$ orthogonal to v and such that the vectors v and n form a positively oriented basis, i.e., the rotation from v to n is counterclockwise. Then the vector n is determined uniquely and $\frac{dv(s)}{ds} = k(s)n$. The number $k(s)$

Fig. 1.1 The change of the
sign of oriented curvature

$$k < 0 \qquad\qquad\qquad k > 0$$

is called the *oriented curvature* of the curve γ at the point $\gamma(s)$, and the number $|k(s)|$ is called simply the *curvature*. Under the change of the curve orientation, the oriented curvature changes sign (see Fig. 1.1).

The length of the vector $v(s)$ equals 1; therefore, it has coordinates $(\cos\varphi, \sin\varphi)$. The vector $n(s)$ obtained by rotating $v(s)$ through $90°$ counterclockwise has coordinates $(-\sin\varphi, \cos\varphi)$. Therefore, $\frac{dv}{ds} = \frac{d\varphi}{ds}(-\sin\varphi, \cos\varphi) = \frac{d\varphi}{ds}n$. Thus, $k(s) = \frac{d\varphi}{ds}$, where φ is the angle between the velocity vector $v(s)$ and some constant vector.

Problem 1.2 Prove that the curvature of a circle of radius R equals $1/R$.

Problem 1.2 explains why the quantity $R(s) = 1/|k(s)|$ is called the *radius of curvature* of the curve at the point s.

Theorem 1.1 (Frenet–Serret Formulas) *If s is the arc length parameter, then $\frac{dv}{ds} = kn$ and $\frac{dn}{ds} = -kv$, where k is oriented curvature.*

Proof Let s_0 be a fixed value of the parameter. The frame $v(s)$, $n(s)$ is obtained by rotating $v(s_0)$, $n(s_0)$ through an angle $\varphi(s)$:

$$\begin{pmatrix} v(s) \\ n(s) \end{pmatrix} = \begin{pmatrix} \cos\varphi(s) & \sin\varphi(s) \\ -\sin\varphi(s) & \cos\varphi(s) \end{pmatrix} \begin{pmatrix} v(s_0) \\ n(s_0) \end{pmatrix};$$

here $\varphi(s_0) = 0$. Therefore,

$$\begin{pmatrix} \frac{dv(s)}{ds}(s_0) \\ \frac{dn(s)}{ds}(s_0) \end{pmatrix} = \frac{d\varphi}{ds}(s_0) \begin{pmatrix} -\sin\varphi(s_0) & \cos\varphi(s_0) \\ -\cos\varphi(s_0) & -\sin\varphi(s_0) \end{pmatrix} \begin{pmatrix} v(s_0) \\ n(s_0) \end{pmatrix}$$

$$= k \begin{pmatrix} 0 & 1 \\ -1 & 0 \end{pmatrix} \begin{pmatrix} v(s_0) \\ n(s_0) \end{pmatrix}.$$

$\square$

HISTORICAL COMMENT The Frenet–Serret formulas for curves in three-dimensional space were derived almost simultaneously by Joseph Alfred Serret (1819–1885) and Jean Frédéric Frenet (1816–1900). Their papers were published in different issues of the same journal: Serret's in 1851 and Frenet's in 1852. These formulas are sometimes called the Frenet formulas.

Fig. 1.2 The support line

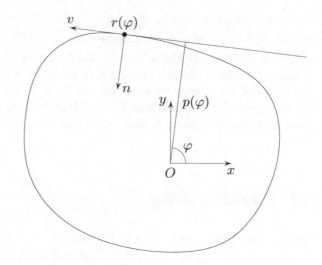

Problem 1.3 Prove that the curvature of a curve $\gamma(t) = \big(x(t), y(t)\big)$ with any parameterization t can be calculated by

$$k^2 = \frac{(x''y' - y''x')^2}{(x'^2 + y'^2)^3}.$$

Problem 1.4 Calculate the curvature of the ellipse $\frac{x^2}{a^2} + \frac{y^2}{b^2} = 1$ at each point.

Problem 1.5 Suppose that a curve γ and a circle of radius R are oriented positively (i.e., counterclockwise), are tangent at a point P, and have the same velocity vector at this point, and the oriented curvature of the curve at P equals k. Prove that if $k > \frac{1}{R}$, then a small neighborhood of the point P on the curve lies inside the circle, and if $k < \frac{1}{R}$, then it lies outside the circle.

Problem 1.6 Suppose that a curve $r(\varphi)$ bounds a convex figure containing the origin O and the parameter φ is the angle between the Ox axis and the perpendicular from O to the support line at the point $r(\varphi)$ (see Fig. 1.2). Let $p(\varphi)$ be the distance between O and the support line. Prove that the radius of curvature at the point $r(\varphi)$ equals $p(\varphi) + p''(\varphi)$.

The Frenet–Serret formulas suggest the following geometric interpretation of the curvature of a curve. Let us draw a normal through each point $\gamma(s)$ of a curve γ and mark off two points at a distance ε (small enough) from the point $\gamma(s)$ on this normal. These points trace two curves $\gamma_{\pm\varepsilon}(s)$. Consider the arcs of these curves corresponding to $s \in [a, b]$. The geometric interpretation of the curvature $k(s_0)$ is as follows: as $a, b \to s_0$, the ratio of the length of the corresponding arc of the curve $\gamma_{\pm\varepsilon}(s)$ to the length of the arc of the initial curve equals $1 + \varepsilon k(s_0)$ or $1 - \varepsilon k(s_0)$ in the first approximation. Indeed, we have $\gamma_{\pm\varepsilon}(s) = \gamma(s) \pm \varepsilon n(s)$, where the vector $n(s)$ is defined by $\frac{dv(s)}{ds} = k(s)n(s)$, s being the arc length parameter. According to the Frenet–Serret formulas, we have $\frac{dn(s)}{ds} = -k(s)v(s)$; therefore, $\frac{d\gamma_{\pm\varepsilon}(s)}{ds} = v(s)\mp\varepsilon k(s)v(s)\pm\varepsilon k'(s)n(s)$. The summand $\varepsilon k'(s)n(s)$ is of order $O(\varepsilon^2)$, because the vectors $v(s)$ and $n(s)$ are orthogonal. Thus, in the first approximation, the length of the velocity vector of the curve $\gamma_{\pm\varepsilon}(s)$ equals $1 \pm \varepsilon k(s)$; therefore, the arc length of this curve equals $\int_a^b (1 \pm \varepsilon k(s))ds$ in the first approximation. This gives the required equation, because the length of the arc of the initial curve equals $b - a$.

1.2   Osculating Circles

Consider two plane curves given by the equations $y = f(x)$ and $y = g(x)$. These curves intersect in a point (x_0, y_0) if $f(x_0) = g(x_0) = y_0$. The curves are tangent (have *contact of order* 1) at the intersection point if $f'(x_0) = g'(x_0)$. For tangent curves, we choose a coordinate system with Ox axis directed along the tangent line to these curves (then $f'(x_0) = g'(x_0) = 0$). Tangent curves have *contact of order n* if $f''(x_0) = g''(x_0)$, ..., $f^{(n)}(x_0) = g^{(n)}(x_0)$. Curves having contact of order 2 are said to be *osculating*.

A point at which a curve and its tangent line have contact of order 2 is called an *inflection point* of this curve. A point $(x_0, f(x_0))$ of a curve $y = f(x)$ is an inflection point if and only if $f''(x_0) = 0$.

If the curvature of a curve is nonzero at some point, then a circle osculating the curve at this point (*osculating circle*) is determined uniquely. The equation of the osculating circle in the case where the curve is given by an equation $y = f(x)$ is derived as follows. Suppose that a circle of radius R centered at (a, b) is locally determined by an implicit function $y = g(x)$, i.e., g satisfies the equation

$$(x - a)^2 + \big(g(x) - b\big)^2 = R^2.$$

Differentiating this equation, we successively obtain

$$2(x - a) + 2g'(x)\big(g(x) - b\big) = 0,$$

$$2 + 2g''(x)\big(g(x) - b\big) + 2\big(g'(x)\big)^2 = 0.$$

If the curve $y = f(x)$ osculates the circle at a point (x_0, y_0), then

$$(x_0 - a)^2 + (y_0 - b)^2 = R^2,$$

$$(x_0 - a) + y_0'(y_0 - b) = 0,$$

$$1 + y_0''(y_0 - b) + (y_0')^2 = 0,$$

where $y_0' = f'(x_0)$ and $y_0'' = f''(x_0)$. Moreover, if this system of equations for a, b, and R has a solution, then the converse is also true: the curve osculates the circle. It is easy to see that if $y_0'' \neq 0$, then this system of equations has a unique solution. Recall that we chose coordinates in which $y_0' = 0$; in these coordinates $a = x_0$, $b = y_0 + \frac{1}{y_0''}$, and $R^2 = \frac{1}{(y_0'')^2}$. According to Problem 1.3, in the situation under consideration, we have $k^2 = \frac{(y_0'')^2}{(1+y_0'^2)^3} = (y_0'')^2$. Thus, the curvature of the osculating circle equals that of the curve at the point of contact.

The center of the osculating circle is called the *center of curvature* of the curve. The center of curvature of a curve γ at a point $\gamma(s)$ lies on the normal to γ at this point at a distance $\pm 1/k(s)$ from it. The center of curvature is the point $\gamma(s) + \frac{1}{k(s)}n(s)$; recall that the direction of the normal vector is chosen so that the vectors $v(s)$ and $n(s)$ form a positively oriented basis.

Problem 1.7 Prove that the center of curvature is the limit position of the intersection point of close normals.

Below we give yet another interpretation of the center of curvature. Let $\gamma(s) = (x(s), y(s))$ be a curve parameterized by arc length. Fix a point $q = (x_0, y_0)$ in the plane and consider the function $F(s) = \|\gamma(s) - q\|^2$ on γ whose value at each point equals the squared distance from this point to the fixed point. Clearly,

$$F'(s) = 2\left(\gamma(s) - q, \frac{d\gamma}{ds}\right),$$

$$F''(s) = 2\left\|\frac{d\gamma}{ds}\right\|^2 + 2\left(\gamma(s) - q, \frac{d^2\gamma}{ds^2}\right) = 2\big(1 + (\gamma(s) - q, kn)\big).$$

Therefore, s_0 is a critical point of the function F if and only if the vector $\overrightarrow{\gamma(s_0)q}$ is orthogonal to the curve γ at $\gamma(s_0)$, and a critical point s_0 is degenerate if and only if $k(s_0) \neq 0$ and $\overrightarrow{\gamma(s_0)q} = \frac{1}{k(s_0)}n$. A point q is said to be a *focal point* of a curve γ if the function $\|\gamma(s) - q\|^2$ has a degenerate critical point. If such a degenerate critical point is s_0, then the point q is uniquely determined by $\overrightarrow{\gamma(s_0)q} = \frac{1}{k(s_0)}n$. Geometrically, this means that the point q is the center of the circle osculating the curve γ at $\gamma(s_0)$. Thus, q is the limit position of the intersection point of the normals to γ at the points $\gamma(s_1)$ and $\gamma(s_2)$ as $s_1 \to s_0$ and $s_2 \to s_0$. In particular, the property of being focal does not depend on the choice of parameterization.

1.3   The Total Curvature of a Closed Plane Curve

The *total oriented curvature* of a closed curve $\gamma : [a, b] \to \mathbb{R}^2$ is defined as $\int_a^b k\, ds$, where s is the arc length parameter and k is oriented curvature of γ. Recall that $k(s) = \frac{d\varphi}{ds}$, where φ is the angle between the velocity vector $v(s)$ and some constant

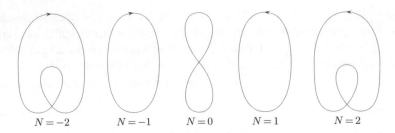

$$N=-2 \qquad N=-1 \qquad N=0 \qquad N=1 \qquad N=2$$

Fig. 1.3 Total oriented curvature

vector (see p. 4). Therefore, $\int_a^b k \, ds = 2\pi N$, where N is the number of full turns (counted with sign) made by the velocity vector as the whole curve is traversed (see Fig. 1.3).

The number N can be calculated as follows. Consider the function $\varphi(s)$, $s \in [a, b]$. Recall that a point s_0 is said to be critical if $\frac{d\varphi}{ds}(s_0) = 0$, and the value of $\varphi(s)$ at a critical point is called a critical value. The measure of the set of critical values is zero. Let us choose a noncritical value φ and consider only those s for which $\varphi(s) = \varphi$. For each of these s, we have either $\varphi'(s) > 0$ or $\varphi'(s) < 0$. In the former case, the velocity vector rotates in the positive direction; we denote the number of such s by $n_+(\varphi)$. In the latter case, the velocity vector rotates in the negative direction; we denote the number of such s by $n_-(\varphi)$. An interval of length $\Delta\varphi$ on which both $n_+(\varphi)$ and $n_-(\varphi)$ are constant contributes $(n_+ - n_-)\Delta\varphi$ to the integral of oriented curvature; therefore, the total oriented curvature is $\int_0^{2\pi}(n_+(\varphi) - n_-(\varphi))d\varphi$.

Remark It can be shown that the difference $n_+(\varphi) - n_-(\varphi)$ does not depend on the chosen noncritical value φ. For example, on passage through a simple critical point, either the numbers n_+ and n_- do not change or one of them increases by 1 and the other decreases by 1. Thus, the number of turns of the velocity vector equals $n_+(\varphi) - n_-(\varphi)$ for any noncritical value φ.

Theorem 1.2 (Umlaufsatz) *The total oriented curvature of a smooth closed curve without self-intersections equals $\pm 2\pi$.*

Proof Let us translate a straight line not intersecting the given curve $\gamma(s)$ until it touches this curve at some point. Note that the curve lies on one side of the line. Thus, we can assume that the coordinate system and the natural parameter $s \in [0, l]$ on the curve $\gamma(s)$ are chosen so that the curve lies in the upper half-plane and $\gamma(0)$ is the origin.

Let T be the triangle consisting of all points (s_1, s_2) whose coordinates satisfy the inequalities $0 \leqslant s_1 \leqslant s_2 \leqslant l$. Consider the map ψ of T to the unit circle $\mathbb{S}^1$ defined as follows: if $s_1 \neq s_2$ and $(s_1, s_2) \neq (0, l)$, then

$$\psi(s_1, s_2) = \frac{\gamma(s_2) - \gamma(s_1)}{\|\gamma(s_2) - \gamma(s_1)\|}; \quad \psi(s, s) = \frac{\gamma'(s)}{\|\gamma'(s)\|}, \quad \psi(0, l) = -\frac{\gamma'(0)}{\|\gamma'(0)\|}.$$

Fig. 1.4 The family of
curves

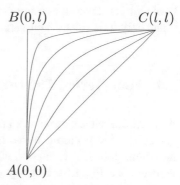

$B(0, l)$ $C(l, l)$

$A(0, 0)$

For a curve without self-intersections, this map is well-defined (does not involve division by zero) and continuous. Its continuity at the point (s, s) is obvious, and to prove continuity at the point $(0, l)$, it suffices to note that

$$\lim_{s \to 0} \frac{\gamma(l) - \gamma(s)}{\|\gamma(l) - \gamma(s)\|} = -\lim_{s \to 0} \frac{\gamma(s) - \gamma(0)}{\|\gamma(s) - \gamma(0)\|}.$$

Let $A(0, 0)$, $B(0, l)$, and $C(l, l)$ be the vertices of T. To the parameter s the restriction of ψ to the side AC assigns the tangent vector $\gamma'(s)$. Therefore, the point $\psi(s, s)$ makes as many turns while traversing AC as the velocity vector does while traversing the curve.

The curve consisting of the segment AC can be connected to that consisting of the segments AB and BC through a continuous family of curves (see Fig. 1.4). Thus, it remains to prove that, while traversing AB and BC, a point on the circle makes one turn (in the positive or negative direction).

As the segment AB is traversed, the vector $\gamma(s) - \gamma(0)$ applied to origin remains in the same half-plane and changes its direction to the opposite one; therefore, it makes a half-turn. While traversing BC, the vector $\gamma(l) - \gamma(s) = \gamma(0) - \gamma(s)$ makes a half-turn in the same direction. As a result, the vector makes one full turn. □

HISTORICAL COMMENT *Umlaufsatz* was proved by Heinz Hopf (1894–1971) in 1933.

Together with the total oriented curvature $\int_a^b k \, ds$ one can consider the *total curvature* $\int_a^b |k| \, ds$. The contributions of all passages through a noncritical value φ to the total curvature are the same, independently of their sign; therefore, the total curvature equals $\int_0^{2\pi} n(\varphi) d\varphi$, where $n(\varphi) = n_+(\varphi) + n_-(\varphi)$ is the number of curve points at which the velocity vector has given direction.

A closed plane curve is said to be *strictly convex* if any straight line intersects it in at most two points. A curve is strictly convex if and only if $n(\varphi) = 1$ for all φ. A closed plane curve is *convex* if it lies on one side of any tangent line. A convex curve differs from a strictly convex curve in that it may have rectilinear fragments. A curve is convex if and only if $n(\varphi) = 1$ for all noncritical values φ.

Problem 1.8 Prove that a smooth closed curve without self-intersections is convex if and only if its oriented curvature does not change sign.

1.4   Four-Vertex Theorem

A *vertex* of a closed curve is a point of local maximum or minimum of its oriented curvature. Every closed curve has at least two vertices—global maximum and minimum points. It is easy to give examples of closed curves with precisely two vertices (see Fig. 1.5). But any curve without self-intersections has at least four vertices. First, we give a simple proof of a weaker statement (for only convex curves, rather than for all non-self-intersecting ones).

Theorem 1.3 *Any smooth convex closed curve has at least four vertices.*

Proof First, we prove that if $\gamma : [a, b] \to \mathbb{R}^2$ is a smooth closed curve, then

$$\int_a^b k'(s)\, ds = 0, \qquad \int_a^b x(s)k'(s)\, ds = 0, \qquad \int_a^b y(s)k'(s)\, ds = 0.$$

The first equation holds because $k(b) = k(a)$. To prove the other two, we apply integration by parts and the second Frenet–Serret formula:

$$\int_a^b k'(s)(x(s), y(s))ds = -\int_a^b k(s)(x'(s), y'(s))ds$$

$$= -\int_a^b kv\, ds = \int_a^b \frac{dn}{ds} ds = 0.$$

Fig. 1.5 Curves with two vertices

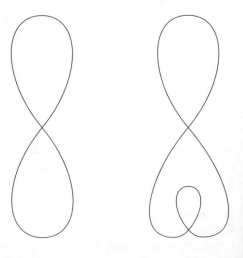

Now suppose that γ is a smooth convex closed curve. At the points of maximum and minimum of its oriented curvature, the derivative k' changes sign. Consider the line $px + qy + r = 0$ joining a point of maximum of the oriented curvature to a point of minimum. Suppose that the derivative k' does not change sign when passing through all other points. Then the product $k'(px + qy + r)$ does not change sign on either of the arcs of the curve γ into which γ is divided by the line $px + qy + r = 0$, because both factors k' and $px + qy + r$ change sign at the points of maximum and minimum and at no other points. The constancy of the sign of $k'(px + qy + r)$ contradicts the relation $\int_a^b k'(px + qy + r) = 0$; hence there is at least one more point at which the derivative k' changes sign.

Thus, the function k' changes sign at the points of maximum and minimum of the function k and at some point or points. It follows from the closedness of the curve that the number of sign changes must be even (if it is finite). Therefore, k' changes sign at at least four points. □

The proof of the four-vertex theorem for closed curves without self-intersections given below was suggested by Osserman [Os]. This proof uses properties of a circle circumscribed about a curve, i.e., a circle of smallest radius enclosing the curve.

Lemma

(a) *Given a smooth closed curve γ, there exists a unique circle C of smallest radius enclosing it (the circle circumscribed about the curve).*

(b) *If the length of an arc of the circle circumscribed about a curve γ is larger than half the length of the circle, then this arc contains at least one point of γ.*

Proof

(a) Given two circles of radius R enclosing the curve γ, we can construct a circle of radius smaller than R also enclosing this curve (see Fig. 1.6).

(b) Suppose that a circle enclosing γ has an arc which is longer than half the length of the circle and does not intersect γ. Then a translate of this circle contains no points of γ; therefore, its radius can be decreased. □

Fig. 1.6 Decreasing the radius of the circle

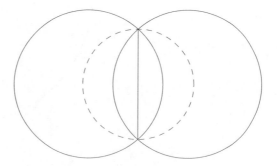

Theorem 1.4 *Let C be the circle circumscribed about a smooth closed curve γ without self-intersections. Then C contains at least two points of γ, and if C contains n points of γ, then γ has at least 2n vertices.*

Proof Let $P_1, \ldots, P_n$ be the common points of the curve γ and the circumscribed circle C numbered so that $P_1 \ldots P_n$ is a convex polygon. Obviously, $n \geqslant 2$. We assume that the curve γ and the circumscribed circle C are oriented positively (counterclockwise). According to Problem 1.5 (see p. 5), we have $k(P_i) \geqslant 1/R$, where R is the radius of C. It suffices to prove that each arc $P_i P_{i+1}$ has a point Q_i for which $k(Q_i) < 1/R$. Indeed, if this is so, then the oriented curvature has a local minimum between the points P_i and P_{i+1} and a local maximum between the points Q_i and Q_{i+1}.

The required point Q_i can be found as follows. Consider the arc γ_i between P_i and P_{i+1}. We can assume that the direction from P_i to P_{i+1} is positive on both the arc γ_i of γ and the corresponding arc C_i of C. According to the lemma, the arc C_i is no longer than half the circle. We assume that the straight line $P_i P_{i+1}$ is vertical and the center of C lies on the left of this line (or on the very line). Choose an interior point Q on the arc γ_i which lies on the right of the line $P_i P_{i+1}$ (see Fig. 1.7). Then the radius R' of the circumcircle C' of the triangle $P_i Q P_{i+1}$ is greater than R.

Fig. 1.7 The choice of the
points Q and Q_i

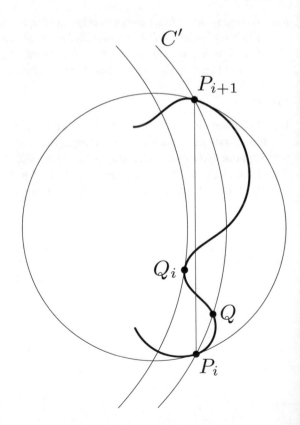

Let us move the circle C' to the left until it takes off the arc γ_i. At the last moment the circle is tangent to the arc at some point Q_i (see Fig. 1.7). The curve γ has no self-intersections; therefore, at the point Q_i the orientation of γ coincides with that of the circle. It follows that $k(Q_i) \leqslant 1/R'$. Indeed, if the oriented curvature of the curve γ at Q_i is positive, then $k(Q_i) \leqslant 1/R'$ according to Problem 1.5, and if $k(Q_i) \leqslant 0$, then $k(Q_i) \leqslant 1/R'$ because R' is positive. Thus, $k(Q_i) \leqslant 1/R' < 1/R$, as required.   □

HISTORICAL COMMENT The four-vertex theorem for convex curves was proved by the Indian mathematician Syamadas Mukhopadhyaya (1866–1937) in 1909.

It can be proved that if a self-intersecting curve bounds a two-dimensional surface immersed in the plane, then this curve has at least four vertices (see [Pi]), and if the immersed surface is not a disk, then it has at least six vertices (see [Ca1] or [Um]).

1.5   The Natural Equation of a Plane Curve

Let $\gamma(s) = (x(s), y(s))$ be a smooth plane curve, where s is the natural parameter. Denoting by $k(s)$ the oriented curvature at s, we obtain the function $k(s)$ corresponding to the curve γ.

Theorem 1.5 *Any smooth function $k(s)$ corresponds to some curve γ, and this curve is determined up to a direct motion of the plane.*

Proof It follows from the Frenet–Serret formulas $v' = kn$ and $n' = -kv$ that both functions $\xi_{1x}(s) = x'(s)$ and $\xi_{1y}(s) = y'(s)$ must satisfy the system of differential equations $\xi_1'(s) = k(s)\xi_2(s)$, $\xi_2'(s) = -k(s)\xi_1(s)$. Let (ξ_{1x}, ξ_{2x}) be its solution with initial condition $(\xi_{1x}(0), \xi_{2x}(0)) = (1, 0)$, and let (ξ_{1y}, ξ_{2y}) be the solution with initial condition $(\xi_{1y}(0), \xi_{2y}(0)) = (0, 1)$. We set

$$x(s) = \int_0^s \xi_{1x}(\tau)d\tau, \quad y(s) = \int_0^s \xi_{1y}(\tau)d\tau$$

and consider the vectors $v(s) = (\xi_{1x}, \xi_{1y})$ and $n(s) = (\xi_{2x}, \xi_{2y})$. Clearly, $v(s) = (x'(s), y'(s))$ and $v'(s) = kn(s)$; therefore, it suffices to check that the vectors $v(s)$ and $n(s)$ have unit length.

Suppose that each of the indices a and b takes the value x or y. Then $(\xi_{1a}\xi_{1b} + \xi_{2a}\xi_{2b})' = k\xi_{2a}\xi_{1b} + k\xi_{1a}\xi_{2b} - k\xi_{1a}\xi_{2b} - k\xi_{2a}\xi_{1b} = 0$. For $s = 0$, the matrix $\begin{pmatrix} \xi_{1x}(s) & \xi_{2x}(s) \\ \xi_{1y}(s) & \xi_{2y}(s) \end{pmatrix}$ equals $\begin{pmatrix} 1 & 0 \\ 0 & 1 \end{pmatrix}$. The rows of this matrix are mutually orthogonal for all s, that is, the matrix is orthogonal, so that its columns are mutually orthogonal as well. This means that v and n are orthogonal vectors of unit length.

Now let us prove the uniqueness of the curve. Let $\gamma(s)$ and $\bar{\gamma}(s)$ be two curves corresponding to the same function $k(s)$. Suppose that at $s = 0$ the vector $v =$

(ξ_{1x}, ξ_{1y}) coincides with $\bar{v} = (\bar{\xi}_{1x}, \bar{\xi}_{1y})$ and the vector $n = (\xi_{2x}, \xi_{2y})$ coincides with $\bar{n} = (\bar{\xi}_{2x}, \bar{\xi}_{2y})$. Since $(\xi_{1a}\bar{\xi}_{1b} + \xi_{2a}\bar{\xi}_{2b})' = 0$, it follows that, for all s, we have $\xi_{1x}\bar{\xi}_{1x} + \xi_{2x}\bar{\xi}_{2x} = 1$ and $\xi_{1y}\bar{\xi}_{1y} + \xi_{2y}\bar{\xi}_{2y} = 1$; thus, $(\xi_{1x} - \bar{\xi}_{1x})^2 + (\xi_{2x} - \bar{\xi}_{2x})^2 + (\xi_{1y} - \bar{\xi}_{1y})^2 + (\xi_{2y} - \bar{\xi}_{2y})^2 = 0$. $\square$

The *natural equation* of a plane curve with natural parameter s is the specification of the curve by the oriented curvature function $k(s)$. Theorem 1.5 shows that a natural equation with smooth function $k(s)$ always has a solution and the natural equation of a curve determines this curve uniquely up to a direct motion.

1.6   Whitney–Graustein Theorem

Let $\gamma_0, \gamma_1 : [0, 1] \to \mathbb{R}^2$ be smooth closed curves. We say that the curves γ_0 and γ_1 are *regularly homotopic* if there exists a family of smooth closed curves γ_t smoothly depending on $t \in [0, 1]$ (and such that $\gamma_t = \gamma_0$ for $t = 0$ and $\gamma_t = \gamma_1$ for $t = 1$). Smooth dependence on t means that the map $(s, t) \mapsto \gamma_t(s)$ is smooth as a map from $[0, 1] \times [0, 1]$ to $\mathbb{R}^2$.

Theorem 1.6 (Whitney–Graustein)   *Curves γ_0 and γ_1 are regularly homotopic if and only if they have equal total oriented curvatures.*

Proof Recall that the total oriented curvature of a plane closed curve equals $2\pi N$, where N is the number of full turns (counted with sign) made by the velocity vector as the whole curve is traversed. Let γ_t be a regular homotopy between the curves γ_0 and γ_1, and let $2\pi N_t$ be the total curvature of γ_t. The integer N_t continuously depends on t. Therefore, it is constant, and hence $N_0 = N_1$.

Now suppose that γ_0 and γ_1 are smooth closed curves of the same total curvature N. Using a regular homotopy, we can replace each of the curves γ_0 and γ_1 by a curve of length 1 with initial point at the origin and initial velocity vector $(1, 0)$. Therefore, we can assume that $s \in [0, 1]$ is the natural parameter, $\gamma_0(0) = \gamma_1(0) = (0, 0)$, and $\gamma_0'(0) = \gamma_1'(0) = (1, 0)$. We can write the velocity vectors of the curves γ_0 and γ_1 in the forms $v_0(s) = e^{i\varphi_0(s)}$ and $v_1(s) = e^{i\varphi_1(s)}$, where $\varphi_0(0) = \varphi_1(0) = 0$ and $\varphi_0(1) = \varphi_1(1) = 2\pi N$. Let us set $\varphi_t(s) = (1 - t)\varphi_0(s) + t\varphi_1(s)$ and consider the curve $\tilde{\gamma}_t$ with velocity vector $v_t(s) = e^{i\varphi_t(s)}$: $\tilde{\gamma}_t(s) = \int_0^s e^{i\varphi_t(\tau)}d\tau$. For $0 < t < 1$, the curve $\tilde{\gamma}_t$ is not necessarily closed, but we can use it to construct the closed curve $\gamma_t(s) = \tilde{\gamma}_t(s) - s\tilde{\gamma}_t(1) = \int_0^s e^{i\varphi_t(\tau)}d\tau - s\int_0^1 e^{i\varphi_t(\tau)}d\tau$. It remains to check that the curve γ_t is smooth, i.e., $\frac{d}{ds}\gamma_t(1) = \frac{d}{ds}\gamma_t(0)$ and $\frac{d}{ds}\gamma_t(s) \neq 0$. Clearly, $\frac{d}{ds}\gamma_t(s) = e^{i\varphi_t(s)} - \int_0^1 e^{i\varphi_t(\tau)}d\tau = v_t(s) - \tilde{\gamma}_t(1)$. The velocities at $s = 0$ and at $s = 1$ are equal because $\varphi_t(0) = 0$ and $\varphi_t(1) = 2\pi N$, whence $v_t(0) = v_t(1)$. To show that $v_t(s) \neq \tilde{\gamma}_t(1)$, it suffices to note that $\|v_t(s)\| = 1$, while $\|\tilde{\gamma}_t(1)\| < 1$, since $\|\tilde{\gamma}_t(1)\| = \left| \int_0^1 e^{i\varphi_t(\tau)}d\tau \right| \leqslant \int_0^1 \left| e^{i\varphi_t(\tau)} \right| d\tau \leqslant 1$ and the function $e^{i\varphi_t(\tau)}$ is not constant. $\square$

HISTORICAL COMMENT The proof of the Whitney–Graustein theorem was published in 1937 in a paper by Hassler Whitney (1907–1989) with a note that the statement and the full proof of this theorem was communicated to him by William Caspar Graustein (1888–1941).

1.7   Tube Area and Steiner's Formula

Let $\gamma : [a, b] \to \mathbb{R}^2$ be a smooth curve without self-intersections. We assume that the parameterization of this curve is natural, i.e., $\|\gamma'(s)\| = 1$. Suppose that, at each point of the curve, the Frenet–Serret frame $v(s) = \gamma'(s)$, $n(s)$ is given. For each ε, we can consider the curve $\gamma_\varepsilon(s) = \gamma(s) + \varepsilon n(s)$. A *tube* of diameter $2r$ is the set of points swept out by the curves γ_ε with $-r \leqslant \varepsilon \leqslant r$. We can also consider the *positive tube* $(0 \leqslant \varepsilon \leqslant r)$ and the *negative tube* $(-r \leqslant \varepsilon \leqslant 0)$.

Theorem 1.7 *For sufficiently small r, the areas of the positive and negative tubes are, respectively,*

$$rL - \frac{r^2}{2} \int_a^b k(s)\,ds \quad and \quad rL + \frac{r^2}{2} \int_a^b k(s)\,ds,$$

where L the length of the curve and k is its oriented curvature.

Proof For sufficiently small r, the curves γ_ε with different $\varepsilon \in [-r, r]$ are pairwise disjoint and have no self-intersections. Therefore, the area of the positive tube equals $\int_0^r L_\varepsilon d\varepsilon$, where L_ε is the length of γ_ε. The Frenet–Serret formulas imply

$$\frac{d\gamma_\varepsilon(s)}{ds} = \frac{d\gamma(s)}{ds} + \varepsilon\frac{dn(s)}{ds} = v(s) - \varepsilon k(s)v(s);$$

therefore, for $|\varepsilon| < |1/k(s)|$,

$$L_\varepsilon = \int_a^b \big(1 - \varepsilon k(s)\big)ds = L - \varepsilon \int_a^b k(s)\,ds.$$

Thus, the area of the positive tube equals

$$\int_0^r \left(L - \varepsilon \int_a^b k(s)\,ds\right) d\varepsilon = rL - \frac{r^2}{2} \int_a^b k(s)\,ds,$$

and the area of the negative tube equals

$$\int_{-r}^0 \left(L - \varepsilon \int_a^b k(s)\,ds\right) d\varepsilon = rL + \frac{r^2}{2} \int_a^b k(s)\,ds.$$

$\square$

Corollary 1 *For sufficiently small r, the area of the tube is exactly* $2rL$.

Corollary 2 (Steiner's Formula) *Let B be a plane figure bounded by a smooth closed curve without self-intersections, and let B_r be the set of points of the plane at a distance of at most r from B. Then, for sufficiently small r, the area of the figure B_r equals $A + rL + \pi r^2$, where A is the area of B and L is the length of the curve bounding B.*

Proof We can assume that the orientation of the curve bounding B is positive. Then the total curvature of this curve equals 2π, and for sufficiently small r, the figure B_r is composed of B and the negative tube.   □

Remark If the figure B is convex, then Steiner's formula holds for all $r \geqslant 0$. Moreover, in the case of a convex figure, it suffices to require that the boundary be only piecewise smooth rather than smooth.

HISTORICAL COMMENT Jakob Steiner (1796–1863) obtained formulas for the area of the r-neighborhood of a convex polygon in the plane and for the length of its boundary, as well as formulas for the volume of the r-neighborhood of a convex polyhedron in three-dimensional space and for the area of its boundary, in 1840.

1.8   The Envelope of a Family of Curves

Consider a family of plane curves C_α depending on a parameter α. The *envelope* of the family of curves C_α is a curve C tangent to each C_α. The envelope may have several connected components. For example, the envelope of the family of circles of fixed radius r centered on a given straight line l consists of the two straight lines parallel to l at a distance r from l.

We assume that the plane is equipped with a Cartesian coordinate system Oxy and the curves C_α are given by equations of the form $f(x, y, \alpha) = 0$, where f is a smooth function.

We will seek an equation for the envelope C in the parametric form $x = \varphi(\alpha)$, $y = \psi(\alpha)$, assuming that C is tangent to each C_α at the point $(\varphi(\alpha), \psi(\alpha))$.

We set $x_0 = \varphi(\alpha_0)$ and $y_0 = \psi(\alpha_0)$. The tangents to the curves C and C_{α_0} at the point (x_0, y_0) are given by the equations

$$\frac{1}{x - x_0} \cdot \frac{d\varphi}{d\alpha} = \frac{1}{y - y_0} \cdot \frac{d\psi}{d\alpha} \text{ and } (x - x_0)\frac{\partial f}{\partial x} + (y - y_0)\frac{\partial f}{\partial y} = 0. \tag{1.2}$$

These lines must coincide, which means that

$$\frac{\partial f}{\partial x} \cdot \frac{d\varphi}{d\alpha} + \frac{\partial f}{\partial y} \cdot \frac{d\psi}{d\alpha} = 0. \tag{1.3}$$

The point $(\varphi(\alpha), \psi(\alpha))$ lies on the curve C_α, whence $f(\varphi(\alpha), \psi(\alpha), \alpha) = 0$ for all α. Differentiating this equation, we obtain

$$\frac{\partial f}{\partial x} \cdot \frac{d\varphi}{d\alpha} + \frac{\partial f}{\partial y} \cdot \frac{d\psi}{d\alpha} + \frac{\partial f}{\partial \alpha} = 0. \tag{1.4}$$

Now, taking into account (1.3), we see that $\frac{\partial f}{\partial \alpha} = 0$. Thus, the envelope (if it exists and is specified parametrically as above) can be found by eliminating the parameter α from the system equations

$$f(x, y, \alpha) = 0, \quad \frac{\partial f}{\partial \alpha}(x, y, \alpha) = 0. \tag{1.5}$$

For equations (1.2) to indeed determine tangent lines, we must require that the derivatives $\frac{\partial f}{\partial x}$ and $\frac{\partial f}{\partial y}$ not vanish simultaneously at the point (x_0, y_0, α_0). If this condition is satisfied, then the argument can be reversed, and the curve found by solving system (1.5) is indeed the envelope.

Example 1.1 The equation for the envelope of the family of curves

$$(x - \alpha)^3 + (y - \alpha)^3 = 3(x - \alpha)(y - \alpha)$$

is satisfied by the line $x = \alpha$, $y = \alpha$, i.e., $x = y$. However, this line is not the envelope; it consists of the self-intersection points of curves in this family.

In many cases, the envelope can be found from the following geometric considerations. Suppose that each pair of curves C_{α_1} and C_{α_2} $(\alpha_1 \neq \alpha_2)$ intersect in one point and the intersection points converge to some point $(x(\alpha), y(\alpha))$ as $\alpha_1 \to \alpha$ and $\alpha_2 \to \alpha$. Then this point $(x(\alpha), y(\alpha))$ satisfies the equations for the envelope. Indeed, if $f(x, y, \alpha_1) = 0$ and $f(x, y, \alpha_2) = 0$, then

$$0 = \frac{f(x, y, \alpha_1) - f(x, y, \alpha_2)}{\alpha_1 - \alpha_2} = \frac{\partial f}{\partial \alpha}(x, y, \alpha^*),$$

where α^* is a point between α_1 and α_2. Therefore, the point $(x(\alpha), y(\alpha))$ satisfies the required equations

$$f(x(\alpha), y(\alpha), \alpha) = 0 \text{ and } \frac{\partial f}{\partial \alpha}(x(\alpha), y(\alpha), \alpha) = 0.$$

Problem 1.9 Find the envelope of the family of lines cutting off a triangle of area $a^2/2$ from a given right angle.

Problem 1.10 Let A and B be fixed points on the arms of an angle with vertex O and suppose that on the segments OA and OB points A_1 and B_1 are chosen so that $OB_1 : B_1B = AA_1 : A_1O$. Prove that the envelope of the family of straight lines A_1B_1 is an arc of a parabola.

Fig. 1.8 Astroid

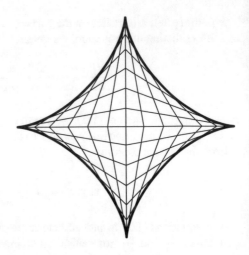

Problem 1.11 Find the envelope of the trajectories of a point particle shot from the origin with a speed of v_0 in a fixed vertical plane.

The envelope in Problem 1.11 is called the *parabola of safety*.

Problem 1.12 Prove that the envelope of the family of straight lines cutting out a segment of constant length l on the coordinate axes (see Fig. 1.8) is given by the equation

$$x^{2/3} + y^{2/3} = l^{2/3}.$$

The envelope in Problem 1.12 is called an *astroid*.

An astroid is a representative of the important class of curves called hypocycloids and epicycloids. A *hypocycloid* is the trajectory of a marked point on a circle of radius r which rolls on an immovable circle of radius R, remaining inside the immovable circle, and an *epicycloid* is the trajectory of a marked point on a circle of radius r which rolls on an immovable circle of radius R, remaining outside this circle.

First, we derive a parametric representation of a hypocycloid, i.e., the trajectory of a marked point on a circle of radius r rolling inside a circle of radius R. This motion of the marked point can be represented as the revolution of the center of the smaller circle along the circle of radius $r_1 = R - r$ at an angular speed of ω_1 and the rotation of the smaller circle of radius $r_2 = r$ at an angular speed of ω_2. The quantities ω_1 and ω_2 have opposite signs and are related by $(r_1 + r_2)\omega_1 = r_2(-\omega_2 + \omega_1)$, which expresses the equality of the arcs of the immovable circle of radius $R = r_1 + r_2$ and the moving circle of radius $r = r_2$ (it rolls without slipping). After reduction this equation takes the form $r_1\omega_1 = -r_2\omega_2$.

The trajectory of the marked point is given parametrically by

$$x = r_1 \cos \omega_1 t + r_2 \cos \omega_2 t,$$
$$y = r_1 \sin \omega_1 t + r_2 \sin \omega_2 t.$$

We set $\omega_1 = 1$ and substitute $r_1 = R - r$, $r_2 = r$, and $\omega_2 = -\frac{r_1}{r_2}\omega_1 = -\frac{R-r}{r}$. As a result, we obtain the following parametric representation of a hypocycloid:

$$x = (R - r) \cos t + r \cos \left(\frac{R - r}{r} t \right),$$

$$y = (R - r) \sin t - r \sin \left(\frac{R - r}{r} t \right).$$

In the same way we obtain a parametric representation of the trajectory in the case where the circle of radius r rolls on the circle of radius R outside. Then the motion of the marked point can be represented as the revolution of the center of the circle of radius $r_2 = r$ along the circle of radius $r_1 = R + r$ at an angular speed of ω_1 and the rotation of the circle of radius $r_2 = r$ at an angular speed of ω_2. The quantities ω_1 and ω_2 have the same sign and are related by $(r_1 - r_2)\omega_1 = r_2(\omega_2 - \omega_1)$. After reduction this equation takes the form $r_1\omega_1 = r_2\omega_2$. As a result, we obtain the following parametric representation of an epicycloid:

$$x = (R + r) \cos t + r \cos \left(\frac{R + r}{r} t \right),$$

$$y = (R + r) \sin t + r \sin \left(\frac{R + r}{r} t \right).$$

In both cases, at $t = 0$, for the marked point we took the point corresponding to the zero angle on both circles. For the hypocycloid, this is the point of tangency of the circles, while for the epicycloid, this is the point farthest from the immovable circle. For the epicycloid, we might also take the point of tangency for the initial point. In that case, the parametric representation of the epicycloid would have the form

$$x = (R + r) \cos t - r \cos \left(\frac{R + r}{r} t \right),$$

$$y = (R + r) \sin t - r \sin \left(\frac{R + r}{r} t \right).$$

If the number r/R is rational, then the corresponding hypo- and epicycloids are closed curves with finitely many singular points (cusps). Some of them have special names. For example, an epicycloid with $R = r$ (one cusp) is called a *cardioid* (because its shape resembles a heart), and an epicycloid with $R = 2r$ (two cusps) is called a *nephroid* (because it resembles a kidney). Problem 1.13 shows that a hypocycloid with four cusps is an astroid.

Problem 1.13 Prove that an astroid is the trajectory of a marked point on a circle of radius $1/4$ which rolls within an immovable circle of radius 1.

Problem 1.14

(a) Choose a number $k \neq 1$ and consider the family of straight lines throught the pairs of points $e^{i\varphi}$ and $e^{ik\varphi}$. Prove that the envelope of this family of lines is a hypo- or epicycloid.
(b) For each integer $k \neq 1$, find the number of cusps of the envelope.

Problem 1.15 Prove that the envelope of the family of straight lines being the reflection of a pencil of parallel rays from a circular mirror is a nephroid (to be more precise, one half of a nephroid: see Fig. 1.9).

Problem 1.16 Consider a circle S on which a point A is chosed. From the point A rays of light emanate and are reflected by the circle. Prove that the envelope of the reflected rays is a cardioid (see Fig. 1.10).

Problem 1.17 Consider a circle S on which a point A is chosen. Prove that the envelope of the family of circles centered on S and passing through A is a cardioid.

Fig. 1.9 Nephroid

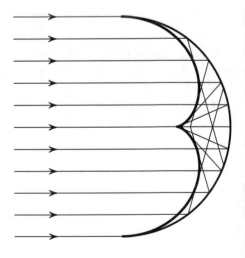

Fig. 1.10 Cardioid

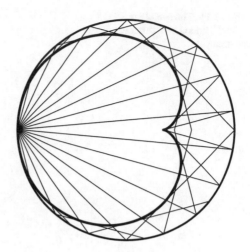

1.9 Evolute and Involute

The *evolute* $E(s)$ of a curve $\gamma(s)$ is the set of all centers of curvature (i.e., focal points) of this curve. Since the center of curvature is the limit position of the intersection points of close normals to the curve, the evolute of a curve can also be defined as the envelope of the family of all normals to this curve.

A curve $\gamma(s)$ with evolute $E(s)$ is called an *involute* of the curve $E(s)$. Each curve has only one evolute but infinitely many involutes. To construct one of the involutes of a curve $E(s)$, we must consider the family of tangents to this curve, choose any point on one of the tangents, and construct its trajectory orthogonal to the tangents.

The extrema of the curvature of a curve correspond to the singular points (cusps) of its evolute (see Fig. 1.11).

The evolute $E(s)$ of a curve $\gamma(s)$ can be defined parametrically by $E(s) = \gamma(s) + R(s)n(s)$, where $R(s)$ is the radius of curvature and $n(s)$ is a normal vector. We assume that the parameterization of the curve $\gamma(s)$ is natural. Then, according to the Frenet–Serret formula, we have $\frac{dn}{ds}(s) = -k(s)v(s) = -\frac{1}{R(s)}v(s)$; therefore,

$$\frac{dE}{ds}(s) = \frac{d\gamma}{ds}(s) + \frac{dR}{ds}(s)n(s) + R(s)\frac{dn}{ds}(s)$$

$$= v(s) + \frac{dR}{ds}(s)n(s) - v(s) = \frac{dR}{ds}(s)n(s).$$

Thus, the length of the arc of $E(s)$ between points with parameters s_1 and s_2 equals

$$\int_{s_1}^{s_2} \left| \frac{dR}{ds} \right| ds.$$

Fig. 1.11 The evolute in a neighborhood of a curvature extremum

If the curvature of $\gamma(s)$ is strictly monotone on this arc, then its length equals

$$|R(s_1) - R(s_2)| = \left| \frac{1}{k(s_1)} - \frac{1}{k(s_1)} \right|.$$

Problem 1.18 Prove that if the curvature of a curve is monotone and does not change sign on an arc of this curve, then the osculating circles corresponding to the points of the arc are disjoint.

HISTORICAL COMMENT The statement formulated in Problem 1.18 was proved by Peter Guthrie Tait (1831–1901) in 1895.

Problem 1.19

(a) Given an ellipse $\frac{x^2}{a^2} + \frac{y^2}{b^2} = 1$, find its focal point corresponding to a point $(a \cos t_0, b \sin t_0)$.
(b) Find an equation of the evolute of an ellipse.

Problem 1.20 Prove that the evolute of a parabola (t, t^2) is the curve $\left(-4t^3, \frac{1+6t^2}{2} \right)$.

A *cycloid* is the curve traced by a marked point of a circle rolling along a fixed straight line.

Problem 1.21 Prove that the evolute of a cycloid is a like cycloid (a translate of the given one).

1.10   Isoperimetric Inequality

The *isoperimetric inequality* is the following inequality between the length L of a closed curve without self-intersections and the area A of the figure which it bounds: $L^2 \geqslant 4\pi A$; the equality is attained only when the given curve is a circle. This inequality has various many-dimensional generalizations, which are often also called isoperimetric inequalities.

The First Proof of the Isoperimetric Inequality Let $\gamma(s) = (x(s), y(s))$ be a given curve with natural parameter $s \in [0, L]$; we will assume that the curve has positive orientation, i.e., is traversed in the counterclockwise direction. Let us draw two parallel support lines of γ, choosing their direction so that γ contain no segments parallel to this direction. In the strip formed by these support lines we inscribe a circle $\bar{\gamma}$ not intersecting γ (see Fig. 1.12). We assume that the coordinate system is chosen so that the origin O coincides with the center of $\bar{\gamma}$ and the Ox axis is perpendicular to the support lines. Let us parameterize the circle $\bar{\gamma}$ by the same parameter s so that each point $\bar{\gamma}(s) = (\bar{x}(s), \bar{y}(s))$ satisfies the equation $\bar{x}(s) = x(s)$ and the circle is traversed counterclockwise (see Fig. 1.12). For the circle, the parameter s is not necessarily natural.

Recall that, according to (1.1) (see p. 3), the area A of the figure bounded by a smooth curve $\gamma : [a, b] \to \mathbb{R}^2$, $\gamma(t) = (x(t), y(t))$, can be calculated by any of the three equivalent formulas

$$A = -\int_a^b yx'dt = \int_a^b xy'dt = \frac{1}{2}\int_a^b (xy' - yx')dt.$$

Thus, for the areas A and $\bar{A}$ of the figures bounded by γ and $\bar{\gamma}$, we have the expressions $A = \int_0^L xy'ds$ and $\bar{A} = -\int_0^L \bar{y}x'ds$. It is also clear that $\bar{A} = \pi r^2$, where r is the radius of the circle $\bar{\gamma}$. Therefore, taking into account the equations $\bar{x}(s) = x(s)$ and $(x')^2 + (y')^2 = 1$ (recall that s is the natural parameter), we obtain

$$A + \pi r^2 = \int_0^L (xy' - \bar{y}x')ds \leqslant \int_0^L \sqrt{(xy' - \bar{y}x')^2}ds$$

$$\leqslant \int_0^L \sqrt{(\bar{x}^2 + \bar{y}^2)((x')^2 + (y')^2)}ds = \int_0^L \sqrt{(\bar{x}^2 + \bar{y}^2)}ds = Lr.$$

Let us combine the inequality proved above and the arithmetic–geometric mean inequality:

$$\sqrt{A} \cdot \sqrt{\pi r^2} \leqslant \frac{1}{2}(A + \pi r^2) \leqslant \frac{1}{2}Lr.$$

This inequality is easily transformed into the required form $4A\pi \leqslant L^2$.

Fig. 1.12 The curve and the circle

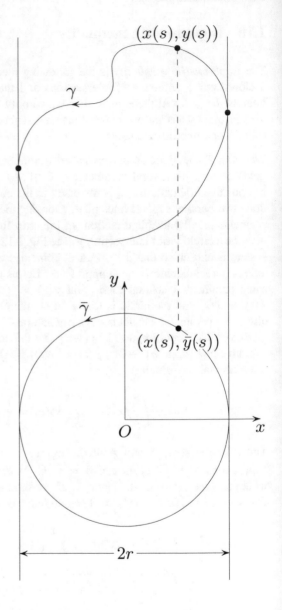

Now let us prove that the equality is attained only in the case where γ is a circle. First, the arithmetic–geometric mean inequality must turn into an equality, so that we must have $A = \pi r^2$, i.e., $L = 2\pi r$. (This implies, in particular, that the radius r does not depend on the direction of the support lines.) Secondly, we must have

$$(xy' - \bar{y}x')^2 = (\bar{x}^2 + \bar{y}^2)((x')^2 + (y')^2),$$

i.e., $(xx' + \bar{y}y')^2 = 0$. Therefore,

$$\frac{x}{y'} = -\frac{\bar{y}}{x'} = \pm\frac{\sqrt{(x^2 + \bar{y}^2)}}{\sqrt{((x')^2 + (y')^2)}} = \pm r,$$

whence $x = \pm ry'$. But the radius r does not depend on the direction of the support lines; therefore, considering the support lines perpendicular to the initial ones (and placing the origin at the intersection point of the axes of symmetry of the two perpendicular strips formed by the support lines), we obtain $y = \pm rx'$. Thus, $x^2 + y^2 = ((x')^2 + (y')^2)r^2 = r^2$.   □

The Second (Hurwitz') Proof of the Isoperimetric Inequality We need Wirtinger's inequality, one of whose simplest proofs is based on the theory of Fourier series.   □

Lemma (Wirtinger's Inequality) *Let $f(t)$ be a 2π-periodic continuously differentiable function. Suppose that $\int_0^{2\pi} f(t)dt = 0$. Then $\int_0^{2\pi}(f'(t))^2 dt \geqslant \int_0^{2\pi}(f(t))^2 dt$, and the equality holds only for $f(t) = a\cos t + b\sin t$.*

Proof From the theory of Fourier series we obtain

$$f(t) = \frac{a_0}{2} + \sum_{n=1}^{\infty}(a_n \cos nt + b_n \sin nt),$$

$$f'(t) = \sum_{n=1}^{\infty}(nb_n \cos nt - na_n \sin nt),$$

where $a_0 = \int_0^{2\pi} f(t)dt = 0$ by assumption. According to Parseval's identity,

$$\int_0^{2\pi}(f(t))^2 dt = \sum_{n=1}^{\infty}(a_n^2 + b_n^2),$$

$$\int_0^{2\pi}(f'(t))^2 dt = \sum_{n=1}^{\infty} n^2(a_n^2 + b_n^2);$$

therefore,

$$\int_0^{2\pi}(f'(t))^2 dt - \int_0^{2\pi}(f(t))^2 dt = \sum_{n=1}^{\infty}(n^2 - 1)(a_n^2 + b_n^2) \geqslant 0,$$

and the equality holds only if $a_n = b_n = 0$ for $n > 1$.   □

We proceed to the proof of the isoperimetric inequality itself. It suffices to consider the case of $L = 2\pi$. In this case, it is required to prove that $A \leqslant \pi$.

Choose a coordinate system so that the center of mass of the curve lies on the Oy axis, i.e., $\int_0^{2\pi} x(s)ds = 0$ (s is the natural parameter). We again use the area formula which we have already used in the first proof: $A = \int_0^{2\pi} xy'ds$. It follows from the identity $(x')^2 + (y')^2 = 1$ that $2\pi = \int_0^{2\pi} ((x')^2 + (y')^2)ds$. Therefore,

$$2(\pi - A) = \int_0^{2\pi} ((x')^2 + (y')^2 - 2xy')ds$$

$$= \int_0^{2\pi} ((x')^2 - x^2)ds + \int_0^{2\pi} (x - y')^2ds.$$

According to Wirtinger's inequality, the first term is nonnegative, and hence $A \leqslant \pi$, as required.

The equality holds when $x(s) = a\cos s + b\sin s$ and $x - y' = 0$, i.e., $y(s) = a\sin s - b\cos s + c$. These equations determine a circle.

1.11   Affine Unimodular Differential Geometry

The length, curvature, and normal of a curve in the plane do not change under the motions of this plane. For a smooth curve, similar notions can be defined so that they remain invariant not only under motions but also under any unimodular affine transformations of the plane. (An affine transformation is said to be *unimodular* if it preserves area.) Such a "length" cannot be defined for a straight line; therefore, the smooth curves under consideration must not only have nonvanishing velocity vector but also possess a certain additional nondegeneracy property; namely, at each point, the velocity and acceleration vectors must be linearly independent. In affine differential geometry such curves are called *nondegenerate*.

Problem 1.22 Prove that the linear independence of the velocity and acceleration vectors does not depend on the choice of the parameterization of a smooth curve. Does the vanishing of the acceleration vector depend on the parameterization?

Problem 1.23 Prove that the points at which the velocity and acceleration vectors are linearly dependent are precisely the inflection points.

In affine unimodular geometry the simplest curve (an analogue of a straight line in Euclidean geometry) is a parabola. First, we define affine length for a parabola, and then generalize this definition to an arbitrary nondegenerate curve.

The definition of affine length for a parabola is based on the following property of a parabola. Suppose that the tangents to a parabola at points α, β, and γ form a triangle ABC (see Fig. 1.13). Then the areas of the three triangles each of which is formed by two tangents and the straight line joining the points of tangency are related by

$$\sqrt[3]{A_{\alpha\beta C}} + \sqrt[3]{A_{\beta\gamma A}} = \sqrt[3]{A_{\gamma\alpha B}}.$$

Fig. 1.13 Tangents to a parabola

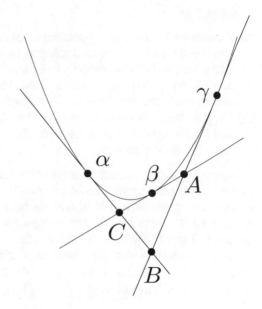

Let us prove this relation. A parabola can be defined parametrically as $r(t) = a + bt + \frac{1}{2}ct^2$, where a, b, and c are fixed vectors. Consider the triangle bounded by the line $r(t_1)r(t_2)$, the line passing through the point $r(t_1)$ in the direction of the vector $r'(t_1) = b + ct_1$, and the line passing through $r(t_2)$ in the direction of $r'(t_2) = b + ct_2$.

Recall that, given two vectors x and y in the plane, their *pseudoscalar product* $x \wedge y$ equals the oriented area of the parallelogram spanned by these vectors. Pseudoscalar product can also be defined in a different way as follows. Consider the three-dimensional space containing the given plane; then the vector product of any two vectors in this plane is perpendicular to the plane. Therefore, the vector product of vectors x and y in a given plane can be regarded as the number $x \wedge y$ (with a sign).

The oriented area of the triangle under consideration is expressed in terms of pseudoscalar product as follows:

$$\frac{1}{2} \cdot \frac{(r(t_2) - r(t_1)) \wedge r'(t_1) \cdot (r(t_2) - r(t_1)) \wedge r'(t_2)}{r'(t_1) \wedge r'(t_2)}.$$

It is easy to check that $(r(t_2) - r(t_1)) \wedge r'(t_1) = \frac{1}{2}(c \wedge b)(t_2 - t_1)^2$ and $r'(t_1) \wedge r'(t_2) = (b \wedge c)(t_2 - t_1)$. Therefore, the (signed) area in question equals $\frac{1}{8}(b \wedge c)(t_2 - t_1)^3$. Let the points α, β, and γ correspond to the parameters t_1, t_2, and t_3. Then the numbers $\sqrt[3]{A_{\alpha\beta C}}$, $\sqrt[3]{A_{\beta\gamma A}}$, and $\sqrt[3]{A_{\gamma\alpha B}}$ are proportional to $t_2 - t_1$, $t_3 - t_2$, and $t_3 - t_1$, respectively, whence the required formula.

Problem 1.24

(a) Consider a smooth curve bounding a convex figure and a not too large arc of
    this curve (that is, an arc containing no two points at which the tangents to the
    curve are parallel). We define the distance between points A and B of this arc
    as $\rho(A, B) = \sqrt[3]{A_{ABX}}$, where X is the intersection point of the tangents at A
    and B. Prove that the distance thus defined satisfies the triangle inequality.
(b) Suppose that, whenever a point B of this arc lies between two other points A
    and C, the equality $\rho(A, B) + \rho(B, C) = \rho(A, C)$ holds. Prove that then the
    arc is an arc of a parabola.

Now we specify a parameterization of the parabola $r(t) = a + bt + \frac{1}{2}ct^2$ similar
to the natural parameterization of a curve as follows. Let us define a "distance"
between two points A and B of this parabola as the cube root of the area of the
triangle ABX, where X is the intersection point of the tangents to the parabola at
the points A and B. We have already seen that, up to proportionality, such a "natural
parameter" coincides with the parameter t. Now we must choose an appropriate
proportionality coefficient. Let us replace the parameter t by $s = \lambda t$, where λ is a
constant number. Then $\frac{dr}{dt} = \frac{dr}{ds} \cdot \frac{ds}{dt} = \lambda \frac{dr}{ds}$ and $\frac{d^2r}{dt^2} = \lambda^2 \frac{d^2r}{ds^2}$. Therefore,

$$b \wedge c = \frac{dr}{dt} \wedge \frac{d^2r}{dt^2} = \lambda^3 \frac{dr}{ds} \wedge \frac{d^2r}{ds^2}.$$

Thus, setting $\lambda = \sqrt[3]{b \wedge c} = \sqrt[3]{\frac{dr}{dt} \wedge \frac{d^2r}{dt^2}}$, we obtain $\frac{dr}{ds} \wedge \frac{d^2r}{ds^2} = 1$. The condition
$\frac{dr}{ds} \wedge \frac{d^2r}{ds^2} = 1$ is invariant with respect to unimodular affine transformations; hence
the parameter s is the required analogue of the natural parameter. The absolute value
of the difference of the parameters corresponding to two points of the parabola is
called the *affine length* of the arc between these points.

Affine length can be defined in a similar way for any curve $r(t)$. We replace the
parameter t by

$$s = \int \sqrt[3]{\frac{dr}{dt} \wedge \frac{d^2r}{dt^2}}\, dt.$$

Let us check that this change of parameter yields $\frac{dr}{ds} \wedge \frac{d^2r}{ds^2} = 1$. Clearly, $\frac{dr}{ds} = \frac{dr}{dt} \cdot \frac{dt}{ds}$
and $\frac{d^2r}{ds^2} = \frac{d^2r}{dt^2}\left(\frac{dt}{ds}\right)^2 + \frac{dr}{dt} \cdot \frac{d^2t}{ds^2}$. Therefore, $\frac{dr}{ds} \wedge \frac{d^2r}{ds^2} = \frac{dr}{dt} \cdot \frac{dt}{ds} \wedge \frac{d^2r}{dt^2}\left(\frac{dt}{ds}\right)^2 + \frac{dr}{dt} \cdot$
$\frac{dt}{ds} \wedge \frac{dr}{dt} \cdot \frac{d^2t}{ds^2} = \frac{dr}{dt} \cdot \frac{dt}{ds} \wedge \frac{d^2r}{dt^2}\left(\frac{dt}{ds}\right)^2 = \frac{dr}{dt} \wedge \frac{d^2r}{dt^2}\left(\frac{dt}{ds}\right)^3$. It remains to note that
$\left(\frac{dt}{ds}\right)^3 = \left(\frac{dr}{dt} \wedge \frac{d^2r}{dt^2}\right)^{-1}$.

We could take the equality $\frac{dr}{ds} \wedge \frac{d^2r}{ds^2} = 1$ for the definition of the parameterization
s, but then we would have to prove the existence of such a parameterization (its
invariance with respect to unimodular affine transformations is obvious).

Problem 1.25 Let Δs be the affine length of an arc AB, and let X be the intersection point of the tangents at the points A and B. Prove that as A tends to B, the ratio $\sqrt[3]{A_{ABX}}/\Delta s$ tends to $1/2$.

The parameter s determining the affine length of an arc is similar to the natural parameter. An affine normal can be defined by using this parameter in precisely the same way as the usual normal is defined by using the natural parameter. Namely, the *affine normal* is the straight line parallel to the vector $r'' = \frac{d^2 r}{ds^2}$. In particular, the affine normal to a parabola at any point is parallel to the axis of this parabola.

Problem 1.26 Take a point A on a curve and draw a chord parallel to the tangent at this point. Prove that the affine normal is the tangent at A to the curve formed by the midpoints of all such chords.

The *affine curvature* of a curve is defined as follows. The parameterization s is chosen so that $r' \wedge r'' = 1$. Differentiating this equation, we obtain $r' \wedge r''' = 0$, because $r'' \wedge r'' = 0$. Thus, the vector r''' is proportional to r'. The proportionality coefficient with the opposite sign is the affine curvature $k(s)$: $r'''(s) = -k(s)r'(s)$. In particular, the affine curvature of a parabola identically vanishes, because the vector r'' is constant.

Problem 1.27 Prove that $k = -r''' \wedge r''$. (It is assumed that the parameterization is chosen so that $r' \wedge r'' = 1$.)

Problem 1.28

(a) Prove that an ellipse is a curve of constant positive affine curvature and a hyperbola is a curve of constant negative affine curvature.
(b) Prove that the affine normals to an ellipse and to a hyperbola pass through their centers.

Problem 1.29 Given a curve $y = f(x)$, prove that $ds = (f'')^{1/3} dx$ and

$$k = \frac{5}{9}(f'')^{-8/3}(f''')^2 - \frac{1}{3}(f'')^{-5/3}f''''.$$

1.12   Projective Differential Geometry

Projective length can be defined for a smooth nondegenerate curve in the projective plane. The notion of a nondegenerate curve is the same as in affine differential geometry: a curve is nondegenerate if the velocity and acceleration vectors are linearly independent at each point. But defining projective curvature requires an additional nondegeneracy condition: the point at which curvature is defined must not be sextatic (see p. 33).

Before proceeding to curves in the projective plane, we discuss curves on the projective line, because they are a nontrivial object in projective differential geometry. A smooth curve on the projective line is a map $\gamma : \mathbb{R} \rightarrow \mathbb{RP}^1$. The smoothness of the curve presumes also that the velocity vector vanishes nowhere. For a curve given in homogeneous coordinates $(x_1(t) : x_2(t))$, this means that $x_1' x_2 - x_2' x_1 \neq 0$. Indeed, we can take x_1/x_2 or x_2/x_1 for the affine coordinate, and in both cases, the derivative of the affine coordinate with respect to t is a fraction with numerator $\pm(x_1' x_2 - x_2' x_1)$.

Curves γ_1 and γ_2 are considered equivalent if $\gamma_2(t) = g(\gamma_1(t))$, where g is a projective transformation.

There is a convenient description of equivalence classes of smooth curves on the projective line in terms of second-order linear differential equations. Consider a smooth curve $\gamma(t) = (x_1(t) : x_2(t))$ on the projective line. The functions $x(t) = x_1(t)$ and $x(t) = x_2(t)$ satisfy the differential equation

$$\begin{vmatrix} x'' & x' & x \\ x_1'' & x_1' & x_1 \\ x_2'' & x_2' & x_2 \end{vmatrix} = 0.$$

By condition, $x_1' x_2 - x_2' x_1 \neq 0$; therefore, this equation can be written in the form

$$x'' + p_1 x' + p_2 x = 0.$$

The solution space of this equation is two-dimensional. Let y_1 and y_2 be two independent solutions. Then $y_i = a_i^j x_j$ for some matrix (a_i^j) with nonzero determinant. Hence the curve $(y_1(t) : y_2(t))$ is equivalent to the curve $(x_1(t) : x_2(t))$. Choosing various bases in the solution space, we obtain all curves equivalent to the given one.

Note that to the same curve different differential equations correspond. A point of the projective line does not change when both of its homogeneous coordinates are multiplied by the same nonzero number. We set $y_i(t) = \lambda(t) x_i(t)$, where λ is a nowhere vanishing function. If $y = \lambda x$, then $y' = \lambda' x + \lambda x'$ and $y'' = \lambda'' x + 2\lambda' x' + \lambda x''$. Therefore,

$$y'' + q_1 y' + q_2 y = \lambda x'' + (2\lambda' + q_1 \lambda)x' + (\lambda'' + q_1 \lambda' + q_2 \lambda)x.$$

If we choose q_1 and q_2 so that $2\lambda' + q_1 \lambda = \lambda p_1$ and $\lambda'' + q_1 \lambda' + q_2 \lambda = \lambda p_2$, i.e., $q_1 = p_1 - \frac{2\lambda'}{\lambda}$ and

$$q_2 = -\frac{\lambda''}{\lambda} + 2\frac{\lambda'^2}{\lambda^2} - p_1 \frac{\lambda'}{\lambda} + p_2,$$

then y will satisfy the equation $y'' + q_1 y' + q_2 y = 0$.

The equations $x'' + p_1 x' + p_2 x = 0$ and $y'' + q_1 y' + q_2 y = 0$ determine the same curve. To eliminate this ambiguity, we choose λ so that $p_1 - \frac{2\lambda'}{\lambda} = 0$, i.e., $q_1 = 0$. Then

$$q_2 = -\frac{\lambda''}{\lambda} + p_2 = p_2 - \frac{1}{4}p_1^2 - \frac{1}{2}p_1',$$

and the differential equation takes the form

$$y'' + Q(t)y = 0, \quad \text{where} \quad Q = p_2 - \frac{1}{4}p_1^2 - \frac{1}{2}p_1'.$$

Consider a curve of the form $(\lambda(t)f(t) : \lambda(t))$. Let us express Q for this curve in terms of f. First, we choose λ so that both functions $x_1(t) = \lambda(t)f(t)$ and $x_2(t) = \lambda(t)$ satisfy the equation $x'' + Qx = 0$. Then $(\lambda f)'' + Q(\lambda f) = 0$ and $\lambda'' + Q\lambda = 0$, whence $2\lambda' f' + \lambda f'' = 0$. Therefore, $\ln \lambda = -\frac{1}{2} \ln f' + C$, i.e., $\lambda = C(f')^{-1/2}$. Thus,

$$Q = -\frac{\lambda''}{\lambda} = \frac{1}{2}\frac{f'''}{f'} - \frac{3}{4}\left(\frac{f''}{f'}\right)^2 = \frac{1}{2}\left(\frac{f''}{f'}\right)' - \frac{1}{4}\left(\frac{f''}{f'}\right)^2.$$

The quantity $2Q$ is called the *Schwarzian derivative* of the function f and denoted by $S(f)$.

The functions $f(t)$ and $g(t) = \frac{af(t)+b}{cf(t)+d}$ determine equivalent curves on the projective line; hence $S(g) = S(f)$. This is one of the main properties of Schwarzian derivative.

Now we pass from curves on the projective line to curves in the projective plane. We begin with defining projective length and projective curvature for such curves. Consider a smooth curve $\gamma(t) = (x_1(t) : x_2(t) : x_3(t))$ in the projective plane. The functions $x_1(t)$, $x_2(t)$, and $x_3(t)$ satisfy the third-order differential equation

$$\begin{vmatrix} x''' & x'' & x' & x \\ x_1''' & x_1'' & x_1' & x_1 \\ x_2''' & x_2'' & x_2' & x_2 \\ x_3''' & x_3'' & x_3' & x_3 \end{vmatrix} = 0.$$

We assume that the coefficient of x''' does not vanish, i.e.,

$$\begin{vmatrix} x_1'' & x_1' & x_1 \\ x_2'' & x_2' & x_2 \\ x_3'' & x_3' & x_3 \end{vmatrix} \neq 0.$$

This condition means that the curve has no inflection points. Indeed, e.g., in the affine coordinates $(x_1 : x_2 : 1)$ we obtain the condition $\begin{vmatrix} x_1'' & x_1' \\ x_2'' & x_2' \end{vmatrix} \neq 0$. If the curve has no inflection points, then the corresponding equation can be written in the form

$$x''' + p_1 x'' + p_2 x' + p_3 x = 0.$$

Choosing various sets of independent solutions, we obtain all curves equivalent to the given one.

Now we must take into account the coincidence of points with proportional homogeneous coordinates. Let us replace the homogeneous coordinate x by $y = \lambda^{-1} x$. A simple calculation shows that y satisfies the equation

$$\lambda y''' + (3\lambda' + p_1\lambda) + (3\lambda'' + 2p_1\lambda' + p_2\lambda)y'' + (\lambda''' + p_1\lambda'' + p_2\lambda' + p_3\lambda)y = 0.$$

If we choose λ so that $3\lambda' + p_1\lambda = 0$, then the equation will take the form (we again write x instead of y)

$$x''' + P_2 x' + P_3 x = 0, \tag{1.6}$$

where $P_2 = p_2 - p' - \frac{1}{3}p_1^2$ and $P_3 = p_3 - \frac{1}{3}p_1'' + \frac{2}{27}p_1^3 - \frac{1}{3}p_1 p_2$.

Now consider how Eq. (1.6) transforms under the change of variables

$$(t, x) \mapsto (s = f(t), y = g(t)^{-1} x).$$

We denote differentiation with respect to s by a dot:

$$x' = g'y + gf'\dot{y},$$
$$x'' = g''y + (2g'f' + gf'')\dot{y} + g(f')^2\ddot{y},$$
$$x''' = g'''y + (3g''f' + 3g'f'' + gf''')\dot{y} + 3(g'(f')^2 + gf'f'')\ddot{y} + \dddot{y}.$$

We want the coefficient of $\ddot{y}$ to vanish; hence we must require that $g'(f')^2 + gf'f'' = 0$, i.e., $g = c/f'$. Then y will satisfy the equation

$$(f')^2\dddot{y} + (P_2 - 2S(f))\dot{y} + [P_3/f' - f''P_2/(f')^2 - f'''/(f')^2 + 2((f'')^2/(f')^3)']y = 0.$$

Thus, if f is a solution of the equation

$$S(f) = \frac{1}{2}P_2$$

(and if we again replace y by x and s by t), then we obtain

$$x''' + Rx = 0,$$

where

$$R = \left(P_3 - \frac{1}{2}P_2'\right) \Big/ (f')^3.$$

For what follows, we need the quantity

$$P = P_3 - \frac{1}{2}P_2'.$$

It is used to define the *projective arc length* element $ds = \sqrt[3]{P}dt$. For a curve parameterized by s, P identically equals 1.

Problem 1.30 Prove that if $P = 0$ at all points of some curve, then this curve is a conic.

The points of a curve at which P vanishes are called *sextatic*. The order of contact of an osculating conic at a sextatic point is higher than that at a nonsextatic point.

In a neighborhood of a point at which $P \neq 0$, we can take the projective length s for a parameter. Let us write Eq. (1.6) in the form

$$\frac{d^3}{ds^3}x + 2k\frac{d}{ds}x + hx,$$

i.e., put $P_2 = 2k$ and $P_3 = h$. For the chosen parameterization, we have $P = 1$, i.e., $1 = P_3 - \frac{1}{2}P_2' = h - k'$, whence $h = 1 + k'$. Thus, the equation has the form

$$x''' + 2kx' + (1 + k')x = 0.$$

The quantity k is called the *projective curvature* of the curve at the given point.

1.13   The Measure of the Set of Lines Intersecting a Given Curve

The topic which we will now discuss refers to integral, rather than differential, geometry, but it is very closely related to the considerations of this chapter. We will obtain a formula relating the measure of the set of straight lines intersecting a given closed curve to the length of this curve (in doing so, we have to take into account the multiplicity of each line, that is, the number of its intersection points with the curve). Before deriving the formula, we must define a measure on the set of lines in the plane.

To each straight line in the plane not passing through the origin we can assign the following two parameters p and θ. From the origin O we draw the perpendicular OP to the line; the parameter p is the length of the segment OP, and θ is the angle of rotation from the ray Ox to the ray OP. The line with given parameters p and θ is determined by the equation $x \cos \theta + y \sin \theta = p$.

Thus, we have to define a measure on sets of points in the plane with coordinates (p, θ), where $p > 0$ and $0 \leqslant \theta < 2\pi$. Of course, we are interested in an invariant measure, which is preserved by the motions of the plane. First, we discuss how invariant measures in the plane look like. We will calculate the area $A(S)$ of a set S by the formula

$$A(S) = \iint_S f(x, y) dx\, dy,$$

where $f(x, y)$ is some function. For what functions f is the area thus defined preserved under motions? Let us show that only for constants. Consider a motion $(\bar{x}, \bar{y}) \mapsto (x, y)$ given by

$$x = a + \bar{x} \cos \varphi - \bar{y} \sin \varphi, \quad y = b + \bar{x} \sin \varphi + \bar{y} \cos \varphi \qquad (1.7)$$

(any direct motion of the plane can be specified in this way). Suppose that this motion takes a figure $\bar{S}$ to a figure S. We are interested in those functions f for which

$$\iint_{\bar{S}} f(\bar{x}, \bar{y}) d\bar{x}\, d\bar{y} = \iint_S f(x, y) dx\, dy,$$

whatever the motion and the figure. By the change of variables formula for multiple integrals

$$\iint_S f(x, y) dx\, dy = \iint_{\bar{S}} f(x(\bar{x}, \bar{y}), y(\bar{x}, \bar{y})) \frac{\partial(x, y)}{\partial(\bar{x}, \bar{y})} d\bar{x}\, d\bar{y},$$

where

$$\frac{\partial(x, y)}{\partial(\bar{x}, \bar{y})} = \begin{vmatrix} \frac{\partial x}{\partial \bar{x}} & \frac{\partial x}{\partial \bar{y}} \\ \frac{\partial y}{\partial \bar{x}} & \frac{\partial y}{\partial \bar{y}} \end{vmatrix} = \begin{vmatrix} \cos \varphi & -\sin \varphi \\ \sin \varphi & \cos \varphi \end{vmatrix} = 1.$$

Thus,

$$\iint_{\bar{S}} f(x(\bar{x}, \bar{y}), y(\bar{x}, \bar{y})) \frac{\partial(x, y)}{\partial(\bar{x}, \bar{y})} d\bar{x}\, d\bar{y} = \iint_{\bar{S}} f(x(\bar{x}, \bar{y}), y(\bar{x}, \bar{y})) d\bar{x}\, d\bar{y};$$

therefore, we are interested in functions for which

$$\iint_{\bar{S}} f(\bar{x}, \bar{y}) d\bar{x}\, d\bar{y} = \iint_{\bar{S}} f(x(\bar{x}, \bar{y}), y(\bar{x}, \bar{y})) d\bar{x}\, d\bar{y}$$

whatever the figure S and the motion. Since the above equality holds for any figure, it follows that, for any motion, we have

$$f(x(\bar{x}, \bar{y}), y(\bar{x}, \bar{y})) = f(\bar{x}, \bar{y});$$

so $f(x, y) = \text{const}$, because any point of the plane can be mapped to any other point by a motion.

Now we can introduce a measure on the set of lines. The motion (1.7) takes each line $x \cos \theta + y \sin \theta = p$ to the line

$$\bar{x} \cos(\theta - \varphi) + \bar{y} \sin(\theta - \varphi) = p - a \cos \varphi - b \sin \varphi.$$

Thus, if we associate points with coordinates (p, θ) to lines, then the motion (1.7) takes each point (p, θ) to the point $(p - a \cos \varphi - b \sin \varphi, \theta - \varphi)$. It is easy to check that the Jacobian of this map equals 1 and any point corresponding to a line can be taken to any other point corresponding to a line by such a map. Therefore, it is natural to define the measure of a set S of lines as $\iint_S dp \, d\theta$. The lines passing through the origin can be ignored, because these lines form a set of measure zero.

Theorem 1.8 *Let γ be a smooth curve of length l. Then the measure of the set of lines intersecting the curve γ (with multiplicity taken into account) equals $2l$.*

Proof First, consider the case where γ is a straight line segment of length l. Since the measure is invariant with respect to motions, we can assume that the midpoint of this segment is at the origin and the segment itself lies on the Ox axis. Then the measure of the set of lines intersecting γ equals

$$\iint dp \, d\theta = \int_0^{2\pi} \left(\int_0^{(l/2)|\cos \theta|} dp \right) d\theta = \int_0^{2\pi} \frac{l}{2} |\cos \theta| \, d\theta = 2l.$$

Next, consider the case where γ is a polygonal chain of length l with i links; let l_i denote the length of the ith link. In this case, a line may intersect several links, and such a line must be counted with weight $n = n(p, \theta)$ equal to the number of intersection points. (A line intersecting the chain in a vertex intersects two links, but such lines form a set of measure zero.) Applying the formula derived above to all links of the chain and summing, we obtain

$$\iint n \, dp \, d\theta = 2 \sum_i l_i = 2l.$$

The general case is proved by passing to the limit. □

HISTORICAL COMMENT The first formulas of integral geometry were obtained by Morgan William Crofton (1826–1915).

1.14   Solutions of Problems

1.1 If the endpoints of the chord have coordinates (x_1, y_1) and (x_2, y_2), then the point M has coordinates

$$x = \frac{bx_1 + ax_2}{a + b}, \quad y = \frac{by_1 + ay_2}{a + b}.$$

Suppose that, as t varies from t_0 to t_1, each of the points $(x_1(t), y_1(t))$ and $(x_2(t), y_2(t))$ traces the whole curve γ. According to formula (1.1), the area A bounded by γ equals $\int_{t_0}^{t_1} y_1 dx_1 = \int_{t_0}^{t_1} y_2 dx_2$. Therefore,

$$\int_{t_0}^{t_1} \frac{(ab + b^2)y_1 dx_1 + (a^2 + ab)y_2 dx_2}{(a + b)^2} = \frac{(ab + b^2)A + (a^2 + ab)A}{(a + b)^2} = A.$$

Let us subtract the area A' bounded by γ' from this expression for the area A. Clearly,

$$A' = \int_{t_0}^{t_1} \frac{(by_1 + ay_2)(b\, dx_1 + a\, dx_2)}{(a + b)^2}.$$

Therefore, the desired area $A - A'$ equals

$$\frac{ab}{(a + b)^2} \int_{t_0}^{t_1} (y_2 - y_1)(dx_2 - dx_1).$$

Let us check that $\int_{t_0}^{t_1} (y_2 - y_1)(dx_2 - dx_1) = \pi(a + b)^2$. This integral equals the area bounded by the curve which is traced by the point with coordinates $(x_2 - x_1, y_2 - y_1)$. But this figure is a disk of radius $a + b$, because the distance from the point with these coordinates to the origin equals the length of the given chord.

1.2 A circle of radius R can be specified by the parametric equations $x(t) = R\cos \omega t$, $y(t) = R \sin \omega t$. Then $v(t) = (-\omega R \sin \omega t, \omega R \cos \omega t)$. For $\omega = 1/R$, we have $\|v(t)\| = 1$, i.e., t is the natural parameter. Moreover, $\frac{dv}{dt} = -\omega(\cos \omega t, \sin \omega t)$ and $k = \|\frac{dv}{dt}\| = \omega = 1/R$.

1.3 If s is the natural parameter, then $\frac{d\gamma}{dt} = \frac{d\gamma}{ds} \cdot \frac{ds}{dt}$ and $\|\frac{d\gamma}{ds}\| = 1$. Therefore, $\left(\frac{ds}{dt}\right)^2 = \left\|\frac{d\gamma}{dt}\right\|^2 = \gamma'^2$ and $\frac{d\gamma}{ds} = \gamma' \cdot (\gamma'^2)^{-1/2}$. Differentiating with respect to t, we obtain

$$\frac{d^2\gamma}{ds^2} \cdot \frac{ds}{dt} = \frac{\gamma''}{\sqrt{\gamma'^2}} - \frac{(\gamma', \gamma'')\gamma'}{(\sqrt{\gamma'^2})^3} = \frac{\gamma''\gamma'^2 - (\gamma', \gamma'')\gamma'}{(\sqrt{\gamma'^2})^3}.$$

Let us raise this equation to the second power. Taking into account the equations $\left\|\frac{d^2\gamma}{ds^2}\right\|^2 = k^2$ and $\left(\frac{ds}{dt}\right)^2 = \gamma'^2$, we obtain

$$k^2 = \frac{\gamma''^2\gamma'^2 - (\gamma', \gamma'')^2}{(\gamma'^2)^3} = \frac{(x''^2 + y''^2)(x'^2 + y'^2) - (x''x' + y''y')^2}{(x'^2 + y'^2)^3}.$$

This expression is easy to reduce to the desired form.

1.4 Consider the following parameterization of an ellipse: $x(t) = a\cos t$, $y(t) = b\sin t$. We have

$$x''y' - y''x' = -ab(\cos^2 t + \sin^2 t) = -ab,$$
$$x'^2 + y'^2 = a^2\sin^2 t + b^2\cos^2 t.$$

Therefore, according to Problem 1.3,

$$k^2 = \frac{a^2 b^2}{(a^2\sin^2 t + b^2\cos^2 t)^3}.$$

1.5 We can assume that the point of tangency is the origin and the common tangent vector has coordinates $(1, 0)$. Then the center of the circle is point $(0, R)$, so that the points (x, y) inside the circle are those for which $x^2 + (y - R)^2 < R^2$, i.e., $x^2 + y^2 < 2Ry$, and the points outside the circle are those for which $x^2 + y^2 > 2Ry$.

Let $\gamma(t) = (x(t), y(t))$, and let the point of tangency correspond to $t = 0$. Then, by assumption, $(x'(0), y'(0)) = (1, 0)$, whence $x(t) = t + \frac{1}{2}x''t^2 + O(t^3)$ and $y(t) = \frac{1}{2}y''t^2 + O(t^3)$; here $x'' = x''(0)$ and $y'' = y''(0)$. Therefore, $x^2(t) + y^2(t) = t^2 + O(t^3)$ and $2Ry(t) = y''t^2 + O(t^3)$.

Applying the formula of Problem 1.3, we obtain $k^2 = (y'')^2$. In view of the direction of the tangent vector, we have $k = y''$ rather than $k = -y''$ (for small t, the curve lies in the upper half-plane if $k > 0$ and in the lower one if $k < 0$). Thus, $2Ry(t) = kRt^2 + O(t^3)$. Therefore, if $kR > 1$, then, for small t, the inequality $2Ry(t) > x^2(t) + y^2(t)$ holds and the point $(x(t), y(t))$ is inside circle, and if $kR < 1$, then this point is outside the circle.

1.6 The velocity vector is directed along the support line. Let $v = (-\sin\varphi, \cos\varphi)$ be the unit velocity vector, and let $n = v' = (-\cos\varphi, -\sin\varphi)$ be the unit normal vector (see Fig. 1.2). Clearly, $n' = -v$.

Let s be the natural parameter. Since $\frac{dv}{d\varphi} = n$ and $\frac{dv}{ds} = kn$, where k is the curvature, it follows that $\frac{ds}{d\varphi} = \frac{1}{k} = R$, where R is the radius of curvature.

Differentiating the equation $p = -(n, r)$, we obtain $p' = (-n', r) - (n, r') = (v, r) - (n, r')$. The vector r' is parallel to v, so that $(n, r') = 0$ and hence $p' = (v, r)$. Differentiating, we obtain $p'' = (v', r) + (v, r') = (n, r) + \left(v, \frac{dr}{d\varphi}\right) =$

$-p + \left(v, \frac{dr}{ds} \cdot \frac{ds}{d\varphi}\right)$. Here $\frac{dr}{ds} = v$ and $\frac{ds}{d\varphi} = \frac{1}{k} = R$. Therefore, $p'' = -p + R$, i.e., $R = p + p''$.

1.7 We can assume that the curve is given by an equation $y = f(x)$ and $f'(0) = 0$. We are interested in the center of curvature at the point $(0, f(0))$. The normal to the curve at a point $(\varepsilon, f(\varepsilon))$ is determined by the equation $(x - \varepsilon) + f'(\varepsilon)(y - f(\varepsilon)) = 0$, and the normal at the point $(0, f(0))$ is the coordinate axis Oy. These normals intersect in $\left(0, f(\varepsilon) + \frac{\varepsilon}{f'(\varepsilon)}\right)$. Letting ε tend to 0, we obtain the point $\left(0, f(0) + \frac{1}{f''(0)}\right)$. This is precisely the center of curvature.

1.8 For a curve without self-intersections, we have $\int_a^b k \, ds = \pm 2\pi$ by virtue of *Umlaufsatz*. If the curve is convex, then $\int_a^b |k| \, ds = 2\pi = \pm \int_a^b k \, ds$. Hence either $\int_a^b (|k| + k) \, ds = 0$ or $\int_a^b (|k| - k) \, ds = 0$. But $|k| \pm k \geqslant 0$. Therefore, in the former case, we have $k = -|k|$ for all points of the curve, and in the latter, we have $k = |k|$; thus, the sign of the oriented curvature k is constant.

Conversely, suppose that the oriented curvature k of a closed curve without self-intersections has constant sign. Then $\int_a^b |k| \, ds = \pm \int_a^b k \, ds = 2\pi$, i.e., $\int_0^{2\pi} n(\varphi) d\varphi = 2\pi$. The velocity vector of a curve without self-intersections makes a full turn; therefore, $n(\varphi) \geqslant 1$ for all φ. Thus, $n(\varphi) = 1$ for almost all φ. But it the curve is nonconvex, then $n(\varphi) > 1$ for φ in a set of nonzero measure.

1.9 Let us introduce the coordinate system whose axes coincide with sides of the given right angle. The straight lines in question intersect the coordinates axes at the points $(\alpha a, 0)$ and $(0, a/\alpha)$, where $\alpha > 0$. The line corresponding to a parameter α is given by the equation $x + \alpha^2 y = \alpha a$. Two lines with parameters α_1 and α_2 intersect in the point with coordinates $\left(\frac{a\alpha_1\alpha_2}{\alpha_1 + \alpha_2}, \frac{a}{\alpha_1 + \alpha_2}\right)$. As $\alpha_1 \to \alpha$ and $\alpha_2 \to \alpha$, this point tends to $\left(\frac{a\alpha}{2}, \frac{a}{2\alpha}\right)$. Such points lie on the hyperbola $xy = a^2/4$.

1.10 We can assume that O is the origin, $A = (1, 1)$, and $B = (-1, 1)$. Then $A_1 = (1 - \alpha, 1 - \alpha)$ and $B_1 = (-\alpha, \alpha)$ for some $\alpha \in [0, 1]$. The line passing through the points A_1 and B_1 is determined by the equation $(2\alpha - 1)x + y = 2\alpha(1 - \alpha)$. It is easy to check that if (x_0, y_0) is the intersection point of the lines corresponding to parameters α_1 and α_2, then $x_0 = 1 - \alpha_1 - \alpha_2$. As $\alpha_1 \to \alpha$ and $\alpha_2 \to \alpha$, the point x_0 tends to $1 - 2\alpha$. Substituting the expression $\alpha = \frac{1-x}{2}$ into the equation of the line, we see that the envelope is determined by the equation $-x^2 + y = \frac{1}{2}(1 - x^2)$, i.e., $y = \frac{1+x^2}{2}$.

1.11 Take the coordinate system with axis Oy pointing vertically upward and axis Ox directed along the horizontal component of the velocity v_0. Then, at a moment of time t, the particle has coordinates $x(t) = v_0 t \cos \alpha$, $y(t) = v_0 t \sin \alpha - \frac{gt^2}{2}$, where α is the angle at which it was shot. For example, if the particle was shot vertically upward, then $y(t) = v_0 t - \frac{gt^2}{2}$, and hence $y(t)$ is maximum at $t = \frac{v_0}{g}$

and $y(t) = \frac{v_0^2}{2g}$. If the envelope is a parabola, then it must be defined by the equation $y = \frac{v_0^2}{2g} - kx^2$. To find k, we calculate the maximum possible coordinate of the intersection point of a trajectory with the Ox axis (shot range). If $y(t_0) = 0$ and $t_0 \neq 0$, then $t_0 = \frac{2v_0 \sin \alpha}{g}$, and $x(t_0) = \frac{2v_0^2 \sin \alpha \cos \alpha}{g} = \frac{v_0^2 \sin 2\alpha}{g}$. Therefore, the maximum range of shot equals $\frac{v_0^2}{g}$. Accordingly, for k, we obtain the equation $y = \frac{v_0^2}{2g} - k \left(\frac{v_0^2}{g} \right)^2 = 0$, whence we find $k = \frac{g}{2v_0^2}$.

Now let us prove that the parabola $y = \frac{v_0^2}{2g} - \frac{g}{2v_0^2}x^2$ is indeed the envelope of the family of trajectories. To this end, it suffices to show that any trajectory lies below this parabola and shares one point with it, i.e., for all t, the inequality

$$\frac{v_0^2}{2g} - \frac{g \cos^2 \alpha}{2}t^2 \geq v_0 t \sin \alpha - \frac{gt^2}{2}$$

holds, and it turns into an equality at some t. Replacing $\cos^2 \alpha$ by $1 - \sin^2 \alpha$, we obtain the inequality

$$\frac{v_0^2}{2g} - v_0 t \sin \alpha + \frac{g \sin^2 \alpha}{2}t^2 \geq 0,$$

i.e., $\frac{g}{2} \left(\frac{v_0}{g} - t \sin \alpha \right)^2 \geq 0$. It turns into an equality at $t = \frac{v_0}{g \sin \alpha}$.

1.12 Consider a segment with endpoints $(a_1, 0)$ and $(0, b_1)$ of length l and a segment with endpoints $(a_2, 0)$ and $(0, b_2)$ of the same length l. If $a = \frac{a_1 + a_2}{2}$ and $b = \frac{b_1 + b_2}{2}$, then $a_1 = a + \alpha$, $b_1 = b - \beta$, $a_2 = a - \alpha$, and $b_2 = b + \beta$ for some α and β. The equation $(a + \alpha)^2 + (b - \beta)^2 = (a - \alpha)^2 + (b + \beta)^2$ gives $a\alpha = b\beta$.

Let us find the coordinates of the intersection point of the segments under consideration (we assume that a_1 and a_2, as well as b_1 and b_2, have the same sign). The straight lines containing these segments are determined by the equations

$$\frac{x}{a + \alpha} + \frac{y}{b - \beta} = 1, \qquad \frac{x}{a - \alpha} + \frac{y}{b + \beta} = 1.$$

Considering the sum and the difference of these equations, we obtain

$$\frac{ax}{a^2 - \alpha^2} + \frac{by}{b^2 - \beta^2} = 1, \qquad \frac{\alpha x}{a^2 - \alpha^2} = \frac{\beta y}{b^2 - \beta^2}.$$

Taking into account the relation $a\alpha = b\beta$, we can rewrite the last equation in the form $\frac{x}{a(a^2-\alpha^2)} = \frac{y}{b(b^2-\beta^2)}$. Substituting it into the first equation, we see that

$$x = \frac{a(a^2 - \alpha^2)}{a^2 + b^2}, \quad y = \frac{b(b^2 - \beta^2)}{a^2 + b^2}.$$

We are interested in the limit as $\alpha \to 0$ and $\beta \to 0$. Clearly, the limit expressions are $x = a^3/l^2$ and $y = b^3/l^2$. Thus,

$$x^{2/3} + y^{2/3} = (a^2 + b^2)/l^{4/3} = l^2/l^{4/3} = l^{2/3}.$$

1.13 In the case where $R = 4r$, we obtain the following parametric representation of the hypocycloid:

$$x = 3r \cos t + r \cos 3t = 4r \cos^3 t,$$

$$y = 3r \sin t - r \sin 3t = 4r \sin^3 t,$$

whence $x^{2/3} + y^{2/3} = (4r)^{2/3} = R^{2/3}$.

1.14

(a) Consider $A = e^{i\varphi}$, $B = e^{ik\varphi}$, $A' = e^{i(\varphi+\alpha)}$, and $B' = e^{ik(\varphi+\alpha)}$. Let C be the limit position of the intersection point of the straight lines AB and $A'B'$ as $\alpha \to 0$. Clearly, if $k > 0$, then the point C lies on the segment AB, and if $k < 0$, then it lies outside this segment. Let us show that $AC : CB = 1 : |k|$. Indeed, $AC : CB' = \sin B' : \sin A = \pm \sin \alpha : \sin k\alpha \to 1 : \pm k$ and $CB' : CB = \sin B : \sin B' \to 1$. We conclude that each point C on the envelope has coordinates

$$\frac{e^{ik\varphi} + ke^{i\varphi}}{1+k} = \frac{1}{1+k}(\cos k\varphi + k\cos\varphi, \ \sin k\varphi + k\sin\varphi).$$

It is seen from the parametric representation of a hypo- and an epicycloid (see p. 19) that the points with such coordinates form a hypo- or epicycloid.

(b) A n s w e r: $|k - 1|$. The cusps correspond to the pairs of diametrically opposite points $e^{i\varphi}$ and $e^{ik\varphi}$, for which $e^{i\varphi} + e^{ik\varphi} = 0$. Canceling out $e^{i\varphi}$, we obtain the equation $e^{i(k-1)\varphi} = -1$. It has $|k - 1|$ solutions.

1.15 We will assume that the rays are parallel to the axis Ox and that they are reflected from the unit circle. Consider the ray falling to a point $A = e^{i\psi}$. After reflection this ray falls to a point A_1, which is obtained as follows. Let A' be the starting point of this ray of light, i.e., the point symmetric to A with respect to the axis Oy. Then A_1 is symmetric to A' with respect to the diameter AB. A simple calculation of angles shows that $A_1 = e^{i(3\psi+\pi)}$. Let $\psi = \varphi + \alpha$. Then $3\psi + \pi =$

Fig. 1.14 Cardioid

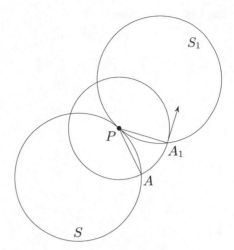

$3\varphi + 3\alpha + \pi = 3\varphi + \alpha$ for $\alpha = -\frac{\pi}{2}$. As a result, we find ourselves in the situation of Problem 1.14 with $k = 3$, which corresponds to $\frac{R+r}{r} = 3$, i.e., $R = 2r$.

1.16 We assume that the circle is unit and $A = (-1, 0)$. Then the ray falling to a point $e^{i\psi}$ falls to the point $e^{i(2\psi+\pi)}$ after reflection. Let $\psi = \varphi + \alpha$. Then $2\psi + \pi = 2\varphi + 2\alpha + \pi = 2\varphi + \alpha$ for $\alpha = -\pi$. As a result, we find ourselves in the situation of Problem 1.14 with $k = 2$, which corresponds to an epicycloid with one cusp.

1.17 Consider the circle S_1 equal to the given circle S and tangent to it at the point A. We mark the point coinciding with A on the circle S_1 and roll S_1 on the circle S, keeping the latter fixed. Suppose that, at some moment of time, the circles are tangent at a point P and the marked point of S_1 is A_1 (Fig. 1.14). Then the arcs PA and PA_1 are equal, and hence so are the segments PA and PA_1. At the moment under consideration the motion of S_1 is the rotation about the point P; therefore, the velocity of the point A_1 is perpendicular to the straight line PA_1, i.e., the velocity vector of the curve traced by the point A_1 is tangent at A_1 to the circle of radius PA centered at P. The point A_1 traces a cardioid, and this cardioid is tangent to all circles in the family under consideration.

1.18 Suppose for definiteness that the curvature radius of a curve $\gamma(s)$ is positive and decreases from a point s_1 to a point s_2. Consider the points $E(s_1)$ and $E(s_2)$ being the centers of curvature corresponding to the parameters s_1 and s_2 (see Fig. 1.15). The length of the arc of the evolute between $E(s_1)$ and $E(s_2)$ equals $R(s_1) - R(s_2)$. Thus, the distance between the centers of osculating circles is smaller than the difference of their radii. Therefore, the circle of smaller radius lies strictly inside the circle of greater radius.

Fig. 1.15 Tait's theorem

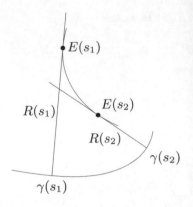

1.19

(a) Given a point (x_0, y_0), consider the function

$$F(t) = (x_0 - a\cos t)^2 + (y_0 - b\sin t)^2$$

on the ellipse. We are interested in (x_0, y_0) for which the function $F(t)$ has degenerate critical point t_0. At such a point, we have $F'(t_0) = 0$ and $F''(t_0) = 0$, i.e.,

$$x_0 a \sin t_0 - y_0 b \cos t_0 + (b^2 - a^2)\sin t_0 \cos t_0 = 0,$$

$$x_0 a \cos t_0 + y_0 b \sin t_0 + (b^2 - a^2)(\cos^2 t_0 - \sin^2 t_0) = 0.$$

Solving this system of equations, we obtain

$$x_0 = \frac{a^2 - b^2}{a}\cos^3 t_0, \quad y_0 = -\frac{a^2 - b^2}{a}\sin^3 t_0.$$

(b) It is seen from the expressions for focal points that these points lie on the curve

$$(ax)^{2/3} + (by)^{2/3} = (a^2 - b^2)^{2/3}.$$

1.20 Given a point (x_0, y_0), consider the function $F(t) = (x_0 - t)^2 + (y_0 - t^2)^2$ on the parabola. We are interested in (x_0, y_0) for which $F'(t_0) = 0$ and $F''(t_0) = 0$, i.e.,

$$4t_0^3 + 2t_0(1 - 2y_0) - 2x_0 = 0,$$

$$12t_0^2 + 2(1 - 2y_0) = 0.$$

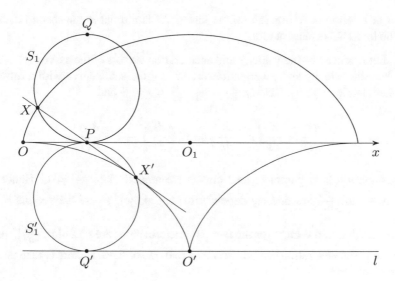

Fig. 1.16 Cycloid

Solving this system of equations, we obtain

$$y_0 = \frac{1 + 6t_0^2}{2}, \qquad x_0 = -4t_0^3.$$

1.21 Suppose that the fixed line is the Ox axis and the marked point coincides with the origin O at the initial moment. We also assume that the circle has radius 1 and remains in the upper half-plane. Suppose that, at some moment of time, the marked point is at a point X; then the rolling circle S_1 is tangent to the Ox axis at a point P (see Fig. 1.16), and the length of the segment OP equals that of the arc PX. Let PQ be the diameter of the circle S_1, and let O_1 be the point $(\pi, 0)$, i.e., the midpoint of the segment with endpoints at the first and second positions of the marked point on the Ox axis. Then the length of the segment O_1P equals that of the arc QX. Therefore, if S_1' is the circle symmetric to S_1 with respect to P and X' and Q' are its points symmetric to X and Q, then the length of the arc $Q'X'$ equals that of the segment O_1P. Consider the straight line l through Q' parallel to the given fixed line. Let O' be the projection of the point O_1 on l. Then the length of the segment O_1P equals that of the segment $Q'O'$. Therefore, the length of the arc $Q'X'$ equals that of the segment $Q'O'$, i.e., the point X' lies on the cycloid traced by a marked point on a circle of radius 1 rolling on the line l toward the initial circle (this marked point coincides with O' at the initial moment).

Let us show that the line XX' is simultaneously a normal to the first cycloid and a tangent to the second. The point P is the instant center of rotation of the point X; therefore, the line PX is normal to the first cycloid. The point Q' is the instant center of rotation of X'; therefore, the line $Q'X'$ is normal to the second cycloid, and the line PX' is tangent to it.

1.22 Let r' and r'' be the velocity and acceleration vectors of the curve $r(t)$. Let us calculate the velocity and acceleration vectors for the same curve with a different parameterization $r(t(s))$. Clearly, $\frac{dr}{ds} = \frac{dr}{dt} \cdot \frac{dt}{ds} = r' \cdot \frac{dt}{ds}$ and

$$\frac{d^2r}{ds^2} = \frac{d}{ds}\left(r' \cdot \frac{dt}{ds}\right) = r''\left(\frac{dt}{ds}\right)^2 + r' \cdot \frac{d^2t}{ds^2}.$$

By assumption both parameterized curves are smooth, i.e., $\frac{dt}{ds} \neq 0$. Hence the vectors $\frac{dr}{ds}$ and $\frac{d^2r}{ds^2}$ are linearly dependent if and only if so are the vectors r' and r''.

The acceleration vectors for the two parameterizations are r'' and $r''\left(\frac{dt}{ds}\right)^2 + r' \cdot \frac{d^2t}{ds^2}$. The parameterization can be chosen so that one of these vectors is zero and the other is nonzero.

1.23 Choose a coordinate system and a parameterization so that the curve has the form $r(t) = (t, f(t))$. Then $r' = (1, f')$ and $r'' = (0, f'')$. The vectors r' and r'' at a point $(t_0, f(t_0))$ are linearly dependent if and only if $f''(t_0) = 0$, i.e., this point is a point of inflection.

1.24 We use the notation of Fig. 1.13 (assuming that it shows an arbitrary convex arc). Let

$$\frac{\alpha C}{CB} = \frac{a}{1-a}, \qquad \frac{C\beta}{\beta A} = \frac{b}{1-b}, \qquad \frac{BA}{A\gamma} = \frac{c}{1-c}.$$

Then $A_{\alpha\beta C} = \frac{ab}{1-a} A_{ABC} = \frac{ab}{1-a} c(1-a) A_{\gamma\alpha B} = abc A_{\gamma\alpha B}$ and $A_{\beta\gamma A} = (1-a)(1-b)(1-c) A_{\gamma\alpha B}$. Thus, we must prove that

$$\sqrt[3]{abc} + \sqrt[3]{(1-a)(1-b)(1-c)} \leqslant 1.$$

This easily follows from the arithmetic–geometric mean inequality:

$$\sqrt[3]{abc} + \sqrt[3]{(1-a)(1-b)(1-c)} \leqslant \frac{a+b+c}{3} + \frac{1-a+1-b+1-c}{3} = 1.$$

(b) It is seen from the solution of problem (a) that $\rho(A, B) + \rho(B, C) = \rho(A, C)$ if and only if $a = b = c$. For us it is only important that $a = c$. Indeed, according to Problem 1.10 (see p. 17), given a fixed angle $\alpha B\gamma$, the envelope of the corresponding family of segments AC is an arc of a parabola.

1.25 Let $A = r(s)$, and let $B = r(s + \Delta s) = r + r'\Delta s + r''\frac{(\Delta s)^2}{2} + \ldots$, where $r = r(s), r' = r'(s)$, and $r'' = r''(s)$. Then $X = r + \lambda r' = r(s + \Delta s) + \mu r'(s + \Delta s)$

for some numbers λ and μ. Clearly, $r'(s+\Delta s) = r'+r''\Delta s+\dots$. Therefore, $\lambda r' = r'\Delta s+r''\frac{(\Delta s)^2}{2}+\mu r'+\mu r''\Delta s+\dots$. Taking into account the linear independence of the vectors r' and r'', we obtain $\lambda \approx \mu+\Delta s$ and $\mu+\frac{\Delta s}{2} \approx 0$. Thus, $\lambda \approx \frac{\Delta s}{2}$, and hence the area of the triangle ABX approximately equals the area of the triangle with vertices r, $r+r'\frac{\Delta s}{2}$, and $r+r'\Delta s+r''\frac{(\Delta s)^2}{2}$. The area of this triangle is

$$\frac{1}{2}r'\frac{\Delta s}{2} \wedge \left(r'\Delta s+r''\frac{(\Delta s)^2}{2}\right) = r' \wedge r''\frac{(\Delta s)^3}{8} = \frac{(\Delta s)^3}{8},$$

because the parameter s is chosen so that $r' \wedge r'' = 1$.

1.26   Choose the parameter s determining the affine arc length so that $A = r(0)$. Let Oxy be the coordinate system with origin at A with respect to which $r'(0) = (1, 0)$ and $r'' = (0, 1)$. Then in a neighborhood of A the given curve is defined by the equations $x(s) = s + a_1s^3 + a_2s^4 + \dots$ and $y(s) = \frac{1}{2}s^2 + b_1s^3 + b_2s^4 + \dots$. The first equation gives $s = x+c_1x^3+c_2x^4+\dots$; therefore, $y = \frac{1}{2}x^2+d_1x^3+d_2x^4+\dots$. Let us compare the curve $y = \frac{1}{2}x^2 + d_1x^3 + d_2x^4 + \dots$ with the curve $y = \frac{1}{2}x^2$. The distance between the points of these curves in the vertical direction is of order x^3; hence the distance in the horizontal direction is of order x^3 as well. Therefore, the curve formed by the midpoints of the chords under consideration deviates from the Oy axis by a distance of order $x^3 \sim y^{3/2}$. Such a curve is tangent to the Oy axis, which is precisely the affine normal, because it is directed along r''.

1.27   The pseudoscalar multiplication of both sides of the equation $r''' = -kr'$ by r'' on the right yields $r''' \wedge r'' = -kr' \wedge r'' = -k$.

1.28

(a)   An ellipse can be represented in the parametric form $r(t) = a\cos(\lambda t) + b\sin(\lambda t)+c$, where a, b, and c are constant vectors and λ is an arbitrary nonzero number. A simple calculation shows that $r' \wedge r'' = \lambda^3 a \wedge b$ and $r''' = -\lambda^2 r'$. Choosing λ^2 so that $\lambda^3 a \wedge b = 1$, we obtain the required parameterization. Moreover, $k(s) = \lambda^2$ is a positive constant.

   For a hyperbola, we consider the parameterization $r(t) = a\cosh(\lambda t) + b\sinh(\lambda t) + c$, choose λ so that $\lambda^3 a \wedge b = 1$, and obtain $k(s) = -\lambda^2$.

(b)   The midpoints of the chords parallel to a diameter in an ellipse or a hyperbola lie on the conjugate diameter. Applying Problem 1.26, we obtain the required result.

1.29   First, we take x for a parameter. Then $r(x) = (x, f(x))$, $r'(x) = (1, f'(x))$, and $r''(x) = (0, f''(x))$; therefore, $r' \wedge r'' = f''(x)$. By definition $ds = (r' \wedge r'')^{1/3}dx$, whence $ds = (f'')^{1/3}dx$.

   To calculate k, we must pass to the parameterization by s. The formula $\frac{d^3r}{ds^3} = -k\frac{dr}{ds}$ shows that it suffices to calculate the first coordinate of the vectors $\frac{d^3r}{ds^3}$ and $\frac{dr}{ds}$. Clearly, $\frac{dr}{ds} = r'\frac{dx}{ds} = (f'')^{-1/3}r'$; therefore, the first coordinate of $\frac{dr}{ds}$ is $(f'')^{-1/3}$.

Next, the first coordinate of $\frac{d^2r}{ds^2}$ equals

$$\frac{d}{ds}((f'')^{-1/3}) = (f'')^{-1/3}((f'')^{-1/3})'$$

$$= -\frac{1}{3}(f'')^{-1/3}(f'')^{-4/3}f''' = -\frac{1}{3}(f'')^{-5/3}f'''.$$

Finally, the first coordinate of the vector $\frac{d^3r}{ds^3}$ equals

$$\frac{d}{ds}(-\frac{1}{3}(f'')^{-5/3}f''') = -\frac{1}{3}(f'')^{-1/3}((f'')^{-5/3}f''')'$$

$$= -\frac{1}{3}(f'')^{-1/3}[-\frac{5}{3}(f'')^{-8/3}(f''')^2 + (f'')^{-5/3}f''''].$$

Dividing the first coordinate of $\frac{d^3r}{ds^3}$ by the first coordinate of $\frac{dr}{ds}$, we obtain the required result.

1.30 For a certain parameterization t of the curve under consideration, the curve is given by the equation $x''' = 0$. This equation has the three independent solutions 1, t, and t^2. Hence the curve is projectively equivalent to the conic $(1 : t : t^2)$.

Chapter 2
Curves in Space

The next simplest object of differential geometry after a plane curve is a curve in space, three- or many-dimensional. In addition to curvature, a curve in three-dimensional space has one more characteristic, torsion. A space curve lies in one plane if and only if its torsion identically vanishes. A curve in n-dimensional space is characterized by numbers $\varkappa_1, \ldots, \varkappa_{n-1}$, which generalize curvature and torsion. Again, a curve lies in one hyperplane if and only if $\varkappa_{n-1} = 0$ at all points of this curve.

2.1 Curvature and Torsion: The Frenet–Serret Formulas

Let $\gamma : [a, b] \to \mathbb{R}^3$ be a smooth curve, i.e., $\gamma(t) = \big(x(t), y(t), z(t)\big)$, where x, y, and z are smooth functions and $v(t) = \frac{d\gamma(t)}{dt} \neq 0$ for all $t \in [a, b]$. It is convenient to replace the parameter t by the *natural parameter* $s = s(t) = \int_0^t \|v(\tau)\| \, d\tau$, for which $\|v(s)\| = \left\| \frac{d\gamma}{ds} \right\| = 1$.

Suppose that $\frac{dv}{ds} \neq 0$ at each point of γ. Then to each point of γ we can assign the orthonormal frame $e_1(s)$, $e_2(s)$, $e_3(s)$ defined as follows. First, $e_1(s) = v(s)$. The vector $e_2(s)$ is uniquely determined by the condition $\frac{dv}{ds}(s) = k(s)e_2(s)$, where $k(s) > 0$. The head of the vector $e_1(s)$ moves on the unit sphere, so that $e_2(s)$ is orthogonal to $e_1(s)$. The pair e_1, e_2 of orthogonal vectors can be extended in a unique way to a positively oriented orthonormal basis by adding a third vector, and e_3 is this third vector. The frame thus obtained is called the *Frenet–Serret frame*. The vector e_2 is called the *principal normal* to the curve.

In what follows, we usually consider *regular space curves* γ, which satisfy the condition $\frac{dv}{ds} \neq 0$ at each point, because the Frenet–Serret frame exists only for regular curves.

© The Author(s), under exclusive license to Springer Nature Switzerland AG 2022
V. V. Prasolov, *Differential Geometry*, Moscow Lectures 8,
https://doi.org/10.1007/978-3-030-92249-8_2

Clearly, $\begin{pmatrix} e_1(s) \\ e_2(s) \\ e_3(s) \end{pmatrix} = A(s) \begin{pmatrix} e_1(s_0) \\ e_2(s_0) \\ e_3(s_0) \end{pmatrix}$, where $A(s)$ is an orthogonal matrix. Let us

show that the matrix $A'(s_0) = \frac{dA}{ds}(s_0)$ is skew-symmetric. Indeed, $A(s)A(s)^T = I$ is the identity matrix. Therefore, $A'(s)A(s)^T + A(s)A'(s)^T = 0$. Moreover, $A(s_0) = I$, whence $A'(s_0) + A'(s_0)^T = 0$. Thus,

$$\begin{pmatrix} e_1' \\ e_2' \\ e_3' \end{pmatrix} = \begin{pmatrix} 0 & a_{12} & a_{13} \\ -a_{12} & 0 & a_{23} \\ -a_{13} & -a_{23} & 0 \end{pmatrix} \begin{pmatrix} e_1 \\ e_2 \\ e_3 \end{pmatrix}. \tag{2.1}$$

But in view of the choice of the vector e_2, we have $e_1' = ke_2$. Therefore, $a_{12} = k$ and $a_{13} = 0$. The number k is called the *curvature* of the given space curve, and the number $\varkappa = a_{23}$ is called the *torsion* of this curve. Curvature is defined for any smooth curves, while torsion is defined only for regular curves.

HISTORICAL COMMENT The curvature and torsion of a space curve were introduced by Michel Ange Lancret in his 1806 paper *Memoir on Curves with Double Curvature*. He defined them in terms of infinitesimal angles of rotation of the normal and osculating planes. In the nineteenth century the name 'curves of double curvature' was used for space curves.

Formulas (2.1) can be written in the form

$$e_1' = ke_2,$$

$$e_2' = -ke_1 + \varkappa e_3,$$

$$e_3' = -\varkappa e_2.$$

These formulas are known as the *Frenet–Serret formulas*.

HISTORICAL COMMENT Recall that Serret and Frenet derived these formulas for curves in three-dimensional space almost simultaneously.

Problem 2.1 Suppose that all normal planes to a curve $\gamma(s)$ pass through a fixed point $x_0 \in \mathbb{R}^3$. Prove that this curve lies on a fixed sphere centered at x_0.

Problem 2.2 Calculate, up to sign, the curvature and the torsion of the *helix* $\gamma(t) = (R\cos t, R\sin t, ht)$.

Problem 2.3

(a) Prove that if the curvature of a smooth curve identically vanishes, then this curve lies on a straight line.
(b) Prove that if the torsion of a regular curve identically vanishes, then this curve lies in a plane.

Problem 2.4 Prove that the curvature of a curve $\gamma(t) = \big(x(t), y(t), z(t)\big)$ with any parameterization t is calculated by the formula

$$k^2 = \frac{(\gamma' \times \gamma'')^2}{(\gamma'^2)^3}.$$

Problem 2.5

(a) Prove that the torsion of a curve $\gamma(s) = \big(x(s), y(s), z(s)\big)$ with natural parameterization s is calculated by the formula

$$\varkappa = \pm \frac{(\gamma', \gamma'', \gamma''')}{k^2},$$

where $(\gamma', \gamma'', \gamma''') = (\gamma', \gamma'' \times \gamma''')$ is the scalar triple product of three vectors.

(b) Prove that the torsion of a curve $\gamma(t) = \big(x(t), y(t), z(t)\big)$ with any parameterization t is calculated by the formula

$$\varkappa = \pm \frac{(\gamma', \gamma'', \gamma''')}{(\gamma' \times \gamma'')^2}.$$

Problem 2.6 Let $\gamma(s)$ be a curve with natural parameterization on the sphere of radius R centered at x_0. Prove that $k \geqslant 1/R$ and if $\varkappa \neq 0$, then $\gamma(s) = x_0 - \rho e_2 - \rho'\sigma e_3$, where $\rho = 1/k$ and $\sigma = 1/\varkappa$. (In particular, $R^2 = \rho^2 + (\rho'\sigma)^2$.)

Problem 2.7 Let $\gamma(s)$ be a curve with natural parameterization for which $\rho = 1/k$ and $\sigma = 1/\varkappa$ are defined. Prove that this curve lies entirely on a fixed sphere if and only if

$$\frac{\rho}{\sigma} + \frac{d}{ds}\left(\sigma \frac{d\rho}{ds}\right) = 0.$$

Problem 2.8 A curve lies on a sphere and has constant curvature. Prove that this curve is a circle.

In Problem 2.7 the torsion $\varkappa$ is required to be nonvanishing. The following two problems give criteria for a curve to lie on a sphere which do not assume that $\varkappa \neq 0$. Recall that, according to Problem 2.6, the curvature k of a curve on a sphere does not vanish.

Problem 2.9 [Wo2] Let $\gamma(s)$, $s \in [0, l]$, be a smooth curve (with natural parameterization) whose curvature nowhere vanishes. Prove that this curve lies on a fixed sphere if and only if there exists a smooth function $f(s)$, $s \in [0, l]$, for which

$$f\varkappa = \frac{d}{ds}\left(\frac{1}{k}\right) \quad \text{and} \quad \frac{df}{ds} + \frac{\varkappa}{k} = 0.$$

Prove also that the radius of the sphere in this case equals $\sqrt{\frac{1}{k^2} + f^2}$.

Problem 2.10 ([Br1, Wo2]) Prove that a smooth curve $\gamma(s)$, $s \in [0, l]$, with natural parameterization lies on a sphere if and only if

$$\left(A \cos \int_0^s \varkappa \, ds + B \sin \int_0^s \varkappa \, ds \right) k(s) = 1$$

for some constants A and B, and the radius of the sphere then equals $\sqrt{A^2 + B^2}$.

A curve $\gamma(s)$ is said to be of *constant slope* if there exists a fixed unit vector e for which $(e, e_1) = \text{const}$; here $e_1 = \frac{d\gamma}{ds}$ is the unit velocity vector.

Problem 2.11 Prove that a curve with $k \neq 0$ is of constant slope if and only if $\varkappa = ck$ for some constant c.

Let $\gamma(s)$ be a smooth curve, where s is the natural parameter. Suppose that $k(0) \neq 0$. Then at the point $\gamma(0)$ the Frenet–Serret frame e_1, e_2, e_3 is uniquely defined. We have

$$\gamma'(0) = e_1,$$
$$\gamma''(0) = ke_2,$$
$$\gamma'''(0) = k'e_2 + ke_2' = -k^2e_1 + k'e_2 - k\varkappa e_3,$$

$$\cdots \quad \cdots$$

Calculating the derivatives of γ can be continued, because the derivatives of the vectors e_1, e_2, and e_3 can be expressed in terms of the functions $k = k(s)$ and $\varkappa = \varkappa(s)$.

If a curve $\gamma(s) = \left(x^1(s), x^2(s), x^3(s) \right)$ is defined by not merely smooth but analytic functions x^1, x^2, and x^3, then the above argument shows that the functions $k(s)$ and $\varkappa(s)$ determine the curve $\gamma(s)$ uniquely up to a motion of Euclidean space. The solution of Problem 2.12 shows that this is true not only for analytic functions.

Problem 2.12 Prove that, for any two differentiable functions $k(s)$ and $\varkappa(s)$, where $k(s) > 0$, there exists a unique (up to a motion) smooth curve $\gamma(s) = \left(x^1(s), x^2(s), x^3(s) \right)$ with curvature $k(s)$ and torsion $\varkappa(s)$.

Suppose in addition that $\varkappa(0) \neq 0$ and choose coordinates x^1, x^2, x^3 associated with the Frenet–Serret frame at the point $\gamma(0)$. It follows from the equations for

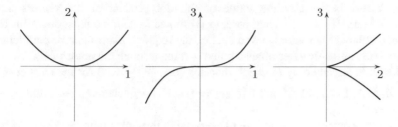

Fig. 2.1 The projections of the curve

the derivatives of $\gamma(s)$ at zero (see p. 50) that, in these coordinates, $\gamma(s) = \left(x^1(s), x^2(s), x^3(s)\right)$, where

$$x^1(s) = s - \frac{k^2 s^3}{6} + \ldots ,$$

$$x^2(s) = \frac{1}{2}ks^2 + \frac{k' s^3}{6} + \ldots ,$$

$$x^3(s) = -\frac{k\varkappa s^3}{6} + \ldots .$$

Thus, the projection of the curve $\gamma(s)$ on the coordinate planes associated with the Frenet–Serret frame have the form shown in Fig. 2.1.

2.2 An Osculating Plane

Let $\gamma(s)$ be a regular space curve with natural parameterization. Take a plane passing through the point $\gamma(s_0)$. We denote the distance between the points $\gamma(s_0)$ and $\gamma(s_0 + \Delta s)$ by $d(\Delta s)$ and the distance from the point $\gamma(s_0 + \Delta s)$ to the chosen plane by $h(\Delta s)$. Let us try to choose the plane so that the curve would move away from it as slowly as possible, or, more precisely, so as to maximize the greatest positive integer m for which $\lim_{\Delta s \to 0} \frac{h(\Delta s)}{d^m(\Delta s)} = 0$. Below we show how to choose a plane for which $m = 2$ and prove its uniqueness (essentially using the regularity of the curve).

Let n be the unit normal vector to the chosen plane. Clearly,

$$\gamma(s_0 + \Delta s) - \gamma(s_0) = \gamma'(s_0)\Delta s + \frac{\gamma''(s_0)}{2}(\Delta s)^2 + \cdots = \Delta s e_1 + \frac{k}{2}(\Delta s)^2 e_2 + \ldots .$$

Thus, $h(\Delta s)$ is the absolute value of the dot product of the vectors n and $\Delta s e_1 + \frac{k}{2}(\Delta s)^2 e_2 + \ldots$. Therefore, to minimize $h(\Delta s)$, we must choose n to be perpendicular to the vectors e_1 and e_2, i.e., the required plane must be parallel to e_1 and e_2. This uniquely determines a plane passing through the point $\gamma(s_0)$.

Now it is already easy to check that $\lim_{\Delta s \to 0} \frac{h(\Delta s)}{d^2(\Delta s)} = 0$ for the chosen plane. Indeed, $h(\Delta s) = \varepsilon_1 (\Delta s)^2$ and $d^2(\Delta s) = (\Delta s)^2 + \varepsilon_2$, where $\varepsilon_1 \to 0$ and $\varepsilon_2 \to 0$ as $\Delta s \to 0$.

Given a regular curve $\gamma(s)$, the plane passing through a point $\gamma(s_0)$ and parallel to the first two vectors e_1 and e_2 of the Frenet–Serret frame at this point is called an *osculating plane*.

An osculating plane can also be found as follows. The plane passing through a point $\gamma(s_0)$ and perpendicular to a vector n is determined by the equation $(x - \gamma(s_0), n) = 0$. Let us substitute $x = \gamma(s)$ into the left-hand side and consider the function $f(s) = (\gamma(s) - \gamma(s_0), n)$ thus obtained. It is required to choose the vector n so as to maximize the number of derivatives of $f(s)$ vanishing at s_0. The equation $f'(s_0) = 0$ means that the vector n is perpendicular to e_1, and $f''(s_0) = 0$ means that n is perpendicular to e_2.

In a similar way, given a curve $\gamma(s)$, we can construct the osculating sphere at a point $\gamma(s_0)$. But for this purpose it is required not only that the curve be regular (i.e., the curvature k be nonvanishing) but also that the torsion $\varkappa$ be nonvanishing as well. The sphere centered at a and passing through a point $\gamma(s_0)$ is determined by the equation $\|x - a\|^2 - R^2 = 0$, where $R = \|\gamma(s_0) - a\|$. We want to choose the center a of this sphere so as to maximize the number of derivatives of the function $f(s) = \|\gamma(s) - a\|^2$ vanishing at s_0 (we throw away the constant, because it vanishes under differentiation). Clearly,

$$f' = 2(\gamma - a, \gamma') = 2(\gamma - a, e_1),$$
$$f'' = 2(\gamma - a, \gamma'') + 2\|\gamma'\|^2 = 2(\gamma - a, ke_2) + 2.$$

Then, taking into account the relations $e_1 \perp e_2$ and $(ke_2)' = k'e_2 + k(e_2)' = k'e_2 - k^2 e_1 + k\varkappa e_3$, we obtain

$$f''' = 2(\gamma - a, k'e_2 - k^2 e_1 + k\varkappa e_3).$$

Let $a = \gamma(s_0) + pe_1 + qe_2 + re_3$. The condition $f'(s_0) = 0$ means that $p = 0$, and the condition $f''(s_0) = 0$ means that $-kq + 1 = 0$, i.e., $q = 1/k$. Finally, for $a = \frac{1}{k}e_2 + re_3$, the condition $f'''(s_0) = 0$ means that $\frac{k'}{k} + rk\varkappa = 0$. As a result, we obtain that the required sphere is centered at

$$\gamma(s_0) + \frac{1}{k(s_0)}e_2(s_0) - \frac{k'(s_0)}{\varkappa(s_0)k^2(s_0)}e_3(s_0),$$

and passes through the point $\gamma(s_0)$. It is called an *osculating sphere*.

2.3 Total Curvature of a Closed Curve

Let $\gamma(s)$ be a smooth closed curve in $\mathbb{R}^3$ with natural parameterization. We will assume that $s \in [0, l]$, where l is the length of γ. The numbers $\int_0^l k(s)\, ds$ and $\int_0^l \varkappa(s)\, ds$ are called, respectively, the *total curvature* and the *total torsion* of the curve γ.

To a curve γ we can assign its *velocity indicatrix* by taking the velocity vector $\frac{d\gamma}{ds}$ at each point of the curve and treating this vector as a point of the unit sphere. Thus, the velocity indicatrix is a closed curve on the unit sphere. The total curvature of a curve γ equals the length of its velocity indicatrix, because $k(s)\, ds = \|dv\|$.

Theorem 2.1 (Fenchel [Fe1]) *The total curvature of a closed curve γ is at least 2π.*

First Proof First, we prove the following auxiliary assertion.

Lemma *If any plane through the center of a unit sphere intersects a curve[1] ω lying on this sphere in at least n points, then the length of ω is at least $n\pi$.*

Proof To each point of ω we assign the set of all planes which pass through this point and the center of the sphere, and to every such plane we assign the pair of points in which the sphere intersects the normal to the plane through the center of the sphere. As a result, an infinitesimal arc $d\alpha$ will be assigned a spherical lune of angle $d\alpha$ (that is, the part of the sphere captured between two planes which make an angle of $d\alpha$). Its area equals $4d\alpha$ (the area of a spherical lune of angle π equals that of the entire sphere, i.e., 4π). Thus, the area of the figure assigned to a curve of length L equals $4L$, and the area of a domain covered m-fold is counted m times. By assumption each point of the sphere is covered at least n-fold. Therefore, $4L \geqslant 4\pi n$, i.e., $L \geqslant \pi n$.   □

Now let us prove that any plane through the center of the sphere intersects the velocity indicatrix in at least two points (or in one double point, but the set of such points has measure zero). Indeed, the curve γ can be enclosed by two planes Π_1 and Π_2 parallel to the given plane in such a way that the planes Π_1 and Π_2 are tangent to γ at points $\gamma(s_1)$ and $\gamma(s_2)$. Then the velocity vectors at these points are parallel to the given plane. This means that the velocity indicatrix intersects the given plane in $s = s_1$ and $s = s_2$.

Applying the lemma with $n = 2$, we see that the length of the indicatrix is at least 2π.   □

Second Proof Fix a positive integer n and consider the spatial n-gon with vertices $A_1 = \gamma(l/n)$, $A_2 = \gamma(2l/n)$, ..., $A_n = \gamma(l) = \gamma(0)$. Let B_i $(i = 1, \ldots, n)$ be the points in which the unit sphere intersects the translate of the ray $A_i A_{i+1}$ with origin at the center of the sphere. If n is large enough, then the points B_i and B_{i+1}

---

[1] This curve is not required to be closed or connected.

are not diametrically opposite, so that the spherical polygon $B_1 \ldots B_n$ is uniquely determined. For large n, this polygon is little different from the velocity indicatrix of γ. Therefore, it suffices to prove that the sum of the lengths of sides of the polygon $B_1 \ldots B_n$ is at least 2π. The sum of the lengths of sides of the polygon $B_1 \ldots B_n$ equals the sum of the exterior angles of the polygon $A_1 \ldots A_n$. Let φ_i be the angle of $A_1 \ldots A_n$ at the vertex A_i. We are interested in the sum $\sum(\pi - \varphi_i) = n\pi - \sum \varphi_i$.

Let us join the vertex A_n with the vertices A_2, A_3, $\ldots$, A_{n-2} by straight line segments. We obtain $n - 2$ triangles. The angles φ_1 and φ_{n-1} are angles of these triangles, and each of the remaining angles but φ_n does not exceed the sum of two angles of these triangles (the angle φ_n does not exceed the sum of several ones). Therefore, the sum of φ_i does not exceed the sum of the angles of $n - 2$ triangles, i.e., $\sum \varphi_i \leqslant (n - 2)\pi$. Thus, $\sum(\pi - \varphi_i) = n\pi - \sum \varphi_i \geqslant 2\pi$, as required. ☐

Remark The total curvature of a closed curve γ equals 2π only if γ is a plane convex curve.

HISTORICAL COMMENT Werner Fenchel (1905–1988) proved the theorem about the total curvature of a closed curve in 1929.

Problem 2.13

(a) Prove that if the length of a (connected) closed curve ω on the unit sphere is less than $2\delta \leqslant 2\pi$, then the curve lies in a spherical disk[2] of spherical diameter δ.
(b) Find a proof of Fenchel's theorem which uses assertion (a).

HISTORICAL COMMENT In his 1948 paper [Bo] Karol Borsuk (1905–1982) conjectured that the total curvature of a *knotted* curve γ is at least 4π. In 1949 this conjecture was proved independently first by István Fáry (1922–1984) [Fa2] and then by John Milnor (born 1931) [Mi2].

Theorem 2.2 (Fáry–Milnor) *The total curvature of a knotted curve in $\mathbb{R}^3$ is at least 4π.*

Proof It suffices to prove that a plane in general position passing through the center of the sphere intersects the velocity indicatrix of a knotted curve in at least four points. (After that, the lemma in the first proof of Fenchel's theorem can be applied.) A plane in general position intersects a closed curve in an even number of points; therefore, it suffices to prove that the number of intersection points is greater than two. Suppose that a curve γ is enclosed by two parallel planes Π_1 and Π_2 and these planes Π_1 and Π_2 are tangent to γ at points $\gamma(s_1)$ and $\gamma(s_2)$. Suppose also that no other plane parallel to Π_1 and Π_2 is tangent to γ. Then the points $\gamma(s_1)$ and $\gamma(s_2)$ divide the curve γ into two arcs, and as a point moves on any of these arcs, its distance to Π_1 changes monotonically. This means that the curve γ is unknotted. ☐

---

[2] A spherical disk of diameter δ is the set of points of a sphere whose spherical distance to a certain fixed point of the sphere (the center of the disk) is at most $\delta/2$.

Problem 2.14 Prove that if the maximum curvature of a closed space curve equals k, then the length l of this curve satisfies the inequality $l \geqslant 2\pi/k$.

2.4 Bertrand Curves

Curves $\gamma(s)$ and $\gamma^*(s)$ are called *Bertrand curves* if, for each s, the principal normal to the curve γ at the point $\gamma(s)$ coincides with the principal normal to γ^* at $\gamma^*(s)$. We assume that s is the natural parameter for the curve γ; for γ^*, this parameter is already not necessarily natural.

It is seen directly from the definition that $\gamma^*(s) = \gamma(s) + ae_2$. Therefore,

$$\frac{d\gamma^*}{ds} = \frac{d\gamma}{ds} + a'e_2 + a\frac{de_2}{ds}$$

$$= (1 - ak)e_1 + a'e_2 + a\varkappa e_3.$$

The tangent vector $\frac{d\gamma^*}{ds}$ must be orthogonal to the principal normal vector e_2, whence $a' = 0$, i.e., the coefficient a is constant. This means that the distance between the corresponding points of the Bertrand curves is constant.

Let s^* be the natural parameter for the curve γ^*, and let e_1^*, e_2^*, e_3^* be the Frenet–Serret frame for this curve. Then the vector e_1^* is orthogonal to $e_2^* = \pm e_2$, i.e., $e_1^* = \cos\omega\, e_1 + \sin\omega\, e_3$. Hence

$$k^* e_2^* \frac{ds^*}{ds} = \frac{de_1^*}{ds^*} \cdot \frac{ds^*}{ds} = \frac{de_1^*}{ds}$$

$$= e_2(k\cos\omega - \varkappa\sin\omega) + \frac{d\cos\omega}{ds}e_1 + \frac{d\sin\omega}{ds}e_3.$$

By condition, the normal vectors e_2^* and e_2 are collinear; therefore, the angle ω must be constant. Moreover, the vectors $\frac{d\gamma^*}{ds} = (1 - ak)e_1 + a\varkappa e_3$ and $e_1^* = \cos\omega\, e_1 + \sin\omega\, e_3$ are collinear, which implies $\begin{vmatrix} 1 - ak & a\varkappa \\ \cos\omega & \sin\omega \end{vmatrix} = 0$, i.e.,

$$ak\sin\omega + a\varkappa\cos\omega = \sin\omega. \tag{2.2}$$

If $\sin\omega = 0$ and $a \neq 0$, then $\varkappa = 0$, which corresponds to plane curves. But if $\sin\omega \neq 0$, then, setting $b = a\cot\omega$, we can rewrite Eq. (2.2) in the form

$$ak + b\varkappa = 1. \tag{2.3}$$

It is possible to reverse this argument and show that condition (2.3) is not only necessary but also sufficient for the curves γ and γ^* to be Bertrand curves.

HISTORICAL COMMENT Bertrand curves were studied by Joseph Bertrand (1822–1900) in 1850.

2.5   The Frenet–Serret Formulas in Many-Dimensional Space

We say that a curve $\gamma(s) = (x_1(s), \ldots, x_n(s))$ in n-space is *regular* if the vectors $\gamma'(s), \gamma''(s), \ldots, \gamma^{(n-1)}(s)$ are linearly independent at each point s of this curve. At every point of a regular curve the *Frenet–Serret frame* $e_1(s), \ldots, e_n(s)$ can be defined; it is constructed as follows. The vectors $e_1(s), \ldots, e_{n-1}(s)$ are obtained by applying the Gram–Schmidt orthonormalization process to the vectors $\gamma'(s)$, $\gamma''(s), \ldots, \gamma^{(n-1)}(s)$, and the vector $e_n(s)$ completes these vectors to a positively oriented orthonormal basis. Recall that the Gram–Schmidt orthonormalization process generates the vectors $e_1, \ldots, e_{n-1}$ by induction; to obtain a vector e_k, we take the vector $\lambda_1 e_1 + \cdots + \lambda_{k-1} e_{k-1} + \gamma^{(k)}$, where $\lambda_i = -(e_i, \gamma^{(k)})$, and divide it by its length. Such a choice of the coefficients λ_i ensures the orthogonality of e_k to the vectors $e_1, \ldots, e_{k-1}$.

The vector e_i is a linear combination of $\gamma', \gamma'', \ldots, \gamma^{(i)}$; therefore, the vector e_i' is a linear combination of $\gamma'', \gamma''', \ldots, \gamma^{(i+1)}$ and hence of $e_1, e_2, \ldots, e_{i+1}$.

Consider the numbers $\varkappa_i = (e_i', e_{i+1})$ for $i = 1, \ldots, n-1$. If $i \leqslant n-2$, then the number (e_i', e_{i+1}) is of the same sign as $(\gamma^{(i+1)}, e_{i+1}) > 0$, so that $\varkappa_i$ is positive (the number $\varkappa_{n-1}$ can have any sign and even be vanishing).

Theorem 2.3 *The derivatives of the vectors the Frenet–Serret basis vectors are expressed in terms of these vectors themselves as follows:*

$$\begin{pmatrix} e_1' \\ \vdots \\ e_n' \end{pmatrix} = K \begin{pmatrix} e_1 \\ \vdots \\ e_n \end{pmatrix}, \text{ where } K = \begin{pmatrix} 0 & \varkappa_1 & 0 & \ldots & 0 & 0 \\ \varkappa_1 & 0 & \varkappa_2 & \ldots & 0 & 0 \\ 0 & -\varkappa_2 & 0 & \ldots & 0 & 0 \\ & & \ldots\ldots\ldots\ldots & & \\ 0 & 0 & 0 & \ldots & 0 & \varkappa_{n-1} \\ 0 & 0 & 0 & \ldots & -\varkappa_{n-1} & 0 \end{pmatrix}.$$

Proof Let $e_i' = k_{i1}e_1 + \cdots + k_{in}e_n$. Then $(e_i', e_j) = k_{ij}$. From the identity $(e_i, e_j) = \delta_{ij}$ it follows that $(e_i', e_j) + (e_i, e_j') = 0$, i.e., $k_{ij} + k_{ji} = 0$. Thus, the matrix K is skew-symmetric. The vector e_i' is a linear combination of $e_1, \ldots, e_{i+1}$; therefore, $k_{ij} = 0$ for $j \geqslant i+2$. Since the matrix K is skew-symmetric, it follows that $k_{ij} = 0$ for $i \geqslant j+2$.   □

Corollary *A curve lies entirely in a hyperplane if and only if $\varkappa_{n-1} = 0$ at all points of this curve.*

Proof The identity $\varkappa_{n-1} \equiv 0$ is equivalent to the constancy of the vector e_n. If e_n is constant, then the vectors $e_1, \ldots, e_{n-1}$ lie in the hyperplane perpendicular to e_n. For our purposes, it is only essential that the velocity vector e_1 of the curve $\gamma(s)$ lies in this hyperplane. Indeed, this implies that γ lies in the hyperplane perpendicular to the vector e_n and passes through the initial point $\gamma(s_0)$.

Now suppose that a curve γ lies in a hyperplane. Then all vectors $e_1, \ldots, e_{n-1}$ are parallel to this hyperplane. Therefore, the vector e_n is perpendicular to it, i.e., constant. $\qquad\square$

The quantity $\varkappa_{n-1}$ is called the *torsion* of the curve γ in n-space.

Theorem 2.4 *For any positive functions $\varkappa_1(s), \ldots, \varkappa_{n-2}(s)$ and any nonvanishing function $\varkappa_{n-1}(s)$, there exists a unique (up to a motion) curve with natural parameterization in n-dimensional space for which the given functions are precisely the quantities through which the derivatives of the vectors in the Frenet–Serret basis are expressed.*

Proof Let $e_1(s), \ldots, e_n(s)$ be the Frenet–Serret frame. Consider the matrix $F(s) = (e_1(s), \ldots, e_n(s))^T$. The Frenet–Serret equations can be written in the form $F'(s) = K(s)F(s)$, where $K(s)$ is the matrix in the statement of Theorem 2.3.

Now let us construct the matrix $K(s)$ from the given functions $\varkappa_1(s), \ldots, \varkappa_{n-2}(s)$ and find a solution of the matrix differential equation $F'(s) = K(s)F(s)$ with an arbitrary initial orthogonal matrix $F(s_0)$. Let us show that all matrices $F(s)$ are then orthogonal as well. We have $(FF^T)' = F'F^T + F(F')^T = K(FF^T) + (FF^T)K^T$. Let $X = FF^T$. The matrix differential equation $X' = KX + XK^T$ with initial condition $X = I$ has a unique solution. Therefore, $X \equiv I$ is the solution of this equation, because $K + K^T = 0$. It follows from the uniqueness of the solution that $F(s)F(s)^T = I$ for all s, i.e., all matrices $F(s)$ are orthogonal.

The matrix $F(s)$ determines, in particular, the unit vector $e_1(s)$. Let $\gamma(s_0) = q_0$ be any initial point. Consider the curve $\gamma(s) = q_0 + \int_{s_0}^{s} e_1(t)dt$. We have $\gamma' = e_1$. Therefore, it follows from $F' = KF$ that $\gamma'' = e_1' = \varkappa_1 e_2$ and $\gamma''' = (\varkappa_1 e_2)' = \varkappa_1' e_2 + \varkappa_1 e_2' = \varkappa_1' e_2 + \varkappa_1(-\varkappa_1 e_1 + \varkappa_2 e_3) = (-\varkappa_1^2 e_1 + \varkappa_1' e_2) + \varkappa_1 \varkappa_2 e_3$. Similar considerations show that the vector $\gamma^{(k)}$ equals the sum of some linear combination of the vectors $e_1, \ldots, e_{k-1}$ and the vector $\varkappa_1 \varkappa_2 \ldots \varkappa_{k-1} e_k$. Note that we always have $\varkappa_1 \varkappa_2 \ldots \varkappa_{k-1} \neq 0$, and hence the vector e_k is determined uniquely. Therefore, $e_1, \ldots, e_n$ is the Serret-Frenet basis, and the given functions are precisely the quantities through which the derivatives of the Frenet–Serret basis vectors are expressed. $\qquad\square$

HISTORICAL COMMENT An analogue of the Frenet–Serret formulas for a curve in n-space was derived by Camille Jordan (1838–1922) in 1874.

2.6 Solutions of Problems

2.1 A normal plane is orthogonal to the vector e_1, which means that $\big(x_0 - \gamma(s), e_1\big) = 0$. Therefore, $\frac{d}{ds}\|x_0 - \gamma(s)\|^2 = -2\big(x_0 - \gamma(s), e_1\big) = 0$.

2.2 It is easy to check that $\left\|\frac{d\gamma}{dt}\right\| = \sqrt{R^2 + h^2}$, whence $t = \frac{s}{\sqrt{R^2+h^2}}$, where s
is the natural parameter. Let $\omega = \frac{1}{\sqrt{R^2+h^2}}$. Then $e_1 = \omega(-R \sin \omega s, R \cos \omega s, h)$
and $e_1' = R\omega^2(-\cos \omega s, -\sin \omega s, 0)$. Therefore, $k = R\omega^2 = \frac{R}{R^2+h^2}$ and $e_2 = (-\cos \omega s, -\sin \omega s, 0)$. Extending the vectors e_1 and e_2 to a positively oriented
orthonormal basis, we obtain $e_3 = (h\omega \sin \omega s, -h\omega \cos \omega s, R\omega)$. To find $\varkappa$, we
write the equation $e_2' = -ke_1 + \varkappa e_3$ for the last coordinate: $0 = -k\omega h + \varkappa R\omega$, i.e.,
$\varkappa = \frac{kh}{R} = \frac{h}{R^2+h^2}$.

2.3

(a) Let $\gamma(s)$ be a curve for which $k \equiv 0$, i.e., $\frac{de_1}{ds} = 0$. Then $\frac{d\gamma}{ds} = e_1 = \text{const}$.
    Therefore, $\gamma(s) = \gamma(0) + se_1$, i.e., the curve γ lies on the line through $\gamma(0)$
    parallel to the vector e_1.
(b) Let $\gamma(s)$ be a curve for which $\varkappa \equiv 0$. Then $e_3 = \text{const}$. The vector e_3 is
    orthogonal to $e_1 = \frac{d\gamma}{ds}$. Therefore, $\frac{d}{ds}(e_3, \gamma) = (e_3, \frac{d\gamma}{ds}) = 0$, i.e., $(e_3, \gamma) = \text{const}$. This means that the curve γ lies in a plane orthogonal to the vector e_3.

2.4 In the same way as in the solution of Problem 1.3 we obtain

$$k^2 = \frac{\gamma''^2 \gamma'^2 - (\gamma', \gamma'')^2}{(\gamma'^2)^3}.$$

Moreover, for any two vectors $a, b \in \mathbb{R}^3$, we have $a^2 b^2 - (a, b)^2 = (a \times b)^2$.

2.5

(a) The equation $e_3' = -\varkappa e_2$ shows that $|\varkappa|^2 = (e_2, e_3')^2$. Moreover, $e_2 = \frac{1}{k}\gamma''$ and
    $e_3 = \frac{1}{k}\gamma' \times \gamma''$. Hence

$$e_3' = \left(\frac{1}{k}\right)' \gamma' \times \gamma'' + \frac{1}{k}\gamma'' \times \gamma'' + \frac{1}{k}\gamma' \times \gamma'''.$$

In this expression for e_3' the first summand is orthogonal e_2 and the second is
zero. Thus, $|\varkappa|^2 = \frac{1}{k^4}(\gamma'', \gamma' \times \gamma''')^2$.
(b) Let us express the derivatives of γ with respect to s in terms of the derivatives
    of γ with respect to t:

$$\frac{d\gamma}{ds} = \frac{d\gamma}{dt} \cdot \frac{dt}{ds},$$

$$\frac{d^2\gamma}{ds^2} = \frac{d^2\gamma}{dt^2} \cdot \left(\frac{dt}{ds}\right)^2 + \frac{d\gamma}{dt} \cdot \frac{d^2t}{ds^2},$$

$$\frac{d^3\gamma}{ds^3} = \frac{d^3\gamma}{dt^3} \cdot \left(\frac{dt}{ds}\right)^3 + \dots,$$

where the dots in the last formula denote a linear combination of the vectors $\frac{d\gamma}{dt}$ and $\frac{d^2\gamma}{dt^2}$. Using the formula for the curvature of a curve with an arbitrary parameterization given in Problem 2.4, we obtain

$$\varkappa = \pm\frac{(\gamma_s', \gamma_s'', \gamma_s''')}{k^2} = \pm\frac{(\gamma_t'^2)^3}{(\gamma_t' \times \gamma_t'')} \cdot (\gamma_t', \gamma_t'', \gamma_t''') \cdot \left(\frac{dt}{ds}\right)^6.$$

We also have $(\gamma_t'^2)^3 = \left(\frac{ds}{dt}\right)^6$, because

$$\left\|\frac{d\gamma}{dt}\right\|^2 = \left\|\frac{d\gamma}{ds}\right\|^2 \cdot \left(\frac{ds}{dt}\right)^2 = \left(\frac{ds}{dt}\right)^2.$$

2.6 By condition $\|\gamma(s) - x_0\|^2 = R^2 = \text{const}$. Differentiating this equation, we obtain $2(\gamma(s) - x_0, e_1) = 0$. Therefore,

$$0 = \frac{d}{ds}(\gamma(s) - x_0, e_1) = (e_1, e_1) + (\gamma(s) - x_0, \frac{d}{ds}e_1)$$

$$= 1 + (\gamma(s) - x_0, ke_2).$$

Thus, $k(\gamma(s) - x_0, e_2) = -1$. The length of the vector $\gamma(s) - x_0$ equals R, whence $1 = k|(\gamma(s) - x_0, e_2)| \leqslant kR$, which implies the required inequality $k \geqslant 1/R$.

Let $\gamma(s) - x_0 = \lambda e_1 + \mu e_2 + \nu e_3$. Then $\lambda = (\gamma(s) - x_0, e_1) = 0$, $\mu = (\gamma(s) - x_0, e_2) = -1/k = -\rho$, and $\nu = (\gamma(s) - x_0, e_3)$. Hence

$$-\rho' = \frac{d}{ds}(\gamma(s) - x_0, e_2) = (e_1, e_2) + (\gamma(s) - x_0, -ke_1 + \varkappa e_3)$$

$$= 0 - k(\gamma(s) - x_0, e_1) + \varkappa(\gamma(s) - x_0, e_3)$$

$$= 0 + \varkappa(\gamma(s) - x_0, e_3).$$

Therefore, $\nu = (\gamma(s) - x_0, e_3) = -\rho'/\varkappa = -\rho'\sigma$.

Since the vectors e_2 and e_3 are mutually orthogonal and have unit length, it follows that $\|\gamma(s) - x_0\|^2 = \| - \rho e_2 - \rho'\sigma e_3\|^2 = \rho^2 + (\rho'\sigma)^2$.

2.7 First, suppose that the curve $\gamma(s)$ lies entirely on a sphere centered at x_0. Then, according to Problem 2.6, we have $x_0 = \gamma(s) + \rho e_2 + \rho'\sigma e_3$. Therefore,

$$0 = \frac{dx_0}{ds} = e_1 + \rho'e_2 + \rho\frac{de_2}{ds} + (\rho'\sigma)'e_3 + \rho'\sigma\frac{de_3}{ds}$$

$$= e_1 + \rho'e_2 + \rho(-ke_1 + \varkappa e_3) + (\rho'\sigma)'e_3 - \rho'\sigma\varkappa e_2$$

$$= \left(\frac{\rho}{\sigma} + (\rho'\sigma)'\right)e_3.$$

Thus, $\frac{\rho}{\sigma} + \frac{d}{ds}\left(\sigma\frac{d\rho}{ds}\right) = 0.$

Now suppose that $\frac{\rho}{\sigma} + \frac{d}{ds}\left(\sigma\frac{d\rho}{ds}\right) = 0$. Consider the point $x_0(s) = \gamma(s) + \rho e_2 + \rho'\sigma e_3$. The calculations performed above show that $\frac{dx_0}{ds} = \left(\frac{\rho}{\sigma} + (\rho'\sigma)'\right)e_3 = 0$, i.e., the point $x_0(s) = x_0$ is fixed. Moreover,

$$\frac{d}{ds}\|x_0 - \gamma(s)\|^2 = \frac{d}{ds}(\rho^2 + (\rho'\sigma)^2) = 2\rho'\sigma\left(\frac{\rho}{\sigma} + (\rho'\sigma)'\right) = 0.$$

This means that all points of the curve $\gamma(s)$ are at the same distance from x_0.

2.8 Let $\gamma(s)$ be a curve with natural parameterization on a sphere centered at x_0. We can assume that $x_0 = 0$. In the solution of Problem 2.6 we obtained the equations $(\gamma, e_2) = -\frac{1}{k}$ and $-\frac{d}{ds}\left(\frac{1}{k}\right) = \varkappa(\gamma, e_3)$. In the case where the curvature k is constant, the second equation takes the form $\varkappa(\gamma, e_3) = 0$. It follows that $\varkappa = 0$ at all points of the curve. Indeed, suppose that $\varkappa \neq 0$ at some point. Then $\varkappa \neq 0$ and hence $(e_3, \gamma) = 0$ in some neighborhood of this point; differentiating the last equation, we obtain

$$0 = (-\varkappa e_2, \gamma) + (e_3, e_1) = -\varkappa(e_2, \gamma).$$

It follows from $(\gamma, e_2) = -\frac{1}{k}$ that $(e_2, \gamma) \neq 0$, whence $\varkappa = 0$.

According to Problem 2.3 (b), a curve with identically vanishing torsion lies in some plane.

2.9 First, suppose that the curve γ lies on a sphere centered at x_0. We can assume that $x_0 = 0$. In the solution of Problem 2.6 we obtained the equations $(\gamma(s), e_2) = -\frac{1}{k}$ and $-\frac{d}{ds}\left(\frac{1}{k}\right) = \varkappa(\gamma(s), e_3)$. Let us introduce the function $f(s) = -(\gamma(s), e_3)$. The second equation is written in its terms as $f\varkappa = \frac{d}{ds}\left(\frac{1}{k}\right)$. We have

$$\frac{df}{ds} = -\left(\frac{d\gamma}{ds}, e_3\right) - \left(\gamma(s), \frac{de_3}{ds}\right) = (\gamma(s), \varkappa e_2) = -\frac{\varkappa}{k};$$

in deriving the last equality, we have used the relation $(\gamma(s), e_2) = -\frac{1}{k}$.

Now suppose that, for the curve γ, there exists a function f for which the above equations hold. We set $x(s) = \gamma(s) + \frac{1}{k}e_2 + fe_3$. Then

$$\frac{dx}{ds} = \frac{d\gamma}{ds} + \frac{d}{ds}\left(\frac{1}{k}\right)e_2 + \frac{1}{k}\frac{de_2}{ds} + \frac{df}{ds}e_3 + f\frac{de_3}{ds}$$

$$= e_1 + \frac{d}{ds}\left(\frac{1}{k}\right)e_2 + \frac{1}{k}(-ke_1 + \varkappa e_3) + \frac{df}{ds}e_3 - f\varkappa e_2 = 0.$$

Therefore, $x(s) = x_0$ is a fixed point. Next,

$$\frac{d}{ds}\|\gamma(s) - x_0\|^2 = \frac{d}{ds}\left(\frac{1}{k^2} + f^2\right) = 2\left(\frac{1}{k}\frac{d}{ds}\left(\frac{1}{k}\right) + f\frac{df}{ds}\right) = 0,$$

and hence the curve $\gamma(s)$ lies on the sphere of radius $\frac{1}{k^2} + f^2$ centered at x_0.

2.10 First, suppose that the specified identity holds for some constants A and B. Then, clearly, $k(s) \neq 0$. Let us write the identity in the form

$$A \cos \int_0^s \varkappa\, ds + B \sin \int_0^s \varkappa\, ds = \frac{1}{k(s)} \tag{2.4}$$

and differentiate it with respect to s. As a result, we obtain

$$\left(-A \sin \int_0^s \varkappa\, ds + B \cos \int_0^s \varkappa\, ds\right)\varkappa(s) = \frac{d}{ds}\left(\frac{1}{k}\right).$$

Consider the function

$$f(s) = -A \sin \int_0^s \varkappa\, ds + B \cos \int_0^s \varkappa\, ds.$$

Obviously, $f\varkappa = \frac{d}{ds}\left(\frac{1}{k}\right)$. It is also easy to check that $\frac{df}{ds} = -\frac{\varkappa}{k}$. Applying Problem 2.9, we obtain the required result.

Now suppose that a smooth curve $\gamma(s)$ lies on a sphere. Then, according to Problem 2.9, there exists a function f for which $f\varkappa = \frac{d}{ds}\left(\frac{1}{k}\right)$ and $\frac{df}{ds} = -\frac{\varkappa}{k}$. Consider the function

$$\theta(s) = \int_0^s \varkappa\, ds,$$

$$g(s) = \frac{1}{k(s)} \cos\theta(s) - f(s)\sin\theta(s), \tag{2.5}$$

$$h(s) = \frac{1}{k(s)} \sin\theta(s) + f(s)\cos\theta(s). \tag{2.6}$$

It is easy to see that $g'(s) = 0$ and $h'(s) = 0$. Thus, $g(s) = A$ and $h(s) = B$, where A and B are constants. Relations (2.5) and (2.6) can be regarded as a system of equations for k and f. Solving this system, we obtain $\frac{1}{k} = A\cos\theta + B\sin\theta$. This coincides with the required identity (2.4).

It remains to prove that if identity (2.4) holds, then the radius of the sphere containing the curve equals $\sqrt{A^2 + B^2}$. According to Problem 2.9, the squared radius of the sphere equals $\frac{1}{k^2} + f^2$. In the case under consideration, $\frac{1}{k} = A\cos\theta + B\sin\theta$ and $f = -A\sin\theta + B\cos\theta$. Therefore, $\frac{1}{k^2} + f^2 = A^2 + B^2$.

2.11 First suppose that γ is a curve of constant slope. Then $(e, e_1) = \cos\alpha$ for some fixed angle α. If $\cos\alpha = \pm 1$, then $e_1 = \pm e$. Hence γ is a straight line, and $k = 0$, which contradicts the assumption. Therefore, $\cos\alpha \neq \pm 1$. We have

$$0 = \frac{d}{ds}(e, e_1) = (e, \frac{d}{ds}e_1) = (e, ke_2) = k(e, e_2),$$

which implies $e_2 \perp e$. Thus, $e = (\cos\alpha)e_1 + (e_3, e)e_3$. Changing the sign of α if necessary, we obtain $e = (\cos\alpha)e_1 + (\sin\alpha)e_3$. Therefore,

$$0 = \frac{d}{ds}e = \cos\alpha\frac{d}{ds}e_1 + \sin\alpha\frac{d}{ds}e_3 = (\cos\alpha)ke_2 - (\sin\alpha)\varkappa e_2.$$

Thus, $\varkappa = (\cot\alpha)k$.

Now suppose that $\varkappa = ck$. If $c = 0$, then $\varkappa \equiv 0$, and hence γ is a plane curve. In this case, for e we can take the vector orthogonal to the plane of the curve. In what follows, we assume that $c \neq 0$. Choose an angle α with $\cot\alpha = c$ and consider the vector $e = (\cos\alpha)e_1 + (\sin\alpha)e_3$. This vector has the required properties, because $(e, e_1) = \cos\alpha = $ const and $\frac{d}{ds}e = (\cos\alpha)ke_2 - (\sin\alpha)\varkappa e_2 = 0$.

2.12 First, suppose given two smooth curves $\gamma(s)$ and $\bar\gamma(s)$ with the same curvature $k(s)$ and torsion $\varkappa(s)$. Let e_1, e_2, e_3 and $\bar e_1, \bar e_2, \bar e_3$ be their Frenet–Serret frames. Then $\frac{d}{ds}(e_1\bar e_1 + e_2\bar e_2 + e_3\bar e_3) = 0$, because

$$e_1'\bar e_1 + e_1\bar e_1' = k(e_2\bar e_1 + e_1\bar e_2),$$

$$e_2'\bar e_2 + e_2\bar e_2' = -k(e_1\bar e_2 + e_2\bar e_1) + \varkappa(e_3\bar e_2 + e_2\bar e_3),$$

$$e_3'\bar e_3 + e_3\bar e_3' = -\varkappa(e_2\bar e_3 + e_3\bar e_2).$$

We can assume that at $s = 0$ we have $e_1\bar e_1 + e_2\bar e_2 + e_3\bar e_3 = 3$, i.e., $e_1 = \bar e_1$, $e_2 = \bar e_2$ and $e_3 = \bar e_3$. Then these relations hold for all s. This proves the uniqueness of a curve γ with given $k(s)$ and $\varkappa(s)$.

To prove the existence of such a curve γ, consider the system of differential equations

$$\frac{du_1}{ds} = ku_2, \quad \frac{du_2}{ds} = -ku_1 + \varkappa u_3, \quad \frac{du_3}{ds} = -\varkappa u_2,$$

where $u_1(s), u_2(s), u_3(s) \in \mathbb{R}^3$. We set $U = \begin{pmatrix} u_1 \\ u_2 \\ u_3 \end{pmatrix} = \begin{pmatrix} u_{11} & u_{12} & u_{13} \\ u_{21} & u_{22} & u_{23} \\ u_{31} & u_{32} & u_{33} \end{pmatrix}$ and

$A = \begin{pmatrix} 0 & k & 0 \\ -k & 0 & \varkappa \\ 0 & -\varkappa & 0 \end{pmatrix}$. Then, we find a solution of the system $\frac{d}{ds}U = AU$ with initial condition $U(0) = I$ (the identity matrix) and put $x^i(s) = \int_0^s u_{1i}(t)\, dt$, so that

$\frac{dx^i}{ds} = u_{1i}$ and $x^i(0) = 0$. Let us check that the curve $\gamma(s) = \left(x^1(s), x^2(s), x^3(s)\right)$ has given curvature $k(s)$ and given torsion $\varkappa(s)$. Since the matrix A is skew-symmetric, we have $\frac{d}{ds}(U^T U) = \left(\frac{d}{ds}U^T\right)U + U^T\left(\frac{d}{ds}U\right) = U^T A^T U + U^T A U = U^T(-A)U + U^T A U = 0$. For $s = 0$, the matrix $U(s)$ is orthogonal, because $U^T U = I$. Therefore, for all s, we have $U(s)^T U(s) = I$, i.e., $U(s)$ is orthogonal everywhere. Thus, the matrix $U(s)$ determines the Frenet–Serret frame of the curve γ.

2.13

(a) First, note that if the length of a connected closed curve on the unit sphere is less than $2\delta \leqslant 2\pi$, then this curve contains no diametrically opposite points of a spherical circle of diameter δ. Indeed, if a connected closed curve passes through two diametrically opposite points, then these points divide it into two arcs, each of length at least δ.

Choose two points A and B on the curve ω so that they divide ω into two arcs of the same length. These points are not diametrically opposite points of the sphere; therefore, the shortest arc AB on the sphere is uniquely determined. Let M be the midpoint of this arc. We claim that the curve ω lies on the spherical disk D of diameter δ centered at M. To show this, first, note that the points A and B lie in the disk D, because they are joined by a curve of length δ and hence the length of the shortest arc of the sphere joining them is at most δ. Now suppose that at least one of the two arcs into which the points A and B divide the curve ω intersects the boundary of D. We denote this arc by ω_1. Consider the curve ω_1' symmetric to ω_1 with respect to the straight line joining M with the center of the sphere. From the curves ω_1 and ω_1' we can compose a closed connected curve. On the one hand, the length of this curve equals that of ω, i.e., is less than 2δ. On the other hand, the arc ω_1 contains a point C of the boundary circle of D, and therefore the arc ω_1' contains the point of this circle diametrically opposite to C. We have arrived at a contradiction.

(b) It follows from assertion (a) that if the length of a connected closed curve on the unit sphere is less than 2π, then this curve lies on one side of a plane passing through the center of the sphere. Precisely the same argument as in the first proof of Fenchel's theorem shows that any plane passing through the center of the sphere intersects the velocity indicatrix in at least two points. If follows that the velocity indicatrix cannot lie on one side of such a plane; therefore, the length of the velocity indicatrix is at least 2π.

2.14 According to Fenchel's theorem, we have $2\pi \leqslant \int_0^l k(s)ds$. Using the Cauchy–Bunyakovsky–Schwarz integral inequality, we see that

$$(2\pi)^2 \leqslant \left(\int_0^l k(s)ds\right)^2 \leqslant \int_0^l ds \cdot \int_0^l k^2(s)ds = l\int_0^l k^2(s)ds \leqslant l^2 k^2,$$

whence $l \geqslant 2\pi/k$.

Chapter 3
Surfaces in Space

One of the most widespread ways of specifying a surface in $\mathbb{R}^3$ is the parametric one, when a surface is given by three functions $x(u, v)$, $y(u, v)$, and $z(u, v)$. In other words, a domain in $\mathbb{R}^2$ with coordinates (u, v) is mapped to $\mathbb{R}^3$. The induced map takes the basis vectors directed along the u- and v-axes to the vectors $e_u = (x_u, y_u, z_u) = \left(\frac{\partial x}{\partial u}, \frac{\partial y}{\partial u}, \frac{\partial z}{\partial u}\right)$ and $e_v = (x_v, y_v, z_v)$ tangent to the surface. We will largely consider *smooth* surfaces, for which the functions x, y, and z are smooth and the vectors e_u and e_v are linearly independent at each point of the surface.

A particular case of the parametric specification of a surface is the graph of a function $f(x, y)$; in this case, $x = u$, $y = v$, and $z = f(u, v) = f(x, y)$. A parametrically specified smooth surface can be locally represented as the graph of some function by drawing the normal line and tangent planes at a given point of the surface and taking coordinates in the tangent plane as x and y and a coordinate on the normal line as z.

For a surface which is the graph of a function $f(x, y)$, tangent vectors are $(1, 0, f_x)$ and $(0, 1, f_y)$.

Surfaces do not always admit a global parameterization; in what follows, we will consider surfaces whose different regions are endowed with different parameterizations. Compact surfaces (without boundary) are called *closed* surfaces. In this chapter we consider surfaces in $\mathbb{R}^3$; this means that, in addition to a surface as a two-dimensional manifold, a map of this manifold to $\mathbb{R}^3$ is given. If this map is a homeomorphism onto its image, then we say that the surface is *embedded*. If the map not subject to any restrictions (except the requirement that the surface must be smooth, i.e., the vectors e_u and e_v must be linearly independent at each point of the surface), then we say that the surface is *immersed*.

With obvious rare exceptions (when the boundary of a surface is mentioned explicitly), we consider surfaces without boundary.

V. V. Prasolov, *Differential Geometry*, Moscow Lectures 8,
https://doi.org/10.1007/978-3-030-92249-8_3

3.1   The First Quadratic Form

The *first quadratic form* of a surface in $\mathbb{R}^3$ is defined on tangent spaces at the points of this surface; to each tangent vector it assigns its inner square (as a vector in $\mathbb{R}^3$).

HISTORICAL COMMENT The first quadratic form was introduced in 1822 by Carl Friedrich Gauss (1777–1855) in the solution of the problem of superposing (isometrically mapping) one surface onto another.

Consider a curve $\gamma(t) = (x(u(t), v(t)), \dots)$ on a parameterized surface. Clearly,

$$\frac{dx}{dt} = \frac{\partial x}{\partial u} \cdot \frac{du}{dt} + \frac{\partial x}{\partial v} \cdot \frac{dv}{dt},$$

so that the velocity vector of this curve is $\gamma' = u'e_u + v'e_v$. The squared length of this vector equals $Eu'^2 + 2Fu'v' + Gv'^2$, where $E = (e_u, e_u)$, $F = (e_u, e_v) = (e_v, e_u)$, and $G = (e_u, e_u)$. In this notation the first quadratic form is often written as $E\,du^2 + 2F du\,dv + G\,dv^2$. The equation $F = 0$ means that the coordinate lines $u = \text{const}$ and $v = \text{const}$ are orthogonal.

Using the first quadratic form, we can calculate the length of a curve and the area of a surface. The length of a curve can be expressed as the integral

$$\int_{t_0}^{t_1} \sqrt{Eu'^2 + 2Fu'v' + Gv'^2}\,dt = \int \sqrt{E\,du^2 + 2F du\,dv + G\,dv^2}.$$

The area element equals the area of the parallelogram with sides $e_u du$ and $e_v dv$, and the area of the parallelogram spanned by the vectors e_u and e_v equals the length of their cross product $e_u \times e_v$. Therefore, the area of a surface can be expressed as the integral

$$\int \|e_u \times e_v\|\,du\,dv.$$

Let us show that $\|e_u \times e_v\| = \sqrt{EG - F^2}$. To this end, we apply the formula

$$(a \times b, c \times d) = (a, c)(b, d) - (a, d)(b, c)$$

to $\|e_u \times e_v\|^2 = (e_u \times e_v, e_u \times e_v)$:

$$\|e_u \times e_v\|^2 = (e_u \times e_v, e_u \times e_v) = (e_u, e_u)(e_v, e_v) - (e_u, e_v)^2 = EG - F^2.$$

Thus, the area of the surface can be expressed as the integral

$$\int \sqrt{EG - F^2}\,du\,dv. \tag{3.1}$$

In differential geometry, even in the two-dimensional case, some formulas are fairly cumbersome, and it is more convenient to use index notation. For this reason, in addition to the notation $E\,du^2 + 2F\,du\,dv + G\,dv^2$, which is convenient in many situations, the notation g_{ij} for the coefficients of the first quadratic form is also used: this is the inner product of the tangent vectors e_i and e_j to the surface that correspond to the basis vectors in the plane $\mathbb{R}^2$. It is convenient to specify the first quadratic form by the coefficient matrix (g_{ij}). For example, if the surface under consideration is the graph of a function $f(x, y)$, then $e_1 = (1, 0, f_x)$ and $e_2 = (0, 1, f_y)$, and hence the matrix of the first quadratic form is $\begin{pmatrix} 1 + f_x^2 & f_x f_y \\ f_x f_y & 1 + f_y^2 \end{pmatrix}$. In this notation formula (3.1) takes the form

$$\int \sqrt{1 + f_x^2 + f_y^2}\,dx\,dy.$$

Example 3.1 For the hemisphere $z = \sqrt{1 - x^2 - y^2}$, where $x^2 + y^2 < 1$, the matrix of the first quadratic form equals

$$\frac{1}{1 - x^2 - y^2} \begin{pmatrix} 1 - y^2 & xy \\ xy & 1 - x^2 \end{pmatrix}.$$

Proof We have

$$e_1 = \left(1, 0, \frac{-x}{\sqrt{1 - x^2 - y^2}}\right) \quad \text{and} \quad e_2 = \left(0, 1, \frac{-y}{\sqrt{1 - x^2 - y^2}}\right).$$

Therefore, e.g.,

$$(e_1, e_1) = 1 + \frac{x^2}{1 - x^2 - y^2} = \frac{1 - y^2}{1 - x^2 - y^2}.$$

$\square$

The first quadratic form on a surface determines an inner product (and, thereby, a metric) in the tangent space to the surface. For this reason, it is also called a *Riemannian metric* or a *metric tensor*. In differential geometry a need for inner product not only in the tangent space but also in the dual (cotangent) space arises. There is a standard construction of inner product in the dual space, which we will now describe. Here and in the sequel, we use the following convention: when in a sum $\sum_i a^i b_i$ the index of summation occurs twice, as a subscript and as a superscript, the summation sign is omitted, i.e., the sum is written in the form $a^i b_i$.

If a vector space V is equipped with an inner product (v, w), then V can be identified with the dual space V^* by associating each vector v with the covector v^* such that $v^*(w) = (v, w)$ for all $w \in V$. After that, an inner product in V^* is defined by $(v^*, w^*) = (v, w)$.

Given a basis $e_1, \ldots, e_n$ in a space V, the corresponding basis in the dual space V^* is $\varepsilon^1, \ldots, \varepsilon^n$, where $\varepsilon^i(e_j) = \delta^i_j$ (here $\delta^i_j = \delta_{ij}$ is the Kronecker delta; we will write it in the tensor form, with sub- and superscripts). Let us show that if the inner product in V is defined by a matrix $G = (g_{ij})$ in the basis $e_1, \ldots, e_n$, then the inner product in the dual space is defined by the matrix $G^{-1} = (g^{ij})$ in the dual basis $\varepsilon^1, \ldots, \varepsilon^n$. First, note that $v^*_j = g_{ij} v^i$. Indeed, on the one hand, $v^*(w) = (v, w) = g_{ij} v^i w^j$, and on the other hand, $v^*(w) = v^*_i \varepsilon^i(w^j e_j) = v^*_i w^j \delta^i_j = v^*_j w^j$. We will represent the vector v as a row of coordinates and the covector v^* as a column of coordinates. Then $v^* = v^T G$, $(v, w) = v^T G w$, and $(v^*, w^*) = v^* X (w^*)^T$, where X is the required matrix. Since $v^* X (w^*)^T = v^T G X G^T w$, it follows that $G X G^T = G$, and hence $X = (G^T)^{-1} = G^{-1}$, because the matrix G is symmetric.

The formula $v^*_j = g_{ij} v^i$ and the similar formula $v^j = g^{ij} v^*_i$ are the coordinate representations of the correspondence between vectors and covectors when the given vector space or its dual space is equipped with inner product.

Example 3.2 For the hemisphere in Example 3.1, the matrix (g^{ij}) equals $\begin{pmatrix} 1 - x^2 & -xy \\ -xy & 1 - y^2 \end{pmatrix}$.

Problem 3.1 Calculate the matrices (g_{ij}) and (g^{ij}) for polar coordinates (r, φ) in the plane.

Problem 3.2 Calculate the matrices (g_{ij}) and (g^{ij}) for the surface of revolution $(f(u) \cos v, f(u) \sin v, u)$.

3.2   The Darboux Frame of a Curve on a Surface

Suppose that a smooth surface $S \subset \mathbb{R}^3$ is *oriented*, i.e., at each point of S a unit normal vector is given. If a curve $\gamma(s)$ lies on the surface S, then, in addition to the Frenet–Serret frame, each point of the curve can be assigned the *Darboux frame* ε_1, $\varepsilon_2, \varepsilon_3$. We assume that the parameterization of the curve is natural. Then ε_1 is, just as in the case of the Frenet–Serret frame, the tangent vector to the curve (the velocity vector), ε_3 is the normal vector to the surface S, and ε_2 is the uniquely determined vector for which $\varepsilon_1, \varepsilon_2, \varepsilon_3$ is an orthonormal positively oriented frame.

HISTORICAL COMMENT The Darboux frame was introduced by Jean Gaston Darboux (1842–1917) in his four-volume 1887–1896 lectures on the general theory of surfaces.

The motion of the Darboux frame is described by the equations

$$\varepsilon'_1 = a\varepsilon_2 + b\varepsilon_3,$$

$$\varepsilon'_2 = -a\varepsilon_1 + c\varepsilon_3,$$

$$\varepsilon'_3 = -b\varepsilon_1 - c\varepsilon_2.$$

The coefficients a, b, and c bear the following names: $a = k_g$ is the *geodesic curvature*, $b = k_n$ is the *normal curvature*, and $c = \varkappa_g$ is the *geodesic torsion*.

HISTORICAL COMMENT In 1830 Ferdinand Gotlibovich Minding (1806–1885) proved that the quantity $\frac{1}{\rho} = \frac{\cos \varphi}{R}$, where $\frac{1}{R}$ is the curvature of a curve on a surface and φ is the angle formed by the osculating plane of the curve and the tangent plane of the surface, belongs to the intrinsic geometry of the surface. Later, in 1848, Pierre Ossian Bonnet (1819–1892) called it geodesic curvature.

The vector ε_1' is the acceleration vector of a curve with natural parameterization (the derivative of the velocity vector). Therefore, the normal curvature is the normal (with respect to the surface) component of the acceleration vector, and the geodesic curvature is the component of the acceleration vector that lies the tangent plane.

If the normal vector and the curve are on the same side of the tangent plane in a neighborhood of some point, then the normal curvature of the curve is positive at this point, and if the normal vector and the curve lie on different sides, then it is negative.

Under the reversal of curve orientation k_n and $\varkappa_g$ remain intact, while k_g changes sign. Indeed, on the change of the curve orientation, the vector ε_3 remains the same, while ds, ε_1, and ε_2 change sign.

Problem 3.3 Prove that the normal curvature k_n and the geodesic torsion $\varkappa_g$ are completely determined by the surface S and the direction of the tangent to the curve.

Problem 3.3 shows that the notion of normal curvature can be defined not only for curves on a surface but also for the surface itself. The *normal curvature* of a surface at a point p in the direction of a tangent vector v is the curvature of the intersection curve of the surface with the plane through p containing the vector v and the normal n to the surface. The acceleration at p of the curve thus obtained is directed along the normal to the surface; therefore, its normal curvature is equal to its curvature. Thus, the normal curvature of the surface at p in the direction v is the normal curvature k_n of any curve on this surface whose tangent vector at p is v.

Problem 3.4 Let e_1, e_2, e_3 be the Frenet–Serret frame of a curve $\gamma(s)$ on a surface S, and let ε_1, ε_2, ε_3 be the Darboux frame. Prove that

$$k_g = k \sin \theta, \quad k_n = k \cos \theta, \quad \text{and} \quad \varkappa_g = \varkappa + \frac{d\theta}{ds},$$

where θ is the angle between the vectors e_2 and ε_3 (that is, between the principal normal to the curve and the normal to the surface).

Problem 3.5 Prove that $k_n^2 + k_g^2 = k^2$.

Problem 3.6

(a) Consider the family of planes passing through a given point of a surface and containing a given tangent vector. Let k be the curvature of the intersection curve of the surface and a normal plane (i.e., a plane containing a normal vector), and let $k(\theta)$ be the curvature of the intersection curve of the surface and the plane making an angle θ with the normal plane. Prove that $k = k(\theta)\cos\theta$.

(b) Prove that the osculating circles of the curves in the family obtained above fill a sphere.

HISTORICAL COMMENT Problems 3.3–3.6 are related to Meusnier's theorem. In 1776 Jean-Baptiste Meusnier (1754–1893) announced the theorem that all curves with common tangent on a surface have the same normal curvature and the osculating circles of these curves fill a sphere (it is known as *Meusnier's theorem*). The equation $k_n = k\cos\theta$ in Problem 3.4 is called *Meusnier's formula*; a different interpretation of Meusnier's formula is given in Problem 3.6 (a). Another proof of Meusnier's formula is given on p. 73.

Problem 3.7 Prove that if a curve lies on a sphere of radius R, then $k_n = \pm\frac{1}{R}$ and $\varkappa_g = 0$.

Problem 3.8 Find the geodesic curvature of a circle of radius r on a sphere of radius R.

3.3   Geodesics

A curve $\gamma(s)$ with natural parameterization on a smooth surface is said to be *geodesic* if its geodesic curvature k_g identically vanishes. A curve with natural parameterization is geodesic if and only if its acceleration vector is orthogonal to the surface at each point. A curve with an arbitrary parameterization is said to be geodesic if its acceleration vector is orthogonal to the surface at each point. The geodesics obtained from a geodesic $\gamma(s)$ by changing the parameterization have the form $\gamma(at+b)$, where a and b are constants, $a \neq 0$. Indeed, we have

$$\frac{d^2\gamma}{dt^2} = \frac{d^2\gamma}{ds^2}\cdot\left(\frac{ds}{dt}\right)^2 + \frac{d\gamma}{ds}\cdot\frac{d^2s}{dt^2},$$

and hence the vector $\frac{d^2\gamma}{dt^2}$ is orthogonal to the surface if and only if $\frac{d^2s}{dt^2} = 0$.

Example 3.3 The curve $\gamma(s) = (\cos s, \sin s, 0)$ on the sphere $x^2 + y^2 + z^2 = 1$ is geodesic.

Proof The velocity vector ε_1 of the curve γ equals $(-\sin s, \cos s, 0)$, whence $\varepsilon_1' = (-\cos s, -\sin s, 0) = -\gamma(s)$. This vector is orthogonal to the surface of the sphere at the point $\gamma(s)$.

Another proof uses Problem 3.8: the geodesic curvature of a circle of radius R on a sphere of radius R equals zero.   □

A shortest curve joining two points is geodesic (see Theorem 3.1 on p. 71). It is easy to see from Example 3.3 that the converse is false: the curve in that example may wind around itself, making arbitrarily many coils. However, the converse is false only globally. Locally it is true: any sufficiently short part of a geodesic is a shortest curve joining given points (see Theorem 3.19 on p. 104).

Problem 3.9 Prove that if a surface contains a straight line, then this line (endowed with the natural parameter s) is geodesic.

Problem 3.10 Suppose given a surface of revolution (i.e., a surface swept out by a plane curve rotating about an axis lying in the plane of the curve).

(a) Prove that a *meridian* (i.e., one of the two symmetric curves in the intersection of the surface and a plane containing the axis of rotation) is a geodesic (the parameterization of the curve is assumed to be natural).
(b) Prove that a *parallel* (i.e., the circle traced by a point A of the rotating curve) is geodesic if and only if the tangent to the meridian at A is parallel to the axis of rotation (the parameterization of the curve is assumed to be natural).

Theorem 3.1 *If a curve $\gamma(s)$ with natural parameterization is shortest among all curves joining points A and B on a given surface, then this curve is geodesic.*

Proof Suppose that $k_g(s_0) \neq 0$ for some s_0. Then $k_g(s)$ does not vanish on some segment $[a, b]$. It suffices to consider the case where $A = \gamma(a)$ and $B = \gamma(b)$. Indeed, any segment of a shortest curve must be shortest among all curves joining its endpoints.

Let $\gamma(s) = r(u^1(s), u^2(s))$, where $r(u^1, u^2)$ is the parameterization of the surface, and let (v^1, v^2) be the coordinates of the vector ε_2 with respect to this parameterization, i.e., $\varepsilon_2 = v^1 \frac{\partial r}{\partial u^1} + v^2 \frac{\partial r}{\partial u^2}$. Consider the family of curves

$$\gamma_t(s) = r(u^1(s) + t\lambda(s)v^1, \ u^2(s) + t\lambda(s)v^2),$$

where $\lambda(s)$ is a smooth function vanishing at the endpoints of the segment $[a, b]$ and positive at the other points of this segment. For each t, the curve $\gamma_t(s)$ joins the points A and B, and at $t = 0$ we obtain the initial (shortest) curve. Therefore, if $L(t)$ is the length of γ_t, then $L'(0) = 0$.

Let $R(s, t) = \gamma_t(s)$. Then $L(t) = \int_a^b \sqrt{\left(\frac{\partial R}{\partial s}, \frac{\partial R}{\partial s} \right)} \, ds$; therefore,

$$L'(t) = \int_a^b \frac{d}{dt} \sqrt{\left(\frac{\partial R}{\partial s}, \frac{\partial R}{\partial s} \right)} \, ds = \int_a^b \frac{\left(\frac{\partial^2 R}{\partial s \, \partial t}, \frac{\partial R}{\partial s} \right)}{\sqrt{\left(\frac{\partial R}{\partial s}, \frac{\partial R}{\partial s} \right)}} \, ds.$$

For $t = 0$, we have $\frac{\partial R}{\partial s} = \frac{d\gamma}{ds}$, and hence $\left(\frac{\partial R}{\partial s}, \frac{\partial R}{\partial s} \right) = \left\| \frac{d\gamma}{ds} \right\| = 1$. Thus, $L'(0) = \int_a^b \left(\frac{\partial^2 R}{\partial s \, \partial t}, \frac{\partial R}{\partial s} \right) \Big|_{t=0} \, ds$. Taking into account the equality

$$\frac{d}{ds} \left(\frac{\partial R}{\partial t}, \frac{\partial R}{\partial s} \right) = \left(\frac{\partial^2 R}{\partial t \, \partial s}, \frac{\partial R}{\partial s} \right) + \left(\frac{\partial R}{\partial t}, \frac{\partial^2 R}{\partial s^2} \right),$$

we obtain

$$L'(0) = \int_a^b \frac{d}{ds}\left(\frac{\partial R}{\partial t}, \frac{\partial R}{\partial s}\right)\Big|_{t=0} ds - \int_a^b \left(\frac{\partial R}{\partial t}, \frac{\partial^2 R}{\partial s^2}\right)\Big|_{t=0} ds.$$

Next, we have $\frac{\partial R}{\partial t}\big|_{t=0} = \sum_i \lambda(s)v^i \frac{\partial r}{\partial u^i} = \lambda(s)\varepsilon_2$ and $\lambda(a) = \lambda(b) = 0$. Therefore,

$$\int_a^b \frac{d}{ds}\left(\frac{\partial R}{\partial t}, \frac{\partial R}{\partial s}\right)\Big|_{t=0} ds = \left(\frac{\partial R}{\partial t}, \frac{\partial R}{\partial s}\right)\Big|_{t=0}\Big|_a^b = 0.$$

Finally, $\dfrac{\partial^2 R}{\partial s^2}\Big|_{t=0} = k_g\varepsilon_2 + k_n\varepsilon_3$. Thus,

$$L'(0) = -\int_a^b (\lambda(s)\varepsilon_2, k_g\varepsilon_2 + k_n\varepsilon_3) ds = -\int_a^b \lambda(s)k_g(s)\,ds \neq 0.$$

We have arrived at a contradiction. $\square$

3.4 The Second Quadratic Form

Let S be a smooth oriented surface in $\mathbb{R}^3$ specified parametrically, which means that the points of S are given in the form $r(u, v)$. Consider a curve $r(u(t), v(t))$ on the surface S. To the tangent vector r' the first quadratic form assigns its inner square (r', r') and the *second quadratic form*, the inner product $(r', -n')$, where n is the unit normal vector to the surface. There is also yet another expression for the second quadratic form: since $(r', n) = 0$ and, therefore, $(r', n') + (r'', n) = 0$, it follows that the second quadratic form assigns the inner product (r'', n) to each tangent vector r'.

At first sight, it may seem that $(r', -n')$ depends on u' and v' linearly rather that quadratically. However, not only r' but also n' depends on u' and v'. From the second expression for the second quadratic form it is clearly seen that this form is quadratic. Indeed, we have $r' = r_u u' + r_v v'$ and $r'' = r_{uu}(u')^2 + 2r_{uv}u'v' + r_{vv}(v')^2$. Taking into account the independence of the normal vector n of u' and v', we see that (r'', n) is a quadratic form in u' and v'. The second quadratic form is often written as $L\,du^2 + 2M\,du\,dv + N\,dv^2$. Here $L = -(r_u, n_u) = (r_{uu}, n)$, $2M = -(r_u, n_v) - (r_v, n_u) = 2(r_{uv}, n)$, and $N = -(r_v, n_v) = (r_{vv}, n)$.

Example 3.4 If a surface is given as the graph of a function $z = f(x, y)$ such that $\frac{\partial f}{\partial x}(0, 0) = 0$ and $\frac{\partial f}{\partial y}(0, 0) = 0$, then at the point $x = y = 0$ the second quadratic form is given by the matrix $\begin{pmatrix} f_{xx} & f_{xy} \\ f_{xy} & f_{yy} \end{pmatrix}$. This matrix is called the *Hessian matrix*, or the *Hessian*, and its determinant is called the *Hessian determinant*.

Proof The given surface can be specified parametrically as $r(x, y) = (x, y, f(x, y))$. At the point under consideration we have $n = (0, 0, 1)$ and

$$r'' = (x'', y'', f_{xx}(x')^2 + 2f_{xy}x'y' + f_{yy}(y')^2).$$

Calculating the inner product of the vectors r'' and n, we obtain the required result.

$\square$

HISTORICAL COMMENT Otto Hesse (1811–1874) introduced Hessian in 1842 in studying curves of the third and the fourth degree. He considered equations in homogeneous coordinates, i.e., the curves were determined by functions of three variables and the Hessian was of order 3.

Consider a curve $r(s)$, where s is the natural parameter, on a surface S. The vector r'' is directed along the principal normal e_2 to the curve and equals the curvature k of the curve in magnitude. Therefore, $(r'', n) = k \cos \theta$, where θ is the angle between the normal $n = \varepsilon_3$ to the surface and the principal normal e_2 to the curve. If the tangent vector r' is fixed, then $(r'', n) = k_0$ is the value of the second quadratic form at the vector r' (which has unit length). Hence

$$k \cos \theta = k_0,$$

where k_0 is the curvature of the intersection curve of the surface and the plane containing the vectors n and r'.

We have obtained another proof of Meusnier's formula, which we already discussed in Problems 3.4 and 3.6 (a). Recall that the number k_0 is called the normal curvature of the surface in the given direction r'. When considering vectors of arbitrary length, rather than only unit ones, we can represent k_0 as the fraction **II/I**, where **II** and **I** are the values of the second and the first quadratic form at the given vector r'.

For any two real symmetric matrices A and B such that A is positive definite, there exists a matrix X for which $X^T A X = I$ is the identity matrix and $X^T B X = D$ is a diagonal matrix. The diagonal entries of D are determined by the condition $\det(B - dA) = 0$. In the case of interest to us, $A = (g_{ij})$ is the matrix of the first quadratic form and B is the matrix of the second quadratic form. The entries k_1 and k_2 of the corresponding diagonal matrix are called the *principal curvatures* of the surface at the given point. The number $H = \frac{1}{2}(k_1 + k_2)$ is called the *mean curvature*, and $K = k_1 k_2$ is called the *Gaussian curvature*.

The reversal of the surface orientation, i.e., the replacement of the vector n by $-n$, changes the sign of the principal curvature.

HISTORICAL COMMENT Gauss introduced Gaussian curvature in 1827 as the ratio of the areas of infinitesimal figures on the sphere and on the given surface corresponding to each other under the spherical Gauss map. Before Gauss, the curvature of a surface was defined in precisely the same way in 1815 by Benjamin

Olinde Rodrigues (1795–1851). Rodrigues also proved that Gaussian curvature equals the product of principal curvatures.

The expression for the Gaussian curvature of a surface of revolution is particularly simple in the case where the parameterization of a meridian is natural. Therefore, in calculating the Gaussian curvature of a surface of revolution, we endow the surface with a parameterization of the form $(f(u)\cos v,\ f(u)\sin v,\ g(u))$. Then the parameterization of a meridian by u is natural if $(f')^2 + (g')^2 = 1$.

Problem 3.11 Calculate the second quadratic form and the Gaussian curvature K of a surface of revolution $(f(u)\cos v,\ f(u)\sin v,\ g(u))$. Show that if the parameterization of a meridian is natural, then $K = -\frac{f''}{f}$.

Problem 3.12 A closed surface embedded in $\mathbb{R}^3$ is contained in a ball of radius R. Prove that the Gaussian curvature of this surface is at least $1/R^2$ at some point.

For a basis $\{e_i\}$ with respect to which the matrix of a quadratic form q is diagonal, the equation $q(\sum \lambda_i e_i) = \sum \lambda_i^2 q(e_i)$ holds. Therefore, if e_1 and e_2 are unit tangent vectors corresponding to the principal curvatures k_1 and k_2, then the unit vector $e_1 \cos \varphi + e_2 \sin \varphi$ corresponds to the normal curvature

$$k_1 \cos^2 \varphi + k_2 \sin^2 \varphi \quad (Euler's\ formula).$$

HISTORICAL COMMENT In 1760 Leonhard Euler (1707–1783) obtained the following expression for the curvature radius r:

$$r = \frac{2fg}{f + g - (f - g)\cos 2\varphi},$$

where f and g are the maximum and minimum curvature radii. The more convenient expression given above was derived by Charles Dupin (1784–1873) in 1813.

Problem 3.13 Let us fix a point q in $\mathbb{R}^3$ and consider the function $F(p) = \|p - q\|^2$ on a surface S.

(a) Prove that a point p is critical if and only if the vector $\overrightarrow{pq}$ is orthogonal to S at p.
(b) Prove that a critical point p is degenerate if and only if the length of the vector $\overrightarrow{pq}$ equals $1/|k_i|$, where k_i is the nonzero principal curvature of the surface at p, and this vector has the corresponding direction.

Problem 3.14 Suppose that a surface S contains entirely a straight line l. Prove that the Gaussian curvature of the surface is nonpositive at each point of this line.

3.5 Gaussian Curvature

In Sect. 3.4 the Gaussian curvature K was defined as the product of principal curvatures, and the mean curvature H was defined as their half-sum. In this section we discuss Gaussian curvature in more detail. First, we obtain an explicit expression for Gaussian and mean curvature in terms of the coefficients of the first and second quadratic forms (see Theorem 3.2). Gaussian curvature differs from mean curvature in that it can be expressed in terms of only the first quadratic form (see Theorem 3.3). Thus, Gaussian curvature is an intrinsic invariant of a surface: it does not depend on the embedding of the surface in space and does not change under bendings, that is, transformations preserving the metric.

Theorem 3.2 *Let A and B be the matrices of the first and second quadratic forms. Then*

$$K = \frac{\det B}{\det A} = \frac{LN - M^2}{EG - F^2} \quad and \quad H = \frac{1}{2} \cdot \frac{LG - 2MF + NE}{EG - F^2}.$$

Proof The principal curvatures are the roots of the equation

$$\begin{vmatrix} L - kE & M - kF \\ M - kF & N - kG \end{vmatrix} = 0,$$

i.e.,

$$(EG - F^2)k^2 - (LG - 2MF + NE)k + LN - M^2 = 0.$$

Therefore, the product of principal curvatures equals $\frac{LN - M^2}{EG - F^2}$, and their half-sum equals $\frac{1}{2} \cdot \frac{LG - 2MF + NE}{EG - F^2}$. □

The expression of Gaussian curvature in terms of the first quadratic form is much more cumbersome, but the calculation itself is fairly simple. In the expression $K = \frac{LN - M^2}{EG - F^2}$ we must get rid of the coefficients of the second quadratic form. For this purpose, it is convenient to use the notation of Sect. 3.4: the surface under consideration is parameterized as $r(u, v)$, n is the unit normal vector to the surface, and r_u and r_v are the partial derivatives of r with respect to u and v. For the coefficients of the second quadratic form we use the expressions $L = (r_{uu}, n)$, $M = (r_{uv}, n)$, and $N = (r_{vv}, n)$. The vector n can be expressed in terms of r_u and r_v as well: the squared length of the vector $r_u \times r_v$ equals $r_u^2 r_v^2 - (r_u, r_v)^2 = EG - F^2$, whence $n = \frac{r_u \times r_v}{\sqrt{EG - F^2}}$.

Theorem 3.3 (Gauss)   *The Gaussian curvature K is expressed in terms of the coefficients of the first quadratic form as*

$$K(EG - F^2)^2 = \begin{vmatrix} -\frac{1}{2}G_{uu} + F_{uv} - \frac{1}{2}E_{vv} & \frac{1}{2}E_u & F_u - \frac{1}{2}E_v \\ F_v - \frac{1}{2}G_u & E & F \\ \frac{1}{2}G_v & F & G \end{vmatrix} - \begin{vmatrix} 0 & \frac{1}{2}E_v & \frac{1}{2}G_u \\ \frac{1}{2}E_v & E & F \\ \frac{1}{2}G_u & F & G \end{vmatrix}$$

Proof   Let us substitute the expressions $L = \left(r_{uu}, \frac{r_u \times r_v}{\sqrt{EG-F^2}}\right)$, $M = \left(r_{uv}, \frac{r_u \times r_v}{\sqrt{EG-F^2}}\right)$, and $N = \left(r_{vv}, \frac{r_u \times r_v}{\sqrt{EG-F^2}}\right)$ for the coefficients of the second quadratic form into the formula $K = \frac{LN - M^2}{EG - F^2}$. We obtain

$$K(EG - F^2)^2 = (r_{uu}, r_u \times r_v) \cdot (r_{vv}, r_u \times r_v) - (r_{uv}, r_u \times r_v)^2$$

$$= \det \begin{pmatrix} r_{uu} \\ r_u \\ r_v \end{pmatrix} \cdot \det \begin{pmatrix} r_{vv} \\ r_u \\ r_v \end{pmatrix} - \det \begin{pmatrix} r_{uv} \\ r_u \\ r_v \end{pmatrix} \cdot \det \begin{pmatrix} r_{uv} \\ r_u \\ r_v \end{pmatrix}.$$

In the last expression the symbols r_{uu}, r_u, etc. in the matrices denote the coordinates of the corresponding vectors written in a row. Let r_{uu}^T, r_u^T, etc. denote the same vectors written in a column. Then

$$K(EG - F^2)^2$$

$$= \det \begin{pmatrix} r_{uu} \\ r_u \\ r_v \end{pmatrix} \cdot \det \begin{pmatrix} r_{vv}^T & r_u^T & r_v^T \end{pmatrix} - \det \begin{pmatrix} r_{uv} \\ r_u \\ r_v \end{pmatrix} \cdot \det \begin{pmatrix} r_{uv}^T & r_u^T & r_v^T \end{pmatrix}$$

$$= \begin{vmatrix} (r_{uu}, r_{vv}) & (r_{uu}, r_u) & (r_{uu}, r_v) \\ (r_{vv}, r_u) & E & F \\ (r_{vv}, r_v) & F & G \end{vmatrix} - \begin{vmatrix} (r_{uv}, r_{uv}) & (r_{uv}, r_u) & (r_{uv}, r_v) \\ (r_{uv}, r_u) & E & F \\ (r_{uv}, r_v) & F & G \end{vmatrix}$$

$$= \begin{vmatrix} (r_{uu}, r_{vv}) - (r_{uv}, r_{uv}) & (r_{uu}, r_u) & (r_{uu}, r_v) \\ (r_{vv}, r_u) & E & F \\ (r_{vv}, r_v) & F & G \end{vmatrix} - \begin{vmatrix} 0 & (r_{uv}, r_u) & (r_{uv}, r_v) \\ (r_{uv}, r_u) & E & F \\ (r_{uv}, r_v) & F & G \end{vmatrix}.$$

(We have used the fact that the elements (r_{uu}, r_{vv}) and (r_{uv}, r_{uv}) occur in the determinants of the matrices under consideration with the same coefficient $EG - F^2$.)

Now we express the inner products in the obtained matrices in terms of the coefficients of the first quadratic form. First, we differentiate the equations $(r_u, r_u) = E$ and $(r_v, r_v) = G$ with respect to u and v. As a result, we obtain

$$(r_{uu}, r_u) = \frac{1}{2}E_u, \quad (r_{uv}, r_u) = \frac{1}{2}E_v, \quad (r_{uv}, r_v) = \frac{1}{2}G_u, \quad (r_{vv}, r_v) = \frac{1}{2}G_v.$$

Then, we differentiate the equation $(r_u, r_v) = F$ with respect to u and v and use the equations already obtained:

$$(r_{uu}, r_v) = F_u - \frac{1}{2}E_v, \quad (r_{vv}, r_u) = F_v - \frac{1}{2}G_u.$$

It remains to calculate $(r_{uu}, r_{vv}) - (r_{uv}, r_{uv})$. For this purpose, we differentiate the equation $(r_{uv}, r_v) = \frac{1}{2}G_u$ with respect to u and the equation $(r_{uu}, r_v) = F_u - \frac{1}{2}E_v$ with respect to v:

$$(r_{uvu}, r_v) + (r_{uv}, r_{vu}) = \frac{1}{2}G_{uu},$$

$$(r_{uuv}, r_v) + (r_{uu}, r_{vv}) = F_{uv} - \frac{1}{2}E_{vv}.$$

It follows that

$$(r_{uu}, r_{vv}) - (r_{uv}, r_{vu}) = -\frac{1}{2}G_{uu} + F_{uv} - \frac{1}{2}E_{vv}.$$

Substituting the found values into the matrices, we obtain the required expression.

$\square$

HISTORICAL COMMENT The theorem that Gaussian curvature can be expressed in terms of the metric and, therefore, does not change under bendings (isometries) of surfaces was proved by Gauss in 1827. Gauss named it *Theorema Egregium* (remarkable theorem).

3.6 Gaussian Curvature and Differential Forms

Here we need only the simplest differential 1- and 2-forms on one- and two-dimensional manifolds, and so we give only the most necessary definitions. A differential 1-form ω is the object dual to a vector field, i.e., a linear function on the vectors of the tangent space. For example, to a function f its differential is assigned, which is the 1-form $df = \frac{\partial f}{\partial x^i}dx^i$ taking the value $\frac{\partial f}{\partial x^i}V^i$ at each vector field V. A differential 2-form is a skew-symmetric bilinear function on pairs of vectors. For example, the form $dx^i \wedge dx^j$ takes the value $V^i W^j - V^j W^i$ at a pair of vectors V, W. In 2-dimensional space the 2-form $dx^1 \wedge dx^2$ is basis, i.e., any 2-form is this form multiplied by some function. The differential of the 1-form $\omega = f_\alpha dx^\alpha$ is the 2-form

$$d\omega = df_\alpha \wedge dx^\alpha = \frac{\partial f_\alpha}{\partial x^i}dx^i \wedge dx^\alpha.$$

The differential of the k-form $\omega = f_{i_1 \ldots i_k} dx^{i_1} \wedge \cdots \wedge dx^{i_k}$ is the $(k+1)$-form

$$d\omega = df_{i_1 \ldots i_k} \wedge dx^{i_1} \wedge \cdots \wedge dx^{i_k}.$$

Using this expression, it is easy to check that $dd\omega = 0$ and $d(\omega_1 \wedge \omega_2) = d\omega_1 \wedge \omega_2 + (-1)^k \omega_1 \wedge d\omega_2$ for a k-form ω_1.

Now we are ready to give another definition of Gaussian curvature, which uses differential forms. Let S be a smooth oriented surface in $\mathbb{R}^3$, and let ε_1, ε_2 be a *moving frame* in some domain of S, that is, a frame whose vectors ε_1 and ε_2 smoothly depend on the point and form an orthonormal frame of the tangent plane at each point. The basis ε_1, ε_2 is associated with the basis ω^1, ω^2 of the dual space (the space of 1-forms) determined by the condition $\omega^i(x^1\varepsilon_1 + x^2\varepsilon_2) = x^i$. The form $\omega^1 \wedge \omega^2$ nowhere vanishes; therefore, $d\omega^1 = \alpha \omega^1 \wedge \omega^2$ and $d\omega^2 = \beta \omega^1 \wedge \omega^2$, where α and β are smooth functions. Consider the 1-form $\omega_1^2 = \alpha \omega^1 + \beta \omega^2$. It satisfies the equations

$$d\omega^1 = \omega_1^2 \wedge \omega^2 \quad \text{and} \quad d\omega^2 = -\omega_1^2 \wedge \omega^1. \tag{3.2}$$

Note that these two equations uniquely determine the 1-form ω_1^2.

The form $d\omega_1^2$ is a 2-form. Therefore, $d\omega_1^2 = -K\omega^1 \wedge \omega^2$, where K is some number; this number is called the *Gaussian curvature* of the surface at the given point. This new definition of Gaussian curvature is equivalent to that given in Sect. 3.4; we will prove this later on (see p. 79).

Theorem 3.4 *The Gaussian curvature K does not depend on the choice of a moving frame.*

Proof The statement to be proved is local, so that we can assume that we work in a small neighborhood homeomorphic to the disk. In this case, any moving frame is obtained from the given one by rotating it through an angle smoothly depending on the point.

We set $\tilde{\omega}^1 = \cos\varphi\,\omega^1 + \sin\varphi\,\omega^2$ and $\tilde{\omega}^2 = -\sin\varphi\,\omega^1 + \cos\varphi\,\omega^2$. Applying (3.2), we obtain

$$d\tilde{\omega}^1 = \cos\varphi\,d\omega^1 + \sin\varphi\,d\omega^2 + d\varphi \wedge (-\sin\varphi\,\omega^1 + \cos\varphi\,\omega^2)$$

$$= \cos\varphi\,\omega_1^2 \wedge \omega^2 - \sin\varphi\,\omega_1^2 \wedge \omega^1 + d\varphi \wedge \tilde{\omega}^2 = (\omega_1^2 + d\varphi) \wedge \tilde{\omega}^2,$$

$$d\tilde{\omega}^2 = -(\omega_1^2 + d\varphi) \wedge \tilde{\omega}^1.$$

Thus, $\tilde{\omega}_1^2 = \omega_1^2 + d\varphi$. Therefore, $d\tilde{\omega}_1^2 = d\omega_1^2$. It is also clear that

$$\tilde{\omega}^1 \wedge \tilde{\omega}^2 = (\omega^1 \cos\varphi + \omega^2 \sin\varphi) \wedge (-\omega^1 \sin\varphi + \omega^2 \cos\varphi) = \omega^1 \wedge \omega^2.$$

$\square$

Remark If $\varepsilon_1 = x_{11}\tilde{\varepsilon}_1 + x_{12}\tilde{\varepsilon}_2$ and $\varepsilon_2 = x_{21}\tilde{\varepsilon}_1 + x_{22}\tilde{\varepsilon}_2$, then $x_{11} = \tilde{\omega}^1(\varepsilon_1) = \cos\varphi$, $x_{12} = \tilde{\omega}^2(\varepsilon_1) = -\sin\varphi$, $x_{21} = \tilde{\omega}^1(\varepsilon_2) = \sin\varphi$, and $x_{22} = \tilde{\omega}^2(\varepsilon_2) = \cos\varphi$. Thus, $\varepsilon_1 = \cos\varphi\tilde{\varepsilon}_1 - \sin\varphi\tilde{\varepsilon}_2$ and $\varepsilon_2 = \sin\varphi\tilde{\varepsilon}_1 + \cos\varphi\tilde{\varepsilon}_2$, i.e., φ is the angle of rotation from the vector ε_1 to $\tilde{\varepsilon}_1$.

The form ω_1^2, which we used to define Gaussian curvature, is related to the geodesic curvature k_g as follows: if ε_1 and ε_2 are the first two vectors in the Darboux frame of a curve $\gamma(s)$, then the restriction of the form ω_1^2 to this curve is the form $k_g ds$. To prove this, consider a moving frame $\varepsilon_1(x)$, $\varepsilon_2(x)$, $\varepsilon_3(x)$ in $\mathbb{R}^3$, where $x \in \mathbb{R}^3$. We have $dx = \omega^i \varepsilon_i$, where ω^i is the form to dual ε_i, and $d\varepsilon_i = \omega_i^j \varepsilon_j$. So far, ω^i and ω_i^j are the mere notations of some 1-forms on $\mathbb{R}^3$. But we will shortly show that the forms ω^1, ω^2, and ω_1^2, which we used to define Gaussian curvature, are the restrictions to the surface of the forms ω^1, ω^2, and ω_1^2 on $\mathbb{R}^3$.

The equations $0 = ddx = d(\omega^i \varepsilon_i) = (d\omega^i)\varepsilon_i - \omega^i \wedge d\varepsilon_i$ show that $d\omega^j = \omega^i \wedge \omega_i^j$. Similarly, the equation $dd\varepsilon_i = 0$ implies $0 = d(\omega_i^j \varepsilon_j) = (d\omega_i^j)\varepsilon_j - \omega_i^j \wedge \omega_j^k \varepsilon_k = (d\omega_i^j)\varepsilon_j - \omega_i^k \wedge \omega_k^j \varepsilon_j$, i.e., $d\omega_i^j = \omega_i^k \wedge \omega_k^j$. Moreover, if the frame is orthonormal, then $(\varepsilon_i, \varepsilon_j) = \delta_{ij}$, and hence $(\varepsilon_i, d\varepsilon_j) + (\varepsilon_j, d\varepsilon_i) = 0$, i.e., $\omega_i^j + \omega_j^i = 0$. In particular, $\omega_i^i = 0$.

The restrictions to the surface of forms on $\mathbb{R}^3$ have zero component along the normal $\varepsilon_3(x)$, and hence $\omega^3 \equiv 0$. Therefore, $d\omega^1 = \omega^1 \wedge \omega_1^1 + \omega^2 \wedge \omega_2^1 + \omega^3 \wedge \omega_3^1 = \omega^2 \wedge \omega_2^1 = -\omega^2 \wedge \omega_1^2$ and $d\omega^2 = \omega^1 \wedge \omega_1^2$. Thus, the forms ω^1, ω^2, and ω_1^2 used to define Gaussian curvature are indeed the restrictions to the surface of the new forms ω^1, ω^2, and ω_1^2 on $\mathbb{R}^3$.

In the new notation, the equations of motion of the Darboux frame for the restrictions of the forms to the curve $\gamma(s)$ are

$$d\varepsilon_1 = \omega_1^2 \varepsilon_2 + \omega_1^3 \varepsilon_3,$$

$$d\varepsilon_2 = -\omega_1^2 \varepsilon_1 + \omega_2^3 \varepsilon_3,$$

$$d\varepsilon_3 = -\omega_1^3 \varepsilon_1 - \omega_2^3 \varepsilon_2;$$

therefore, $\omega_1^2 = k_g ds$, $\omega_1^3 = k_n ds$, and $\omega_2^3 = \varkappa_g ds$.

Problem 3.15 Prove that $d\omega_1^2 = \omega_1^3 \wedge \omega_3^2$, $d\omega_1^3 = \omega_1^2 \wedge \omega_2^3$, and $d\omega_2^3 = \omega_2^1 \wedge \omega_1^3$.

Problem 3.16 Prove that the restrictions of the forms to the surface satisfy the equation $\omega^1 \wedge \omega_1^3 + \omega^2 \wedge \omega_2^3 = 0$.

Let us now prove that the new definition of Gaussian curvature is equivalent to that given in Sect. 3.4. It suffices to check that the curvatures coincide at each point. Let $r = r(u, v)$ be a parameterized surface. Suppose that the matrix of the first quadratic form at a given point is the identity, i.e., $(r_u, r_u) = (r_v, r_v) = 1$ and $(r_u, r_v) = 0$. We set $r_u = \varepsilon_1$, $r_v = \varepsilon_2$, and $n = \varepsilon_3$, where n is the normal vector.

Then $dr = \omega^1 \varepsilon_1 + \omega^2 \varepsilon_2$, because the normal component of dr is zero. Therefore, the value of the second quadratic form at dr equals

$$\mathbf{II}(dr) = (dr, -dn) = (\omega^1 \varepsilon_1 + \omega^2 \varepsilon_2, -\omega_3^j \varepsilon_j).$$

Suppose that $\omega_1^3 = A\omega^1 + B\omega^2$ and $\omega_2^3 = C\omega^1 + D\omega^2$. It follows from the equation $\omega^1 \wedge \omega_1^3 + \omega^2 \wedge \omega_2^3 = 0$ (see Problem 3.16) that $B = C$. Hence, in the notation of Sect. 3.4, we have $dr = \omega^1 \varepsilon_1 + \omega^2 \varepsilon_2 = r_u du + r_v dv$ and $dn = \omega_3^j \varepsilon_j = -\omega_1^3 \varepsilon_1 - \omega_2^3 \varepsilon_2 = -r_u(Adu + Bdv) - r_v(Bdu + Ddv)$. Since the matrix of the first quadratic form is the identity at the given point, if follows that, at this point,

$$\mathbf{II}(dr) = (dr, -dn) = A du^2 + 2B du\, dv + D dv^2,$$

i.e., $L = A$, $M = B$, and $N = D$. Thus,

$$\omega_1^3 \wedge \omega_2^3 = (AD - BC)\omega^1 \wedge \omega^2 = (LN - M^2)\omega^1 \wedge \omega^2 = \det(\mathbf{II})\omega^1 \wedge \omega^2.$$

On the other hand, according to Problem 3.15, we have $\omega_1^3 \wedge \omega_2^3 = -\omega_1^3 \wedge \omega_3^2 = -d\omega_1^2 = K\omega^1 \wedge \omega^2$. Therefore, $\det(\mathbf{II}) = K$. This is precisely what we need, because the matrix of the first quadratic form is the identity at the given point.

Remark As already mentioned, Eq. (3.2) completely determines the form ω_1^2. This means that the forms $d\omega_1^2$ and $\omega^1 \wedge \omega^2$ are completely determined by the metric. Thus, it follows directly from the definition of Gaussian curvature in the language of differential forms that Gaussian curvature is completely determined by the metric.

Problem 3.17 Prove that $-\omega^1 \wedge \omega_2^3 + \omega^2 \wedge \omega_1^3 = 2H\omega^1 \wedge \omega^2$, where H is the mean curvature.

3.7 The Gauss–Bonnet Theorem

The Gauss–Bonnet formula relates the integral of Gaussian curvature over a polygon on an oriented surface to the integral of geodesic curvature over the boundary of this polygon and the sum of exterior angles of the polygon. From this formula an expression for the total Gaussian curvature of a closed oriented surface in terms of the Euler characteristics can be derived, which is known as the Gauss–Bonnet theorem. The Gauss–Bonnet theorem can also be proved without using the Gauss–Bonnet formula. We give both proofs.

The Gaussian curvature K does not depend on the surface orientation; moreover, it can be defined for nonorientable surface. But the Gauss–Bonnet formula and the Gauss–Bonnet theorem involve the integral $\int_G K d\sigma$, where $d\sigma$ is the area form, and the area form exists only on orientable surfaces. In one of the proofs of the Gauss–

Fig. 3.1 The Gauss–Bonnet formula

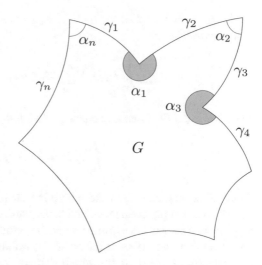

Bonnet theorem we cut the surface into parts and apply the Gauss–Bonnet formula to each part. This can be done only in the case where the surface is orientable.

The Gauss–Bonnet formula applies to a simply connected domain G bounded by a piecewise smooth curve consisting of smooth curves $\gamma_1, \gamma_2, \ldots, \gamma_n$ on a surface M^2; we denote the angles between the curve-pieces at the corners by $\alpha_1, \alpha_2, \ldots, \alpha_n$ (see Fig. 3.1). The angles in the corners can take values from 0 to 2π (in the figure the angles α_1 and α_3 are larger than π), and the exterior angles $\pi - \alpha_k$ can take values from $-\pi$ to π. Recall that a simply connected domain is always orientable.

To prove the Gauss–Bonnet theorem, we need *Umlaufsatz* (Theorem 1.2) in a form more general than that proved above: Given a vector field ε_1 without singularities on a simply connected domain bounded by a smooth positively oriented curve on a surface, the tangent vector $\tilde{\varepsilon}_1$ makes one turn in the positive direction with respect to ε_1 while traversing the curve. *Umlaufsatz* implies this assertion only for a plane domain and a constant vector field ε_1. There is no substantial difference between a constant vector field and a vector field ε_1 without singular points: the index of a vector field without singular points is zero, and hence ε_1 makes zero turns as the curve is traversed. Consider a plane chart parameterizing the given simply connected domain. Let $(\,,\,)_0$ be the inner product of vectors in the plane at some point, and let $(\,,\,)_1$ be the inner product of the corresponding tangent vectors to the surface at the corresponding point. According to *Umlaufsatz*, if angles between vectors are measured by using the inner product $(\,,\,)_0$, then the velocity vector of the curve makes the required number of turns. Therefore, it suffices to check that the number of turns is the same when angles are measured by using the inner product $(\,,\,)_1$. To this end, consider the family of inner products $t(\,,\,)_1 + (1-t)(\,,\,)_0$, $0 \leqslant t \leqslant 1$. Since the number of turns must vary continuously with t, it is constant.

Theorem 3.5 (Gauss–Bonnet Formula) *Let G be a simply connected domain bounded by a piecewise smooth curve consisting of smooth curves $\gamma_1, \gamma_2, \ldots, \gamma_n$*

on a surface M^2, and let $\alpha_1, \alpha_2, \ldots, \alpha_n$ be the angles between these curves at the joint points. Then

$$\sum_k \int_{\gamma_k} k_g ds + \sum_k (\pi - \alpha_k) = 2\pi - \int_G K d\sigma.$$

In particular, if the bounding curve γ is smooth, then

$$\int_\gamma k_g ds = 2\pi - \int_G K d\sigma.$$

Proof First, consider the case where the curve γ is smooth. By assumption the domain G is simply connected, and hence there exists a vector field ε_1 on G without singular points; we can assume that each vector ε_1 has unit length. Let us complete the vector ε_1 to an orthonormal positively oriented basis $\varepsilon_1, \varepsilon_2$ in the tangent plane and construct a form ω_1^2 for which $d\omega_1^2 = -K\omega^1 \wedge \omega^2 = -K d\sigma$. By Stokes' formula $- \int_G K d\sigma = \int_\gamma \omega_1^2$.

In a neighborhood of the curve γ, in addition to the vector field ε_1, we can consider a vector field $\tilde{\varepsilon}_1$ whose restriction to γ is the tangent vector field on γ. Using the vector field $\tilde{\varepsilon}_1$, we construct a form $\tilde{\omega}_1^2$. We have $\omega_1^2 = \tilde{\omega}_1^2 + d\varphi$, where φ the angle of rotation from the vector $\tilde{\varepsilon}_1$ to the vector ε_1 (see the remark after Theorem 3.4). Therefore, $\int_\gamma \omega_1^2 = \int_\gamma \tilde{\omega}_1^2 + \int_\gamma d\varphi = \int_\gamma k_g ds - 2\pi$, because, according to *Umlaufsatz*, as the entire curve is traversed, the angle φ changes by 2π, and the angle of rotation is measured in the opposite direction: from the tangent vector $\tilde{\varepsilon}_1$ to the vector ε_1.

In the general case of a piecewise smooth curve, the exterior angles $\pi - \alpha_k$ at the joint points are included in the integral $\int_\gamma d\varphi$. $\square$

HISTORICAL COMMENT In 1827 Gauss proved that the total curvature of a geodesic triangle (i.e., the integral of Gaussian curvature over a geodesic triangle) equals the sum of its angles measured in radians minus π. The total curvature of a geodesic polygon equals 2π minus the sum of exterior angles. In 1848 Bonnet generalized this statement from geodesic polygons to general closed curves; in the case of curves, the sum of exterior angles of a geodesic polygon is replaced by the integral of geodesic curvature along the curve.

Problem 3.18 Prove that two geodesic lines on a surface which intersect in two points cannot bound a simply connected domain (digon) with nonpositive Gaussian curvature at all points.

Theorem 3.6 (Gauss–Bonnet) *Let M^2 be a closed oriented surface in $\mathbb{R}^3$. Then*

$$\int_{M^2} K d\sigma = 2\pi \cdot \chi(M^2),$$

where $\chi(M^2)$ is the Euler characteristic of the surface M^2.

First Proof We cut the surface M^2 into curvilinear polygons and apply the Gauss–Bonnet formula. Suppose that the number of polygons is f, the number of edges is e, and the number of vertices is v. Let us show that the summation of f Gauss–Bonnet formulas yields

$$2\pi e - 2\pi v = 2\pi f - \int_{M^2} K d\sigma.$$

For the right-hand side, this is obvious; consider the left-hand one. First, the integrals $\int_{\gamma_k} k_g ds$ occur twice with opposite signs for each edge, so that they cancel one another out. Secondly, summing the terms $\sum_k (\pi - \alpha_k)$, we obtain $2\pi e - 2\pi v$. Indeed, the number of terms equal to π coincides with the number of all vertices in all polygons (each vertex is counted as many times as the number of polygons containing it); therefore, it equals the number of all edges in all polygons, which, in turn, equals $2e$, because each edge belongs to precisely two polygons. Finally, the terms α_k at each of the v vertices amount to the full angle 2π.

As a result, we obtain

$$\int_{M^2} K d\sigma = 2\pi(v - e + f) = 2\pi \cdot \chi(M^2).$$

$\square$

Second Proof (Not Using the Gauss–Bonnet Formula) Let us construct a vector field $V(x)$ on M^2 with finitely many singular points. The sum of their indices equals $\chi(M^2)$. Let $\bigsqcup D_{\varepsilon,i}$ be the union of ε-neighborhoods of the singular points; we assume that ε is sufficiently small and these ε-neighborhoods are disjoint. Then

$$\int_{M^2} K d\sigma = \lim_{\varepsilon \to 0} \int_{M^2 \setminus \bigsqcup D_{\varepsilon,i}} K d\sigma.$$

The vector field under consideration has no singular points on $M^2 \setminus \bigsqcup D_{\varepsilon,i}$; therefore, using this field, we can construct a moving frame ε_1, ε_2 and consider the dual frame ω^1, ω^2. From this dual frame we construct a form ω_1^2 for which $d\omega_1^2 = -K\omega^1 \wedge \omega^2$. Applying Stokes' formula, we obtain

$$\int_{M^2 \setminus \bigsqcup D_{\varepsilon,i}} K d\sigma = \int_{M^2 \setminus \bigsqcup D_{\varepsilon,i}} K\omega^1 \wedge \omega^2 = -\int_{M^2 \setminus \bigsqcup D_{\varepsilon,i}} d\omega_1^2 = -\int_{\bigsqcup S_{\varepsilon,i}} \omega_1^2.$$

Together with the initial vector field $V(x)$, consider a "constant" vector field $\tilde{V}(x)$ on $D_{\varepsilon,i}$, that is, a vector field constant in some coordinate system on $D_{\varepsilon,i}$. Using $\tilde{V}(x)$, we construct the form $\tilde{\omega}_1^2$.

In Stokes' formula the orientation of the boundary of a two-dimensional surface is defined as shown in Fig. 3.2. Therefore, the orientations of the circles $S_{\varepsilon,i}$ are negative (i.e., clockwise). We have $\tilde{\omega}_1^2 = \omega_1^2 + d\varphi$, where φ is the angle of rotation

Fig. 3.2 Orientation of the
boundary

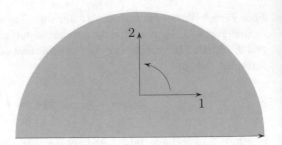

from the vector $V(x)$ to the constant vector $\tilde{V}(x)$. Hence the vector $V(x)$ makes an
angle of $-\varphi$ with the constant direction. Therefore, while traversing the whole circle
in the positive direction, the angle φ changes by -2π ind$_i$, where ind$_i$ is the index
of the given singular point, and while traversing the circle in the negative direction,
φ changes by 2π ind$_i$. Thus,

$$\int_{S_{\varepsilon,i}} \tilde{\omega}_1^2 = \int_{S_{\varepsilon,i}} \omega_1^2 + 2\pi \text{ ind}_i \, .$$

Moreover,

$$\lim_{\varepsilon \to 0} \int_{S_{\varepsilon,i}} \tilde{\omega}_1^2 = \lim_{\varepsilon \to 0} \int_{D_{\varepsilon,i}} d\tilde{\omega}_1^2 = 0.$$

As a result, we obtain

$$\int_{M^2} K d\sigma = 2\pi \sum_i \text{ind}_i = 2\pi \cdot \chi(M^2).$$

$\square$

3.8 Christoffel Symbols

Given a surface $r(u^1, u^2)$, consider the unit normal vector n and the tangent vectors
$r_1 = \frac{\partial r}{\partial u^1}$ and $r_2 = \frac{\partial r}{\partial u^2}$ at each point. Differentiating the equation $(n, n) = 1$, we
obtain $(n_i, n) = 0$, where $n_i = \frac{\partial n}{\partial u^i}$, $i = 1, 2$. Therefore, the vectors n_i are expressed
in terms of r_1 and r_2 as $n_i = c_i^j r_j$. Let us calculate the coefficients c_i^j. On the one
hand, $(n_i, r_k) = -b_{ik}$, where (b_{ik}) is the matrix of the second quadratic form. On
the other hand, $(n_i, r_k) = c_i^j (r_j, r_k) = c_i^j g_{jk}$, where (g_{jk}) is the matrix of the first
quadratic form. Let $b_i^l = b_{ik} g^{kl}$. Then $c_i^l = c_i^j g_{jk} g^{kl} = -b_{ik} g^{kl} = -b_i^l$. Thus,

$$n_i = -b_i^j r_j, \quad \text{where} \quad b_i^j = b_{ik} g^{kj} \quad \text{(Weingarten equations)}.$$

In particular, if the vectors r_1 and r_2 correspond to principal curvatures k_1 and k_2, then $n_i = -\frac{b_{ii}}{g_{ii}} r_i = -k_i r_i$, i.e.,

$$k_i \frac{\partial r}{\partial u^i} + \frac{\partial n}{\partial u^i} = 0 \quad (Rodrigues' \ formula)$$

(the summation with respect to the repeated index i is not performed).

HISTORICAL COMMENT The Weingarten equations were obtained by Julius Weingarten (1836–1910) in 1861. Rodrigues' formula was proved by Benjamin Olinde Rodrigues (1795–1851) in 1815.

Problem 3.19 Prove that if the principal curvatures of a closed surface are equal and different from zero at each point, then the surface is the sphere.

HISTORICAL COMMENT The statement in Problem 3.19 was proved by Meusnier in 1785. The points at which the principal curvatures are equal and different from zero are called *umbilical*.

In each tangent plane to a surface we can consider the linear operator L with matrix (b_i^l). This operator is called the *Weingarten operator*. It is related to the second quadratic form B as $B(u, v) = (Lu, v)$. Geometrically, the operator L is defined as $Lu = -\partial_u n$ (the derivative of n in the direction of u).

Problem 3.20 Prove that $L^2 = 2HL - KI$, where H is the mean curvature, K is the Gaussian curvature, and I is the identity operator.

Now consider the second derivatives $r_{ij} = \frac{\partial^2 r}{\partial u^i \partial u^j}$. The vector r_{ij} is expanded in the basis n, r_1, r_2 as

$$r_{ij} = \Gamma_{ij}^k r_k + a_{ij} n.$$

Let us calculate the coefficients Γ_{ij}^k and a_{ij}. First, note that $(r_k, n) = 0$, and therefore $(r_{ij}, n) = a_{ij}$, i.e., the $a_{ij} = b_{ij}$ are the entries of the matrix of the second quadratic form. The formula

$$r_{ij} = \Gamma_{ij}^k r_k + b_{ij} n \tag{3.3}$$

will often be used in what follows.

Next, we have $(r_{ij}, r_l) = (\Gamma_{ij}^k r_k, r_l) = \Gamma_{ij}^k g_{kl}$, whence $\Gamma_{ij}^p = \Gamma_{ij}^k g_{kl} g^{lp} = (r_{ij}, r_l) g^{lp}$.

The numbers Γ_{ij}^k are called the *Christoffel symbols*, and the formula $r_{ij} = \Gamma_{ij}^k r_k + b_{ij} n$ is known as *Gauss' formula*. It follows from the symmetry of the second derivative $(r_{ij} = r_{ji})$ that the Christoffel symbols are symmetric with respect to the subscripts: $\Gamma_{ij}^k = \Gamma_{ji}^k$.

HISTORICAL COMMENT In 1869 Elwin Bruno Christoffel (1829–1900) considered conditions for the coincidence of Riemann geometries with first quadratic forms $g_{ij}dx_idx_j$ and $h_{ij}dx_idx_j$. This generalized the superposition problem for surfaces to the many-dimensional case. The condition that the geometries coincide was expressed in terms of the Christoffel symbols. In the two-dimensional case, the Christoffel symbols for surfaces were known to Gauss; they are mentioned in his 1827 paper.

Theorem 3.7 (Gauss) *The Christoffel symbols are expressed in terms of the first quadratic form (g_{ij}) as*

$$\Gamma_{ij}^k = \frac{1}{2} g^{kl} \left(\frac{\partial g_{il}}{\partial u^j} - \frac{\partial g_{ij}}{\partial u^l} + \frac{\partial g_{lj}}{\partial u^i} \right).$$

Proof Let $\Gamma_{ijl} = \Gamma_{ij}^k g_{kl} = (r_{ij}, r_l)$. Clearly, $r_{ij} = r_{ji}$; therefore, $\Gamma_{ijl} = \Gamma_{jil}$. Differentiating the equation $g_{ij} = (r_i, r_j)$, we obtain

$$\frac{\partial g_{ij}}{\partial u^l} = (r_{il}, r_j) + (r_i, r_{jl}) = \Gamma_{ilj} + \Gamma_{jli}.$$

Similarly,

$$\frac{\partial g_{il}}{\partial u^j} = \Gamma_{ijl} + \Gamma_{lji}, \qquad \frac{\partial g_{lj}}{\partial u^i} = \Gamma_{lij} + \Gamma_{jil}.$$

Thus,

$$\frac{1}{2} \left(\frac{\partial g_{il}}{\partial u^j} - \frac{\partial g_{ij}}{\partial u^l} + \frac{\partial g_{lj}}{\partial u^i} \right) = \Gamma_{ijl}.$$

It is also clear that $\Gamma_{ij}^k = g^{kl} g_{pl} \Gamma_{ij}^p = g^{kl} \Gamma_{ijl}$. □

Problem 3.21 Calculate the Christoffel symbols for polar coordinates (r, φ) in the plane.

Problem 3.22 Calculate the Christoffel symbols for a surface of revolution $(f(u) \cos v, f(u) \sin v, u)$.

Problem 3.23 Let Γ_{ij}^k be the Christoffel symbols in coordinates (x^1, x^2), and let $\tilde{\Gamma}_{pq}^s$ be the Christoffel symbols in coordinates (y^1, y^2). Prove that

$$\Gamma_{ij}^k = \left(\tilde{\Gamma}_{pq}^s \frac{\partial y^p}{\partial x^i} \cdot \frac{\partial y^q}{\partial x^j} + \frac{\partial^2 y^s}{\partial x^i \partial x^j} \right) \frac{\partial x^k}{\partial y^s}.$$

Problem 3.24 Prove that, at any point of a two-dimensional surface, local coordinates can be chosen so that all Christoffel symbols at this point vanish.

3.9 The Spherical Gauss Map

For the sphere in $\mathbb{R}^3$ considered not only purely topologically but also geometrically, i.e., as a surface with first quadratic form, we use the notation $\mathbb{S}^2$.

Given a smooth oriented surface M^2 in $\mathbb{R}^3$, the *spherical Gauss map* $f\colon M^2 \to \mathbb{S}^2$ takes each point $x \in M^2$ to the head of the unit normal vector $n(x)$ to the surface M^2 at x.

According to Rodrigues' formula, we have $\frac{\partial n}{\partial u^1} = -k_1 \frac{\partial r}{\partial u^1}$ and $\frac{\partial n}{\partial u^2} = -k_2 \frac{\partial r}{\partial u^2}$. Therefore, an infinitesimal parallelogram of area $\left| \frac{\partial r}{\partial u^1} \times \frac{\partial r}{\partial u^2} \right| du^1 du^2$ on the surface M^2 is mapped to an infinitesimal parallelogram of area

$$
\left| \frac{\partial n}{\partial u^1} \times \frac{\partial n}{\partial u^2} \right| du^1 du^2 = \left| k_1 k_2 \frac{\partial r}{\partial u^1} \times \frac{\partial r}{\partial u^2} \right| du^1 du^2 = |K| \left| \frac{\partial r}{\partial u^1} \times \frac{\partial r}{\partial u^2} \right| du^1 du^2
$$

on the sphere. Since the ratio of oriented areas equals K, we obtain the following interpretation of Gaussian curvature:

$$
K = \frac{\text{the oriented area of } f(A) \text{ on } \mathbb{S}^2}{\text{the oriented area of } A \text{ on } M^2},
$$

where A is an infinitesimal figure at a point $x \in M^2$. Thus,

$$
K d\sigma_{M^2} = f^* d\sigma_{\mathbb{S}^2}, \tag{3.4}
$$

where $d\sigma_{M^2}$ and $d\sigma_{\mathbb{S}^2}$ are the area forms on M^2 and on $\mathbb{S}^2$.

Relation (3.4) can also be obtained in the language of differential forms. Recall that $n = \varepsilon_3$ and $K\omega^1 \wedge \omega^2 = -d\omega_1^2 = \omega_1^3 \wedge \omega_2^3$. Here $\omega^1 \wedge \omega^2$ is the area form on the surface M^2. The vectors ε_1 and ε_2 form a basis of both the tangent space to M^2 at the point x and the tangent space to the unit sphere $\mathbb{S}^2$ at the point $f(x)$. For the sphere, the equation $d\varepsilon_3 = \omega_3^1 \varepsilon_1 + \omega_3^2 \varepsilon_2$ is similar to the equation $dx = \omega^1 \varepsilon_1 + \omega^2 \varepsilon_2$ for the surface M^2. Therefore, $\omega_1^3 \wedge \omega_2^3$ is the image of an area element of the surface under the spherical Gauss map.

For any smooth map $g\colon M^n \to N^n$ of manifolds, we have

$$
\int_{M^n} g^* \omega = (\deg g) \int_{N^n} \omega, \tag{3.5}
$$

where $\deg g$ is the degree of g. Indeed, using a partition of unity, we can reduce the proof to the case where the form ω is different from zero only in an arbitrarily small neighborhood. In the case where g maps homeomorphically several open balls to one ball, the task reduces directly to the definition of the degree of a map. The critical values can be ignored, because they constitute a set of measure zero.

Theorem 3.8 *For a closed oriented surface $M^2 \subset \mathbb{R}^3$, the degree of the spherical Gauss map equals $\frac{1}{2}\chi(M^2)$.*

Proof Applying Eq. (3.5) to the spherical Gauss map f, we obtain

$$\int_{M^2} f^* d\sigma_{\mathbb{S}^2} = (\deg f) \int_{\mathbb{S}^2} d\sigma_{\mathbb{S}^2} = 4\pi \cdot (\deg f).$$

According to (3.4), we have $f^* d\sigma_{\mathbb{S}^2} = K d\sigma_{M^2}$. Moreover, by virtue of the Gauss–Bonnet theorem, $\int_{M^2} K d\sigma_{M^2} = 2\pi \cdot \chi(M^2)$. Thus, $4\pi \cdot (\deg f) = 2\pi \cdot \chi(M^2)$, i.e., $\deg f = \frac{1}{2}\chi(M^2)$. $\square$

HISTORICAL COMMENT Gauss introduced the spherical Gauss map in 1827, in studying curved surfaces. He used it to define the Gaussian curvature of a surface. Before Gauss, the curvature of a surface was defined in precisely the same way by Rodrigues in 1815.

Problem 3.25 Prove that the differential of the spherical Gauss map $f: M^2 \to \mathbb{S}^2$ is a self-adjoint operator and the quadratic form associated with this operator is the second quadratic form with the minus sign.

3.10   The Geodesic Equation

Recall that a curve $\gamma(s) = r(u^1(s), u^2(s))$ with natural parameterization on a smooth surface $r(u^1, u^2)$ is said to be geodesic if its geodesic curvature identically vanishes.

Theorem 3.9 *A curve $\gamma(s)$ on a surface $r(u^1, u^2)$ is geodesic if and only if*

$$\frac{d^2 u^k}{ds^2} + \Gamma_{ij}^k \frac{du^i}{ds} \cdot \frac{du^j}{ds} = 0 \text{ for } k = 1, 2.$$

Proof First, suppose that the parameterization of the curve is natural. Consider the velocity vector of the curve $\varepsilon_1 = \frac{du^i}{ds} r_i$, where $r_i = \frac{\partial r}{\partial u^i}$. Clearly,

$$\frac{d\varepsilon_1}{ds} = \frac{du^i}{ds} \cdot \frac{du^j}{ds} r_{ij} + \frac{d^2 u^i}{ds^2} r_i, \text{ where } r_{ij} = \frac{\partial^2 r}{\partial u^i \partial u^j}.$$

Therefore, applying the Gauss formula (Eq. (3.3) on p. 85), we obtain

$$\frac{d\varepsilon_1}{ds} = \left(\Gamma_{ij}^k \frac{du^i}{ds} \cdot \frac{du^j}{ds} + \frac{d^2 u^k}{ds^2} \right) r_k + b_{ij} \frac{du^i}{ds} \cdot \frac{du^j}{ds} n. \tag{3.6}$$

Recall that $\frac{d\varepsilon_1}{ds} = k_g\varepsilon_2 + k_n\varepsilon_3$, where the vector ε_2 lies in the tangent plane and $\varepsilon_3 = n$. The vectors r_1 and r_2 are linearly independent; it follows that $k_g = 0$ if and only if both coefficients multiplying the vectors r_1 and r_2 vanish.

A geodesic with any parameterization has the form $\gamma(as + b)$, where $\gamma(s)$ is a geodesic with natural parameterization and a and b are constants, $a \neq 0$. The curve $\gamma(as + b)$ satisfies the above equation if and only if so does the curve $\gamma(s)$. □

In Problem 3.22 we calculated the Christoffel symbols for a surface of revolution $(f(u)\cos v, f(u)\sin v, u)$. This allows us to write equations for a geodesic on a surface of revolution. One of these two equations has the form

$$\frac{d^2v}{ds^2} + 2\frac{f'}{f} \cdot \frac{du}{ds} \cdot \frac{dv}{ds} = 0.$$

It follows from this equation that, for a geodesic on a surface of revolution, the quantity $f^2\frac{dv}{ds}$ is constant. Indeed,

$$\frac{d}{ds}\left(f^2\frac{dv}{ds}\right) = f^2\left(\frac{d^2v}{ds^2} + 2\frac{f'}{f} \cdot \frac{du}{ds} \cdot \frac{dv}{ds}\right) = 0.$$

A visual geometric interpretation of the constancy of $f^2\frac{dv}{ds}$ is given by Clairault's theorem.

Theorem 3.10 (Clairault) *For a geodesic γ on a surface of revolution, the product $f \sin\alpha$, where f is the distance to the axis of rotation and α is the angle between the velocity vector of the geodesic and the corresponding meridian, is constant.*

Proof The vectors $e_u = (f'\cos v, f'\sin v, 1)$ and $e_v = (-f\sin v, f\cos v, 0)$ are tangent to the meridian and the parallel, respectively. It follows that, on the one hand, $\frac{d\gamma}{ds} = \cos\alpha\frac{e_u}{\|e_u\|} + \sin\alpha\frac{e_v}{f}$ by the definition of the angle α, and on the other hand, a direct calculation shows that $\frac{d\gamma}{ds} = \frac{du}{ds}e_u + \frac{dv}{ds}e_v$. Therefore, $\sin\alpha = f\frac{dv}{ds}$. Thus, $f\sin\alpha = f^2\frac{dv}{ds}$ is constant. □

HISTORICAL COMMENT Alexis Claude Clairault (1713–1765) proved the theorem about a geodesic on a surface of revolution in 1733.

3.11 Parallel Transport Along a Curve

Let $\gamma(t) = r(u^1(t), u^2(t))$, $t \in [a, b]$, be a curve on a surface $r(u^1, u^2)$. Consider a smooth vector field $V(t)$ on the curve γ, i.e., a smooth map $V : [a, b] \to \mathbb{R}^3$. The vector field V is said to be *parallel* along γ if the vector $\frac{dV}{dt}$ is perpendicular to the surface at each point. This definition does not depend on the parameterization of the curve γ, because $\frac{dV}{dt} = \frac{dV}{ds} \cdot \frac{ds}{dt}$, i.e., the vectors $\frac{dV}{dt}$ and $\frac{dV}{ds}$ are proportional.

For an arbitrary (not necessarily tangent to the surface) vector field, the definition of parallel transport along a curve on a surface involves the arbitrariness in the choice of a normal direction. But of greatest interest are vector fields that are tangent to the surface at each point. In what follows, we mainly consider vector fields with this property. A vector field on a surface can be represented in form $V^1 \frac{\partial r}{\partial u^1} + V^2 \frac{\partial r}{\partial u^2}$.

As we will shortly see, a vector field parallel along a curve is uniquely determined by any vector in this field; i.e., given a tangent vector, we can define a uniquely determined parallel transport of this vector along a curve. The geometric meaning of such a transport is as follows. The vector is translated to an infinitely close point of the curve in $\mathbb{R}^3$. The translate is not necessarily tangent to the surface. To obtain a tangent vector, we take the orthogonal projection of the translate on the tangent plane. Such is the geometric interpretation of the perpendicularity of the vector $dV = V(t + dt) - V(t)$ to the tangent plane.

Example 3.5 If a curve γ lies in a plane, then a vector field in the plane is parallel along γ if and only if it is constant (i.e., a vector is translated along γ in the ordinary sense).

Proof We can assume that the plane in which the curve γ lies is given by the equation $x^3 = 0$. Let $V(t) = (V^1(t), V^2(t), 0)$. Then $\frac{dV}{dt} = \left(\frac{dV^1}{dt}, \frac{dV^2}{dt}, 0 \right)$. This vector is orthogonal to the plane $x^3 = 0$ if and only if $\frac{dV^1}{dt} = \frac{dV^2}{dt} = 0$. This means that $V(t) = \text{const}$. $\qquad \square$

Theorem 3.11 *A vector field $V(t) = V^i(t) \frac{\partial r}{\partial u^i}$ is parallel along a curve $\gamma(t)$ if and only if*

$$\frac{dV^k}{dt} + \Gamma^k_{ij} V^i \frac{d\gamma^j}{dt} = 0 \quad for \quad k = 1, 2.$$

Proof First, note that

$$\frac{dV}{dt} = \frac{dV^k}{dt} \cdot \frac{\partial r}{\partial u^k} + V^i \frac{\partial^2 r}{\partial u^i \partial u^j} \cdot \frac{d\gamma^j}{dt}.$$

Next, according to the Gauss formula (Eq. (3.3) on p. 85), we have

$$\frac{\partial^2 r}{\partial u^i \partial u^j} = \Gamma^k_{ij} \frac{\partial r}{\partial u^k} + b_{ij} n.$$

Thus, the orthogonal projection of the vector $\frac{dV}{dt}$ on the tangent plane equals $\left(\frac{dV^k}{dt} + \Gamma^k_{ij} V^i \frac{d\gamma^j}{dt} \right) \frac{\partial r}{\partial u^k}$. Since the vectors $\frac{\partial r}{\partial u^1}$ and $\frac{\partial r}{\partial u^2}$ are linearly independent, it follows that the projection is zero (i.e., the vector field $V(t)$ is parallel along the curve γ) if and only if the equations in the statement of the theorem hold. $\qquad \square$

Corollary 1 *A curve γ is geodesic if and only if it velocity vectors form a vector field parallel along γ.*

Corollary 2 *A vector field parallel along a curve* $\gamma(t)$ *is uniquely determined by any of its vectors. To be more precise, given any vector tangent to a surface at some point of a curve* $\gamma(t)$, *there exists a unique vector field* $V(t)$ *parallel along* $\gamma(t)$ *and containing this vector.*

Proof To construct a vector field $V(t)$ parallel along $\gamma(t)$ and containing the given vector, we must find a solution of a system of differential equations with given initial condition. This can be done in a sufficiently small neighborhood. It remains to cover the interval by a finite number of such neighborhoods.   □

Corollary 2 suggests the definition of the *parallel transport* of a vector along a given curve. This transformation is a map of the tangent plane at an endpoint of the curve to the tangent plane at the other endpoint. It is easy to show that parallel transport preserves inner product. Indeed, if $V(t)$ and $W(t)$ are vector fields parallel along a given curve, then

$$\frac{d}{dt}(V(t), W(t)) = \left(\frac{dV(t)}{dt}, W(t)\right) + \left(V(t), \frac{dW(t)}{dt}\right) = 0,$$

because the vectors $\frac{dV(t)}{dt}$ and $\frac{dW(t)}{dt}$ are orthogonal to all tangent vectors. Since inner product is preserved by parallel transport, it follows that so do the lengths of vectors and the angles between vectors.

HISTORICAL COMMENT The theory of the parallel transport of a vector along a curve on a Riemannian manifold was developed in 1917 by Tullio Levi-Città (1873–1941).

According to Theorem 3.7, the Christoffel symbols are expressed in terms of the metric g_{ij}. Therefore, the parallel transport of a vector along a curve on a cone is the same thing as the parallel transport along the image of this curve on a net of the cone, and the parallel transport of a vector along a plane curve amounts to an ordinary parallel translation.

It is seen directly from the definition of a vector field parallel along a curve γ that if two surfaces are tangent along γ, i.e., the tangent planes to these surfaces coincide at each point of γ, then any vector field parallel along γ with respect to one of the surfaces is also parallel along γ with respect to the other surface. For example, if we consider a cone and a sphere inscribed in it, then a vector field parallel along the circle of tangency γ with respect to the sphere is the same thing as a vector field parallel along γ with respect to the cone. Using this remark, we can describe the parallel transport of a vector along a circle on the sphere.

Problem 3.26 A plane divides a sphere of radius r into two parts of areas S and $4\pi r^2 - S$. Parallel transport along the circle in the intersection of the plane with the sphere takes a vector V_1 to a vector V_2. Prove that the angle between the vectors V_1 and V_2 equals S/r^2.

The notion of parallel transport suggests the following geometric interpretation of geodesic curvature k_g of a curve $\gamma(s)$ on a surface S.

Theorem 3.12 *Consider a parallel vector field $V(s)$ on a curve $\gamma(s)$ with natural parameterization. Let φ be the angle of rotation from the tangent vector $\varepsilon_1 = \frac{d\gamma}{ds}$ of the curve to the vector V. Then $k_g = -\frac{d\varphi}{ds}$.*

Proof Since the length of a vector is preserved by parallel transport, we can assume that $\|V(s)\| = 1$. Then $V = \cos\varphi\varepsilon_1 + \sin\varphi\varepsilon_2$, where ε_2 is the vector in the Darboux frame. Recall that $(\varepsilon_1, \varepsilon_2') = -k_g$ and $(\varepsilon_2, \varepsilon_1') = k_g$. By assumption the vector field V is parallel along γ; hence the vector V' is parallel to the normal vector ε_3 at each point, i.e., V' is perpendicular to the vectors ε_1 and ε_2. Thus,

$$0 = (\varepsilon_1, V') = (\varepsilon_1, \cos\varphi\varepsilon_1' + \sin\varphi\varepsilon_2' - \varphi'\sin\varphi\varepsilon_1 + \varphi'\cos\varphi\varepsilon_2)$$
$$= (\varepsilon_1, \varepsilon_2' - \varphi'\varepsilon_1)\sin\varphi = -(k_g + \varphi')\sin\varphi,$$

because $(\varepsilon_1, \varepsilon_1') = 0$ and $(\varepsilon_1, \varepsilon_2) = 0$. A similar calculation for (ε_2, V') shows that $(k_g + \varphi')\cos\varphi = 0$. One of the numbers $\cos\varphi$ and $\sin\varphi$ is different from zero; therefore, $k_g = -\varphi'$.   □

Corollary *Parallel transport along a closed curve bounding a simply connected domain G rotates each vector through an angle of $\int_G K\,d\sigma$.*

Proof Replacing the geodesic curvature in the Gauss–Bonnet formula

$$\int_\gamma k_g ds = 2\pi - \int_G K\,d\sigma$$

by $-\frac{d\varphi}{ds}$, we see that the angle of rotation of a vector equals $\int_G K\,d\sigma$ up to 2π.   □

3.12 Covariant Differentiation

On any manifold the differentiation of a function in the direction of a vector can be defined. But an attempt to differentiate a vector field in the direction of a vector faces an insurmountable difficulty: we must consider differences of vectors lying in tangent spaces at different points, which requires that these tangent spaces be identified somehow. In the general case, this cannot be done, but in the presence of a metric, such an identification is possible. In fact, we have already performed differentiation in the definition of the parallel transport of a vector along a curve, because a vector field parallel along a curve is a vector field whose derivative in the direction of the tangent vector to the curve is zero. Thus, in fact, we defined what a zero derivative is. Now we give a general definition of the derivative of a vector field in the direction of a vector.

We can do this in two ways. First, we can repeat the construction used in the definition of the parallel transport of a vector along a curve in a more general form. Secondly, we can identify tangent spaces at different points of a curve by applying transport along this curve and use this identification to define differentiation. Below we give both of these definitions and prove their equivalence.

Let f be a function on $\mathbb{R}^n$, and let V and W be vector fields on $\mathbb{R}^n$. Then we can define a function $\partial_W f$ (the derivative of f in the direction of W) by setting its value at a point P to equal the limit

$$\lim_{\varepsilon \to 0} \frac{f(P + \varepsilon W) - f(P)}{\varepsilon} = W^i(P) \frac{\partial f}{\partial x^i}(P).$$

We can also define a vector field $\partial_W V$ (the derivative of V in the direction of W). At a point P this derivative equals the limit

$$\lim_{\varepsilon \to 0} \frac{V(P + \varepsilon W) - V(P)}{\varepsilon} = W^i(P) \frac{\partial V}{\partial x^i}(P).$$

If V and W are vector fields on a surface $S \subset \mathbb{R}^3$, then the vector field $\partial_W V$ is not necessarily tangent to S. To obtain a vector field on S, we take the orthogonal projection of $\partial_W V$ on the tangent space at each point $P \in S$. The vector field thus obtained is called the *covariant derivative* of V in the direction of W and denoted by $\nabla_W V$.

To calculate $\nabla_W V$ for a parameterized surface $r(u^1, u^2)$, we need Eq. (3.3):

$$r_{ij} = \frac{\partial^2 r}{\partial u^i \partial u^j} = \Gamma^k_{ij} r_k + b_{ij} n.$$

If $W = W^j r_j$ and $V = V^i r_i$, then

$$\partial_W V = W^j \frac{\partial}{\partial u^j}(V^i r_i) = W^j \frac{\partial V^i}{\partial u^j} r_i + W^j V^i \frac{\partial r_i}{\partial u^j}$$

$$= W^j \frac{\partial V^i}{\partial u^j} r_i + W^j V^i (\Gamma^k_{ij} r_k + b_{ij} n).$$

To obtain the covariant derivative, it suffices to discard the component parallel to the normal vector n. Thus,

$$\nabla_W V = W^j \frac{\partial V^i}{\partial u^j} r_i + W^j V^i \Gamma^k_{ij} r_k. \tag{3.7}$$

In particular,

$$\nabla_{r_j} r_i = \Gamma^k_{ij} r_k. \tag{3.8}$$

Now let us define covariant derivative by using the parallel transport of a vector along a curve and check that we come to the same definition. Let γ be a curve on a surface S. We set $\frac{d\gamma}{dt} = W$. Given a vector field V, let V_0 and V_ε be its vectors at points $\gamma(0)$ and $\gamma(\varepsilon)$, and let V_0' be V_0 transported along γ to the point $\gamma(\varepsilon)$. Then, at the point $\gamma(0)$, we define

$$\nabla_W V = \lim_{\varepsilon \to 0} \frac{V_\varepsilon - V_0'}{\varepsilon},$$

We have

$$V_\varepsilon = V_0 + \varepsilon W^j \frac{\partial V}{\partial u_j} + o(\varepsilon) = V_0 + \varepsilon W^j \frac{\partial V^i}{\partial u_j} r_i + \varepsilon W^j V^i \frac{\partial r_i}{\partial u^j} + o(\varepsilon).$$

Moreover, it follows from the transport equation

$$\frac{dV^k}{dt} r_k + \Gamma_{ij}^k V^i \frac{d\gamma^j}{dt} r_k = 0$$

(see Theorem 3.11) that

$$\frac{dV}{dt} = \frac{dV^k}{dt} r_k + V^k \frac{dr_k}{dt} = -\Gamma_{ij}^k V^i \frac{d\gamma^j}{dt} r_k + W^j V^i \frac{\partial r_i}{\partial u^j}.$$

Therefore,

$$V_0' = V_0 - \varepsilon W^j V^i \Gamma_{ij}^k r_k + \varepsilon W^j V^i \frac{\partial r_i}{\partial u^j} + o(\varepsilon).$$

Thus,

$$\nabla_W V = W^j \frac{\partial V^i}{\partial u^j} r_i + W^j V^i \Gamma_{ij}^k r_k,$$

which coincides with (3.7).

HISTORICAL COMMENT Covariant differentiation was developed during the decade 1884–1894 by Gregorio Ricci-Curbastro (1853–1925), who named it "absolute differential calculus." Ricci-Curbastro aimed at carrying over the usual procedures of calculus to Riemannian manifolds in such a way that they be invariant with respect to changes of coordinates. The most detailed exposition of this theory is contained in a 1900 joint paper of Ricci and his student Levi-Cività. In 1917 Levi-Cività introduced the notion of parallel transport along a curve on a Riemannian manifold, which had rendered Ricci's formal constructions geometrically evident.

A direct calculation based on the expression for covariant derivative shows that the multiplication of one of the vector fields by a function f results in the transformations

$$\nabla_{fW} V = f \nabla_W V, \quad \nabla_W (fV) = (\partial_W f)V + f \nabla_W V. \tag{3.9}$$

The symmetry of the Christoffel symbols with respect to the subscripts ($\Gamma^k_{ij} = \Gamma^k_{ji}$) implies the symmetry of covariant differentiation:

$$\nabla_W V - \nabla_V W = \partial_W V - \partial_V W.$$

The vector field $\partial_W V - \partial_V W$ is denoted by $[W, V]$ and called the *commutator* of the vector fields W and V. It is easy to check that

$$\partial_{[W,V]} f = \partial_W (\partial_V f) - \partial_V (\partial_W f).$$

Indeed, $\partial_V f = V^i \frac{\partial f}{\partial x^i}$; therefore,

$$\partial_W (\partial_V f) = \partial_W \left(V^i \frac{\partial f}{\partial x^i} \right) = W^j \frac{\partial V^i}{\partial x^j} \frac{\partial f}{\partial x^i} + W^j V^i \frac{\partial^2 f}{\partial x^i \partial x^j}.$$

Hence

$$\partial_W (\partial_V f) - \partial_V (\partial_W f) = \left(W^j \frac{\partial V^i}{\partial x^j} - V^j \frac{\partial W^i}{\partial x^j} \right) \frac{\partial f}{\partial x^i}.$$

This is the derivative of the function f in the direction of the difference of the vector fields $\partial_W V = W^j \frac{\partial V}{\partial x^j}$ and $\partial_V W = V^j \frac{\partial W}{\partial x^j}$.

Thus, the ith coordinate of the vector $[W, V]$ equals $W^j \frac{\partial V^i}{\partial x^j} - V^j \frac{\partial W^i}{\partial x^j}$. It is easy to derive from this expression that, for any function f, we have

$$[fW, V] = f[W, V] - (\partial_V f)W, \quad [W, fV] = f[W, V] + (\partial_W f)V. \tag{3.10}$$

Problem 3.27 Consider the vector fields $r_1 = \frac{\partial r}{\partial u^1}$ and $r_2 = \frac{\partial r}{\partial u^2}$ on a parameterized surface $r(u^1, u^2)$. Prove that $[r_1, r_2] = 0$, i.e., the coordinate vector fields commute.

In applications it is often required to find the covariant derivative of a vector field V given on a curve $\gamma(t)$ in the direction of the velocity vector of this curve. We denote this covariant derivative by $\nabla_{\gamma'} V$. Let us discuss some properties of the covariant derivative $\nabla_{\gamma'} V$.

First, note that the covariant derivative $\nabla_{\gamma'} V$ along the curve γ identically vanishes if and only if the vector field V is parallel along γ. This assertion follows from the definition of covariant derivative, but it can also be proved by comparing the equation of parallel transport along a curve (see Theorem 3.11) and the equation $\nabla_{\gamma'} V = 0$. To this end, we write the parallel transport equation in the form

$$\frac{dV^i}{dt} r_i + \Gamma^k_{ij} V^i \frac{d\gamma^j}{dt} r_k = 0$$

and note that $\frac{dV^i}{dt} = \frac{d\gamma^j}{dt} \cdot \frac{\partial V^i}{\partial u_j}$.

Theorem 3.13 *For vector fields V and W on a curve $\gamma(t)$,*

$$\frac{d}{dt}(V, W) = (\nabla_{\gamma'} V, W) + (V, \nabla_{\gamma'} W).$$

Proof We choose an orthonormal basis e_i in the tangent space at the point $\gamma(0)$ and consider its parallel transport along the curve. At each point of the curve we expand the vectors V and W in the corresponding basis: $V = V^i e_i$ and $W = W^j e_j$. Clearly,

$$\nabla_{\gamma'}(V^i e_i) = \frac{dV^i}{dt} e_i + V^i (\nabla_{\gamma'} e_i) = \frac{dV^i}{dt} e_i,$$

because $\nabla_{\gamma'} e_i = 0$ (the vector field e_i is parallel along the curve). Therefore,

$$(\nabla_{\gamma'} V, W) + (V, \nabla_{\gamma'} W) = (\frac{dV^i}{dt} e_i, W^j e_j) + (V^i e_i, \frac{dW^j}{dt} e_j)$$

$$= \frac{dV^i}{dt} W^i + V^i \frac{dW^i}{dt} = \frac{d}{dt}(V, W).$$

$\square$

Covariant differentiation can also be extended to covector fields. A *covector* is an element of the dual space (a linear function on vectors), and a covector field is a differential 1-form. With each basis r_i in a vector space a dual basis r^j in the covector space is associated; it is defined by the relation $r^j(r_i) = \delta^j_i$. The value of a covector $T = T_j r^j$ at a vector $V = V^i r_i$ equals $TV = T_i V^i$.

The covariant derivative $\nabla_W T$ of a covector field T is defined so as to satisfy the following two conditions:

(1) the Leibniz formula $\nabla_W (TV) = (\nabla_W T)V + T(\nabla_W V)$ holds, in particular,

$$\nabla_{r_j}(TV) = (\nabla_{r_j} T)V + T(\nabla_{r_j} V) = (\nabla_{r_j} T)V + T_i \left(\frac{\partial V^i}{\partial u^j} + \Gamma^i_{kj} V^k \right);$$

(2) $\nabla_W (TV)$ is the usual derivative of the function TV in the direction of W, in particular,

$$\nabla_{r_j}(TV) = \frac{\partial T_i}{\partial u^j} V^i + T_i \frac{\partial V^i}{\partial u^j}.$$

Comparing the two expressions for $\nabla_{r_j}(TV)$, we obtain

$$\left((\nabla_{r_j} T)_i + T_k \Gamma_{ij}^k - \frac{\partial T_i}{\partial u^j} \right) V^i = 0;$$

therefore,

$$(\nabla_{r_j} T)_i = \frac{\partial T_i}{\partial u^j} - T_k \Gamma_{ij}^k.$$

3.13   The Gauss and Codazzi–Mainardi Equations

Let us continue the calculation begun in Sect. 3.8, namely, calculate $r_{ijk} = \frac{\partial^3 r}{\partial u^i \partial u^j \partial u^k}$ and compare the resulting expressions for r_{ijk} and r_{ikj}. Using the notation $b_i^l = b_{ik} g^{kl}$ introduced in Sect. 3.8 and the relations $\frac{\partial n}{\partial u^i} = -b_i^j r_j$ and $r_{ij} = \Gamma_{ij}^l r_l + b_{ij} n$ obtained in the same section, we arrive at

$$
\begin{aligned}
r_{ijk} &= \frac{\partial}{\partial u^k} \left(\Gamma_{ij}^l r_l + b_{ij} n \right) \\
&= \frac{\partial \Gamma_{ij}^l}{\partial u^k} r_l + \Gamma_{ij}^l r_{lk} + \frac{\partial b_{ij}}{\partial u^k} n + b_{ij} \frac{\partial n}{\partial u^k} \\
&= \frac{\partial \Gamma_{ij}^l}{\partial u^k} r_l + \Gamma_{ij}^l \Gamma_{lk}^m r_m + \Gamma_{ij}^l b_{lk} n + \frac{\partial b_{ij}}{\partial u^k} n - b_{ij} b_k^l r_l \\
&= \left(\Gamma_{ij}^l b_{lk} + \frac{\partial b_{ij}}{\partial u^k} \right) n + \left(\frac{\partial \Gamma_{ij}^l}{\partial u^k} + \Gamma_{ij}^p \Gamma_{pk}^l - b_{ij} b_k^l \right) r_l.
\end{aligned}
$$

A similar calculation yields

$$r_{ikj} = \left(\Gamma_{ik}^l b_{lj} + \frac{\partial b_{ik}}{\partial u^j} \right) n + \left(\frac{\partial \Gamma_{ik}^l}{\partial u^j} + \Gamma_{ik}^p \Gamma_{pj}^l - b_{ik} b_j^l \right) r_l.$$

But $r_{ijk} = r_{ikj}$. Moreover, the vectors n and r_l are linearly independent. Therefore,

$$\frac{\partial \Gamma_{ik}^l}{\partial u^j} - \frac{\partial \Gamma_{ij}^l}{\partial u^k} + \Gamma_{ik}^p \Gamma_{pj}^l - \Gamma_{ij}^p \Gamma_{pk}^l = b_{ik} b_j^l - b_{ij} b_k^l \tag{3.11}$$

and

$$\frac{\partial b_{ij}}{\partial u^k} - \frac{\partial b_{ik}}{\partial u^j} = \Gamma_{ik}^l b_{lj} - \Gamma_{ij}^l b_{lk}. \tag{3.12}$$

There are only two independent equations in (3.12):

$$\frac{\partial b_{i1}}{\partial u^2} + \Gamma_{i1}^1 b_{12} + \Gamma_{i1}^2 b_{22} = \frac{\partial b_{i2}}{\partial u^1} + \Gamma_{i2}^1 b_{11} + \Gamma_{i2}^2 b_{21} \tag{3.13}$$

for $i = 1$ and $i = 2$.

HISTORICAL COMMENT Equations (3.11) were obtained by Gauss in 1827. They are known as the *Gauss equations*. Gauss used them to express Gaussian curvature in terms of metric. The history of Eq. (3.13) is more intricate. They were first obtained by Karl Mikhailovich Peterson (1828–1881) in 1853, but his dissertation was published only a hundred years later. In this dissertation, he proved that the first and the second quadratic form determine a surface up to motion. Peterson supplemented the Gauss equations by two more independent equations and proved that if the coefficients of two quadratic forms (the first of them must be positive definite) satisfy these equations, then there exists a surface for which these forms are the first and second quadratic forms, and they determine this surface up to a motion of the space. (The first proof of this theorem was published by Bonnet in 1867; it is usually referred to as Bonnet's theorem.) Later the same equations were obtained by Gaspare Mainardi (1800–1879) (in 1856) and Delfino Codazzi (1824–1873) in 1859. For this reason, in Russia Eq. (3.12) are known as the *Peterson–Codazzi equations* and abroad, as the *Codazzi–Mainardi equations*.

The *Riemann curvature tensor* of a two-dimensional surface is defined as

$$R_{ijk}^l = \frac{\partial \Gamma_{ik}^l}{\partial u^j} - \frac{\partial \Gamma_{ij}^l}{\partial u^k} + \Gamma_{ik}^p \Gamma_{pj}^l - \Gamma_{ij}^p \Gamma_{pk}^l.$$

In terms of this tensor, the Gauss equations can be written in the form

$$R_{ijk}^l = b_{ik} b_j^l - b_{ij} b_k^l. \tag{3.14}$$

Gauss used these equations to prove that Gaussian curvature is an intrinsic characteristic of a surface, i.e., depends only on the metric. Gauss gave the name *Theorema Egregium* ('remarkable theorem') to this statement.

Theorem 3.14 (Gauss)   *The Gaussian curvature K can be expressed in terms of the metric g_{ij} alone.*

Proof Equation (3.14) implies

$$R_{ijk}^l g_{lm} = b_{ik} b_j^l g_{lm} - b_{ij} b_k^l g_{lm}$$

$$= b_{ik} b_{jm} - b_{ij} b_{km}.$$

For $i = k = 1$ and $j = m = 2$, we obtain

$$R^l_{121} g_{l2} = b_{11} b_{22} - b^2_{12} = \det(b_{ij})$$
$$= \det\left((b^k_j)(g_{ik})\right) = K \det(g_{ik}).$$

Thus,

$$K = \frac{1}{\det(g_{ik})} R^l_{121} g_{l2}.$$

It remains to recall that, according to another theorem of Gauss (see Theorem 3.7 on p. 86), the Christoffel symbols can be expressed in terms of the metric g_{ik}.   □

It is easy to check that if the matrix (g_{ij}) is the identity, then

$$R^1_{212} = R^2_{121} = -R^1_{221} = -R^2_{112} = K$$

and all the other components of the Riemann tensor are zero.

3.14   Riemann Curvature Tensor

The Riemann curvature tensor

$$R^l_{ijk} = \frac{\partial \Gamma^l_{ik}}{\partial u^j} - \frac{\partial \Gamma^l_{ij}}{\partial u^k} + \Gamma^p_{ik} \Gamma^l_{pj} - \Gamma^p_{ij} \Gamma^l_{pk}$$

defines a trilinear function $R(X, Y)Z$:

$$R(X^j r_j, Y^k r_k)(Z^i r_i) = X^j Y^k Z^i R^l_{ijk} r_l.$$

Theorem 3.15 *The Riemann curvature tensor is expressed in terms of covariant differentiation as*

$$R(X, Y)Z = \nabla_X \nabla_Y Z - \nabla_Y \nabla_X Z - \nabla_{[X,Y]} Z.$$

Proof First, we check the required equality for the basis vectors. For them, the commutator is zero (see Problem 3.27), so that the last term vanishes and what remains is

$$\nabla_{r_j}(\nabla_{r_k} r_i) - \nabla_{r_k}(\nabla_{r_j} r_i) = \nabla_{r_j}(\Gamma^p_{ik} r_p) - \nabla_{r_k}(\Gamma^p_{ij} r_p)$$

$$= \frac{\partial}{\partial u^j} \Gamma^l_{ik} r_l + \Gamma^p_{ik} \Gamma^l_{pj} r_l - \frac{\partial}{\partial u^k} \Gamma^l_{ij} r_l - \Gamma^p_{ij} \Gamma^l_{pk} r_l$$

$$= \left(\frac{\partial \Gamma^l_{ik}}{\partial u^j} - \frac{\partial \Gamma^l_{ij}}{\partial u^k} + \Gamma^p_{ik} \Gamma^l_{pj} - \Gamma^p_{ij} \Gamma^l_{pk} \right) r_l.$$

Let us check the trilinearity of $\nabla_X \nabla_Y Z - \nabla_Y \nabla_X Z - \nabla_{[X,Y]} Z$. Applying (3.9) to the first two terms and the first equation in (3.10) to the last term, we see that when X is replaced by fX, where f is a function, the terms $\nabla_X \nabla_Y Z$, $-\nabla_Y \nabla_X Z$, and $-\nabla_{[X,Y]} Z$ become, respectively,

$$\nabla_{fX} \nabla_Y Z = f \nabla_X \nabla_Y Z,$$

$$-\nabla_Y \nabla_{fX} Z = -(\partial_Y f) \nabla_X Z - f \nabla_Y \nabla_X Z,$$

$$-\nabla_{[fX,Y]} Z = -\nabla_{f[X,Y] - (\partial_Y f)X} Z = -f \nabla_{[X,Y]} Z + (\partial_Y f) \nabla_X Z.$$

Thus, their sum equals the initial expression multiplied by f. For Y, the calculation is similar. For Z, we have

$$\nabla_X \nabla_Y (fZ) = \nabla_X \big((\partial_Y f) Z + f \nabla_Y Z \big)$$

$$= (\partial_X \partial_Y f) Z + (\partial_Y f) \nabla_X Z + (\partial_X f) \nabla_Y Z + f \nabla_X \nabla_Y Z,$$

$$-\nabla_Y \nabla_X (fZ) = -(\partial_Y \partial_X f) Z - (\partial_X f) \nabla_Y Z - (\partial_Y f) \nabla_X Z - f \nabla_Y \nabla_X Z,$$

$$-\nabla_{[X,Y]} (fZ) = -(\partial_{[X,Y]} f) Z - f \nabla_{[X,Y]} Z.$$

Summing these expressions and taking into account the identity $\partial_X \partial_Y - \partial_Y \partial_X = \partial_{[X,Y]}$, we obtain the required expression. $\square$

3.15 Exponential Map

It follows from the geodesic equation that, for each point p of a surface S, given any tangent vector V at this point and a sufficiently small t, there exists a unique geodesic $\gamma_V(t)$ for which $\frac{d\gamma_V}{dt}(0) = V$ and $\gamma_V(0) = p$. Suppose that $\gamma_V(t)$ is defined for $t = 1$. Then we set $\exp_p(V) = \gamma_V(1)$.

Theorem 3.16 *The differential of the map* $\exp_p$ *at the origin is the identity operator.*

Proof In the tangent space at the point p consider the curve $\alpha(t) = tV$, where V is a fixed vector. For $t = 0$, the velocity vector of this curve equals V. The map $\exp_p$ takes this curve to the curve $\gamma_{tV}(1) = \gamma_V(t)$. For $t = 0$, the velocity vector of $\gamma_V(t)$ on the surface S equals V. $\square$

Corollary *The restriction of the map* $\exp_p$ *to a sufficiently small neighborhood of zero in the tangent space is a diffeomorphism.*

Proof The differential of the map $\exp_p$ at the origin is a nonsingular operator; therefore, by the inverse function theorem, the restriction of the map $\exp_p$ to a sufficiently small neighborhood of zero in the tangent space is a diffeomorphism.

$\square$

For a fixed vector V, the curve $c(\lambda) = \exp_p(\lambda V)$ is geodesic. Indeed, since $\gamma_{\lambda V}(t) = \gamma_V(\lambda t)$, it follows that $\exp_p(\lambda V) = \gamma_{\lambda V}(1) = \gamma_V(\lambda)$.

The length of the velocity vector of the geodesic $\gamma_V(t)$ at each point equals $\|V\|$; therefore, the length of a geodesic arc from the point p to the point $\exp_p(V) = \gamma_V(1)$ equals $\|V\|$. For this reason, the curve $\{\exp_p(V) \mid \|V\| = r\}$ is called the *geodesic circle* of radius r centered at p on the surface S.

For a surface S in $\mathbb{R}^3$, the map $\exp_p$ makes it possible to introduce so-called *normal coordinates* in a small neighborhood of the point $p \in S$. Normal coordinates correspond to the image of a Cartesian coordinate system in the tangent space at the point p under the map $\exp_p$. Let e_1, e_2 be an orthonormal basis, and let $q = \exp_p(x^1 e_1 + x^2 e_2)$. Then (x^1, x^2) are the normal coordinates of the point q.

At the origin the differential of $\exp_p$ is the identity map; therefore, in normal coordinates at the point p the first quadratic form is $(dx^1)^2 + (dx^2)^2$, i.e., $E(p) = G(p) = 1$ and $F(p) = 0$.

In normal coordinates the geodesic lines through the point p are determined by the equations $x^1 = tc^1$ and $x^2 = tc^2$.

Problem 3.28 Prove that in normal coordinates all Christoffel symbols at the point p are zero.

The map $\exp_p$ makes it also possible to introduce *geodesic polar coordinates* in a small neighborhood of a point $p \in S$ as the image of polar coordinates (r, φ) in the tangent plane at the point p under the map $\exp_p$. In geodesic polar coordinates the equation $r = \mathrm{const}$ determines a geodesic circle centered at p, and the equation $\varphi = \mathrm{const}$ determines a geodesic line passing through the point p.

Let us calculate the coefficients of the first quadratic form $E dr^2 + 2F dr\, d\varphi + G d\varphi^2$ in geodesic polar coordinates. The velocity of motion along the coordinate r is 1, whence $E = 1$. The curve with natural parameterization s which is determined by the equation $\varphi = \mathrm{const}$ is geodesic. For this curve, we have $\frac{du^1}{ds} \neq 0$ and $\frac{du^2}{ds} = 0$; therefore, $\Gamma_{11}^2 = 0$.

In the case of surfaces, in which there are only two indices, the equations $(r_{ij}, r_l) = \Gamma_{ij}^k g_{kl}$ (see p. 85) are often more convenient than the general expression for the Christoffel symbols in terms of the metric. Let us write two such equations:

$$\Gamma_{11}^1 E + \Gamma_{11}^2 F = (r_{11}, r_1) = \frac{1}{2} E_1, \quad \Gamma_{11}^1 F + \Gamma_{11}^2 G = (r_{11}, r_2) = F_1 - \frac{1}{2} E_2$$

(here E_1 and F_1 are derivatives with respect to r and E_2 is derivative with respect to φ). Since $E = 1$ and $\Gamma_{11}^2 = 0$, it follows from the first equation that $\Gamma_{11}^1 = 0$ and from the second equation that $F_1 = 0$, i.e., F does not depend on r.

Consider the geodesic circle $\alpha(\varphi)$ centered at p and the geodesic line $\gamma(r)$ passing through p. At their intersection point, we have $F = \left(\frac{d\alpha}{d\varphi}, \frac{d\gamma}{dr} \right)$. Let us fix φ and consider the limit $\lim_{r \to 0} F(r, \varphi)$. The geodesic circle $\alpha(\varphi)$ contracts to the point p; therefore, $\lim_{r \to 0} \left(\frac{d\alpha}{d\varphi}, \frac{d\gamma}{dr} \right) = 0$. But F does not depend of r, whence $F = 0$. Thus, we have proved the following assertion.

Theorem 3.17 (Gauss' Lemma)   *The geodesic* $\exp_p(tV_0)$ *from the center of the geodesic circle* $\{\exp_p(V) \mid \|V\| = r\}$ *is orthogonal to this circle.*

HISTORICAL COMMENT   In 1825 Gauss proved that the set of endpoints of arcs of the same length on geodesics from the same point of a surface is a curve intersecting all these geodesics at right angles.

Thus, in geodesic polar coordinates, the first quadratic form is $dr^2 + G(r, \varphi)d\varphi^2$. It follows by Theorem 3.3 (see p. 76) that the Gaussian curvature K can be calculated by

$$4G^2 K = \left(\frac{\partial G}{\partial r}\right)^2 - 2G\frac{\partial^2 G}{\partial r^2}.$$

A simple calculation shows that this formula is equivalent to

$$K = -\frac{1}{\sqrt{G}} \cdot \frac{\partial^2 \sqrt{G}}{\partial r^2}. \tag{3.15}$$

Let us prove that $\lim_{r\to 0} G = 0$ and $\lim_{r\to 0} \frac{\partial \sqrt{G}}{\partial r} = 1$. Consider the normal coordinates $x^1 = r\cos\varphi$, $x^2 = r\sin\varphi$. Under a change of coordinates the area form is multiplied by the Jacobian determinant of this change; therefore,

$$\sqrt{G} = \sqrt{EG - F^2} = \sqrt{\bar{E}\bar{G} - \bar{F}^2}\frac{\partial(x^1, x^2)}{\partial(r, \varphi)} = r\sqrt{\bar{E}\bar{G} - \bar{F}^2}.$$

At the point p we have $\bar{E} = \bar{G} = 1$ and $\bar{F} = 0$ for the normal coordinates; hence $\lim_{r\to 0} G = 0$ and $\lim_{r\to 0} \frac{\partial \sqrt{G}}{\partial r} = 1$.

Let us write Eq. (3.15) in the form

$$\frac{\partial^2 \sqrt{G}}{\partial r^2} = -K\sqrt{G} \tag{3.16}$$

and differentiate it with respect to r:

$$\frac{\partial^3 \sqrt{G}}{\partial r^3} = -K\frac{\partial \sqrt{G}}{\partial r} - \frac{\partial K}{\partial r}\sqrt{G}.$$

Hence

$$K(p) = \lim_{r\to 0}\frac{\partial^3 \sqrt{G}}{\partial r^3}.$$

Note also that (3.16) implies

$$\lim_{r \to 0} \frac{\partial^2 \sqrt{G}}{\partial r^2} = 0.$$

Thus, we know all coefficients in the decomposition

$$\sqrt{G}(r, \varphi) = \sqrt{G}(0, \varphi) + r \frac{\partial \sqrt{G}}{\partial r}(0, \varphi) + \frac{r^2}{2!} \frac{\partial^2 \sqrt{G}}{\partial r^2}(0, \varphi) + \frac{r^3}{3!} \frac{\partial^3 \sqrt{G}}{\partial r^3}(0, \varphi) + o(r^3)$$

and can write it in the form

$$\sqrt{G}(r, \varphi) = r - \frac{r^3}{3!} K(p) + o(r^3).$$

This formula makes it possible to express the Gaussian curvature at a given point in terms of the lengths of geodesic circles of sufficiently small radius or in terms of the areas of geodesic disks of sufficiently small radius.

Theorem 3.18 (Bertrand–Diguet–Puiseux)

(a) Let $L(r)$ be the length of a geodesic circle of radius r. Then

$$K = \lim_{r \to 0} \frac{3}{\pi} \cdot \frac{2\pi r - L(r)}{r^3}.$$

(b) Let $A(r)$ be the area of a geodesic disk of radius r. Then

$$K = \lim_{r \to 0} \frac{12}{\pi} \cdot \frac{\pi r^2 - A(r)}{r^4}.$$

Proof

(a) Let us introduce geodesic polar coordinates (r, φ) in a neighborhood of a point p. Fix r and consider a geodesic circle $\alpha(\varphi)$ of radius r; the point $\alpha(\varphi)$ has geodesic polar coordinates (r, φ). The length of the velocity vector of this curve at the point (r, φ) equals $\sqrt{G}(r, \varphi)$; therefore,

$$L(r) = \int_0^{2\pi} \sqrt{G}(r, \varphi) d\varphi$$

$$= \int_0^{2\pi} \left(r - \frac{Kr^3}{6} \right) d\varphi + \int_0^{2\pi} o(r^3) d\varphi$$

$$= 2\pi \left(r - \frac{Kr^3}{6} \right) + o(r^3).$$

(b) Clearly,

$$A(r) = \int dA = \int_0^{2\pi} \int_0^r \sqrt{G}(r, \varphi) dr\, d\varphi$$

$$= \int_0^{2\pi} \int_0^r \left(r - \frac{Kr^3}{6} \right) dr\, d\varphi + \int_0^{2\pi} \int_0^r o(r^3) dr\, d\varphi$$

$$= 2\pi \left(\frac{r^2}{2} - \frac{Kr^4}{24} \right) + o(r^4).$$

$\square$

HISTORICAL COMMENT Theorem 3.18 was proved by Bertrand, C. F. Diquet, and Victor Alexandre Puiseux (1820–1883) in 1848.

Geodesic polar coordinates can also be applied to prove that a geodesic locally minimizes arc length.

Theorem 3.19 *Any point p on a surface S has a neighborhood U such that the length of a geodesic γ going from the point p to some point q and lying entirely in the neighborhood U does not exceed the length of any curve α on S joining the points p and q. Moreover, if the lengths of the curves γ and α are equal, then γ and α coincide as nonparameterized curves.*

Proof Consider the image of a neighborhood of the origin in the tangent space at the point p such that the restriction of $\exp_p$ to this neighborhood is a diffeomorphism and let U be an open geodesic disk of radius ε lying entirely in this image. If $q \in U$, then the geodesic circle centered at p and passing through q is contained in U. First, consider the case where the curve $\alpha(t) = (r(t), \varphi(t))$, $t \in [a, b]$, lies inside the geodesic circle. We have

$$L(\alpha) = \int_a^b \sqrt{(r')^2 + (\varphi')^2} dt \geqslant \int_a^b \sqrt{(r')^2} dt \geqslant \int_a^b r' dt = L(\gamma),$$

and the equality is attained if and only if $\varphi = $ const and $r' > 0$. In this case, the nonparameterized curves γ and α coincide.

If the curve α intersects the geodesic circle, then its length is at least the radius of this geodesic circle, and the equality is possible only if the nonparameterized curves γ and α coincide.

$\square$

3.16 Lines of Curvature and Asymptotic Lines

At each point of a smooth curve on of a surface, three quantities are defined: the geodesic curvature k_g, the normal curvature k_n, and the geodesic torsion $\varkappa_g$. Moreover, the normal curvature and the geodesic torsion depend only on the direction of the curve at the given point; they are the same for all curves with common tangent line.

We have already discussed the curves for which $k_g = 0$ at all points (geodesics). The curves for which $k_n = 0$ (asymptotic lines) or $\varkappa_g = 0$ (lines of curvature) are interesting in many respects as well. As we will see, asymptotic lines and lines of curvature differ substantially from geodesics: through any point we can draw a geodesic in any direction, while lines of curvature and asymptotic lines can be drawn in only two directions (moreover, asymptotic lines exist only at points of negative Gaussian curvature). The lines of curvature bisect the angles between asymptotic lines.

A *line of curvature* is a smooth curve on a surface S such that the direction of its tangent at each point corresponds to one of the principal curvatures of the surface at this point. The lines of curvature are integral lines for the two direction fields corresponding to the directions of principal curvatures. At the nonumbilical points (i.e., at those points at which the principal curvatures are different) the lines of curvature intersect at a right angle.

On the sphere and in the plane any curve is a line of curvature, because all points of the sphere and the plane are umbilical and all directions are directions of principal curvature.

Rodrigues' formula shows that the eigendirections of the differential dn, where n is the normal vector to the surface, are the principal curvature directions. It follows that the lines of curvature are the curves consisting of points at which the geodesic torsion $\varkappa_g$ vanishes. Indeed, recall the third formula for the Darboux frame: $\varepsilon_3' = -k_n\varepsilon_1 - \varkappa_g\varepsilon_2$; here $\varepsilon_3 = n$ and ε_1 is a tangent vector to the curve. Thus, a tangent vector to a curve is an eigenvector of the operator dn if and only if $\varkappa_g = 0$.

Problem 3.29 Prove that the parallels and meridians of a surface of revolution are lines of curvature.

Problem 3.30 Prove that if two surfaces intersect along some curve γ at a constant angle and this curve is a line of curvature on one of these surfaces, then it is also a line curvature on the other surface.

If $K(p) < 0$, then there are two directions at the point p for which the normal curvature k_n is zero (in other words, the second quadratic form vanishes in these directions). These directions are said to be asymptotic. They divide the tangent plane into four sectors, in each of which k_n has constant sign. On a surface of negative Gaussian curvature *asymptotic lines* are defined, whose tangents have asymptotic directions. Clearly, if a surface contains a straight line, then this line is asymptotic.

Problem 3.31 Prove that, for an asymptotic line, the Darboux frame coincides with the Frenet–Serret frame (under an appropriate choice of the direction of the normal vector to the surface).

Problem 3.32 Prove that the asymptotic directions at a point p are perpendicular if and only if the mean curvature vanishes at this point.

Both the lines of curvature and the asymptotic lines can be the coordinate lines of some parameterization in a neighborhood of a given point under the natural additional assumption that this point is nonumbilical (for lines of curvature) or is a point of negative Gaussian curvature (for asymptotic lines). Both these assertions about the existence of a parameterization follow from a general theorem, which we will now prove.

Theorem 3.20 *Suppose that on a smooth surface $r(u, v)$ in a neighborhood of a point with coordinates (u_0, v_0) two differential equations*

$$A_1(u, v)du + B_1(u, v)dv = 0, \quad A_2(u, v)du + B_2(u, v)dv = 0$$

are given, and their coefficients satisfy the condition $A_1 B_2 - A_2 B_1 \neq 0$ at the point (u_0, v_0). Then in a neighborhood of this point the surface can be parameterized so that the coordinate lines are integral curves for these equations.

Proof We assume that $A_2 B_1 \neq 0$. Let $v = \varphi(\alpha, u)$ be the solution of the first equation with initial condition $\varphi(\alpha, u_0) = \alpha$, and let $u = \psi(\beta, v)$ be the solution of the second equation with initial condition $\psi(\beta, v_0) = \beta$. The equations $v = \varphi(\alpha, u)$ and $u = \psi(\beta, v)$ are solvable with respect to α and β in a neighborhood of the point (u_0, v_0), because $\frac{\partial \varphi}{\partial \alpha} = 1$ at $u = u_0$ and $\frac{\partial \psi}{\partial \beta} = 1$ at $v = v_0$. Let $\alpha = \alpha(u, v)$ and $\beta = \beta(u, v)$ be their solutions. We claim that $\alpha_u \beta_v - \alpha_v \beta_u \neq 0$ at the point (u_0, v_0). Indeed, the curve $\alpha(u, v) = $ const is an integral of the first equation, and hence the equations $A_1 du + B_1 dv = 0$ and $\alpha_u du + \alpha_v dv = 0$ are proportional, i.e., $A_1 \alpha_v = B_1 \alpha_u$, $\alpha_u \neq 0$. Similarly, $A_2 \beta_v = B_2 \beta_u$, $\beta_v \neq 0$. Suppose that $\alpha_u \beta_v = \alpha_v \beta_u$. Then $A_2 B_1 \alpha_u \beta_v = A_1 B_2 \alpha_v \beta_u = A_1 B_2 \alpha_u \beta_v$, and $\alpha_u \beta_v \neq 0$. Therefore, $A_1 B_2 = A_2 B_1$, which contradicts the assumption of the theorem. Thus, $\alpha_u \beta_v - \alpha_v \beta_u \neq 0$ at the point (u_0, v_0), and hence $\alpha(u, v)$ and $\beta(u, v)$ can be taken for new parameters of the surface. For these parameters, the coordinate lines $\alpha = $ const and $\beta = $ const are integral lines of the equations under consideration.   □

The system of equations in Theorem 3.20 can be given in the form of the product of these equations, i.e., as an equation of the form

$$A \, du^2 + 2B \, du \, dv + C \, dv^2 = 0.$$

The coefficients in this equation must satisfy the inequality $AC - B^2 < 0$.

The asymptotic lines are the solutions of the differential equation

$$L \, du^2 + 2M \, du \, dv + N \, dv^2 = 0,$$

where L, M, and N are the coefficients of the second quadratic form. Moreover, at a point with negative Gaussian curvature, we have $LN - M^2 < 0$.

The direction $(du : dv)$ is the direction of one of the principal curvatures if the vectors

$$\begin{pmatrix} E & F \\ F & G \end{pmatrix} \begin{pmatrix} du \\ dv \end{pmatrix} = \begin{pmatrix} E\,du + F\,dv \\ F\,du + G\,dv \end{pmatrix} \quad \text{and} \quad \begin{pmatrix} L & M \\ M & N \end{pmatrix} \begin{pmatrix} du \\ dv \end{pmatrix} = \begin{pmatrix} L\,du + M\,dv \\ M\,du + N\,dv \end{pmatrix}$$

are proportional, i.e.,

$$\begin{vmatrix} E\,du + F\,dv & F\,du + G\,dv \\ L\,du + M\,dv & M\,du + N\,dv \end{vmatrix} = 0.$$

This equation can be written in the more symmetric form

$$\begin{vmatrix} E & F & G \\ L & M & N \\ du^2 & du\,dv & dv^2 \end{vmatrix} = 0.$$

In a neighborhood of a nonumbilical point it has the form

$$A\,du^2 + 2B\,du\,dv + C\,dv^2 = 0,$$

and through each point of this neighborhood precisely two integral curves pass. Therefore, $AC - B^2 < 0$, and we can apply Theorem 3.20.

In the parameterization $r(u, v)$ whose coordinate lines are lines of curvature many expressions become simpler. For example, in these coordinates the first and second quadratic forms are $E\,du^2 + G\,dv^2$ and $k_1 E\,du^2 + k_2 G\,dv^2$, where k_1 and k_2 are the principal curvatures. Indeed, it follows from the orthogonality of the curvature lines that $F = 0$, and the coefficients of the second quadratic form are obtained by applying Rodrigues' formula. The Codazzi–Mainardi equations become substantially simpler in these coordinates as well, because $M = 0$:

$$L_v = L\Gamma^1_{12} - N\Gamma^2_{11}, \quad N_u = N\Gamma^2_{12} - L\Gamma^1_{22}.$$

Here $L = k_1 E$ and $N = k_2 G$. We also have

$$\Gamma^2_{11} = -\frac{E_v}{2G}, \quad \Gamma^1_{12} = \frac{E_v}{2E}, \quad \Gamma^1_{22} = -\frac{G_u}{2E}, \quad \Gamma^2_{12} = -\frac{G_u}{2G}.$$

Thus, the Codazzi–Mainardi equations take the form

$$(k_1 E)_v = \frac{E_v}{2}(k_1 + k_2), \quad (k_2 G)_u = \frac{G_u}{2}(k_1 + k_2).$$

3.17 Minimal Surfaces

A *minimal surface* $S \subset \mathbb{R}^3$ is a surface with zero mean curvature.

To explain the name, consider the normal variations of the surface. Let D be a bounded domain of the surface, and let $\bar{D}$ be its closure. Given a differentiable function $h\colon \bar{D} \to \mathbb{R}$, the *normal variation* of the surface $r(u, v)$ determined by the function h is the family of surfaces $r_t(u, v) = r(u, v) + th(u, v)n(u, v)$, where $n(u, v)$ is the unit normal vector to the surface at the point $r(u, v)$.

Recall that the area of a surface is given by the integral $\int \sqrt{EG - F^2}\,du\,dv$ (see formula (3.1) on p. 66). Therefore, to evaluate the area of the surface r_t for small t, we calculate the first quadratic form of r_t. To this end, we first calculate the partial derivatives $(r_t)_u = \frac{\partial r_t}{\partial u}$ and $(r_t)_v = \frac{\partial r_t}{\partial v}$:

$$(r_t)_u = r_u + thn_u + th_u n,$$

$$(r_t)_v = r_v + thn_v + th_v n,$$

The coefficients of the first quadratic form of the surface r_t are equal to the pairwise inner products of these partial derivatives:

$$E_t = E + 2th(r_u, n_u) + t^2 h^2(n_u, n_u) + t^2 h_u h_u,$$

$$F_t = F + th[(r_u, n_v) + (r_v, n_u)] + t^2 h^2(n_u, n_v) + t^2 h_u h_v,$$

$$G_t = G + 2th(r_v, n_v) + t^2 h^2(n_v, n_v) + t^2 h_v h_v.$$

If we are interested in the coefficients of the first quadratic form of r_t up to the first order in t, then we can express them in terms of the coefficients of the first and second quadratic forms of r, because

$$(r_u, n_u) = -L, \quad (r_u, n_v) + (r_v, n_u) = -2M, \quad (r_v, n_v) = -N.$$

Therefore,

$$E_t G_t - F_t^2 = EG - F^2 - 2th(LG - 2MF + NE) + o(t).$$

Recall that the mean curvature H is expressed in terms of the coefficients of the first and second quadratic forms as $H = \frac{1}{2} \cdot \frac{LG - 2MF + NE}{EG - F^2}$ (see Theorem 3.2 on p. 75). Hence

$$E_t G_t - F_t^2 = (EG - F^2)(1 - 4thH) + o(t).$$

For the area A_t of a normal variation of the domain $\bar{D}$, we obtain the expression

$$A(t) = \int_{\bar{D}} \sqrt{E_t G_t - F_t^2} \, du \, dv$$

$$= \int_{\bar{D}} \sqrt{1 - 4thH + o(t)} \sqrt{EG - F^2} \, du \, dv.$$

Therefore,

$$A'(0) = -\int_{\bar{D}} 2hH \sqrt{EG - F^2} \, du \, dv.$$

This formula shows that a surface is minimal if and only if $A'(0) = 0$ for any normal variation of any domain of this surface. Indeed, if a surface is minimal, then $H = 0$ at all of its points. Suppose that $H(p) \neq 0$ at some point p and consider a function h vanishing outside a small neighborhood D of p and taking values of the same sign as $h(p) = H(p)$ at those points at which it does not vanish. We have $A'(0) < 0$ for the variation determined by the function h.

HISTORICAL COMMENT The equivalence of the minimality of the area of a surface with given boundary to the vanishing of the mean curvature of this surface was proved by Meusnier in 1776.

In the differential geometry of surfaces it is sometimes convenient to use *isothermal coordinates*, for which the angles between curves on the surface are equal to the angles between the corresponding curves in the coordinate plane, i.e., the map of the coordinate plane to the surface is *conformal* (preserves the angles between curves). For isothermal coordinates, the first quadratic form is $\lambda^2(u, v)(du^2 + dv^2)$.

HISTORICAL COMMENT The existence of isothermal coordinates in the case of a real analytic metric was proved by Gauss in 1822.

In isothermal coordinates with first quadratic form $\lambda^2(du^2 + dv^2)$ the expression for the mean curvature takes the form $H = \frac{L+N}{2\lambda^2}$, where L and N are the coefficients of the second quadratic form. Using this expression for H, we can prove the following stronger statement (from which the above formula is obtained by taking inner product with the normal vector).

Theorem 3.21 *Let $r(u, v)$ be an isothermal parameterization. Then $r_{uu} + r_{vv} = 2\lambda^2 Hn$, where n is the normal vector.*

Proof Differentiating the equation $(r_u, r_u) = (r_v, r_v)$ with respect to u and the equation $(r_u, r_v) = 0$ with respect to v, we obtain

$$(r_{uu}, r_u) = (r_{uv}, r_v) = -(r_u, r_{vv}).$$

Therefore, $(r_{uu} + r_{vv}, r_u) = 0$. Similarly, $(r_{uu} + r_{vv}, r_v) = 0$. Hence the vector $r_{uu} + r_{vv}$ is proportional to n. The proportionality coefficient satisfies the equation $H = \frac{L+N}{2\lambda^2}$, because

$$(n, r_{uu} + r_{vv}) = L + N = 2\lambda^2 H.$$

Thus, $r_{uu} + r_{vv} = 2\lambda^2 H n$.   □

For a function $f(u, v)$, the *Laplace operator* (*Laplacian*) Δf is defined as

$$\Delta f = \frac{\partial^2 f}{\partial u^2} + \frac{\partial^2 f}{\partial v^2}.$$

A function f is said to be *harmonic* if $\Delta f = 0$. Theorem 3.21 readily implies the following description of minimal surfaces.

Corollary Let $(x^1(u, v), x^2(u, v), x^3(u, v))$ *be an isothermal parameterization of a surface. Then this surface is minimal if and only if the functions* $x^1(u, v)$, $x^2(u, v)$, *and* $x^3(u, v)$ *are harmonic.*

Example 3.6 A surface of the form

$$r(u, v) = (a \cosh v \cos u, \ a \cosh v \sin u, \ av)$$

is called a *catenoid*. A catenoid is a minimal surface.

Proof It is easy to check that $E = G = a^2 \cosh^2 v$, $F = 0$, and $r_{uu} + r_{vv} = 0$.   □

Problem 3.33 Prove that any minimal surface of revolution different from the plane is a catenoid.

Example 3.7 A surface of the form

$$r(u, v) = (a \sinh v \cos u, \ a \sinh v \sin u, \ au)$$

is called a *helicoid*. A helicoid is a minimal surface.

Proof It is easy to check that $E = G = a^2 \cosh^2 v$, $F = 0$, and $r_{uu} + r_{vv} = 0$.   □

HISTORICAL COMMENT Catenoids and helicoids were discovered by Meusnier in 1776.

Problem 3.34 Prove that the the spherical Gauss map of a minimal surface with nonvanishing Gaussian curvature is conformal.

3.18   The First Variation Formula

In studying variations of curves on a surface S, it is convenient to consider piecewise smooth curves and their variations. A piecewise smooth curve is a map $\gamma : [0, 1] \to S$ which defines smooth curves on intervals $[t_{i-1}, t_i]$ for a partition $0 = t_0 < t_1 < \cdots < t_k = 1$ of the interval $[0, 1]$. The points $p = \gamma(0)$ and $q = \gamma(1)$ are the endpoints of the curve.

A *variation* (with fixed endpoints) of a curve $\gamma(0, t) = \gamma(t)$ is a family of curves $\gamma(s, t)$, $s \in (-\varepsilon, \varepsilon)$, $t \in [a, b]$, i.e., a map $(-\varepsilon, \varepsilon) \times [0, 1] \to S$. Here s parameterizes the family of curves and t is a parameter on each curve. For a fixed s, we obtain a curve in the family, and for a fixed t, we obtain a curve transverse to the family of curves. The endpoints of the curves are fixed, i.e., $\gamma(s, 0) = p$ and $\gamma(s, 1) = q$ for all s. The restriction of the map to each strip $(-\varepsilon, \varepsilon) \times [t_{i-1}, t_i]$ is smooth.

With each variation two vector fields are associated, which correspond to the partial derivatives with respect to t and with respect to s: $\frac{\partial \gamma(s,t)}{\partial t}$ is a tangent vector to a curve in the family and $\frac{\partial \gamma(s,t)}{\partial s}$ is a tangent vector to a transverse curve, which we will call a *variation vector*.

Problem 3.35 Let $W(t)$ be the restriction to a curve $\gamma(t)$ of some vector field tangent to the surface S. Prove that there exists a variation of γ with variation vector $W(t)$ at each point of γ.

A curve $\gamma(t)$ is said to be *critical* for a functional F if

$$\frac{\partial F(\gamma(s, t))}{\partial s}\bigg|_{s=0} = 0$$

for any variation of this curve.

The geodesics are critical curves for the length functional. For a parameterized curve γ, the *length functional* equals $L(\gamma) = \int_a^b \sqrt{(\gamma', \gamma')}dt$. In some respects the length functional is not as convenient as the *energy functional* $E(\gamma) = \frac{1}{2}\int_a^b (\gamma', \gamma')dt$. The point is that the length functional does not depend of the choice of a parameterization, while the energy functional does. Therefore, for the energy functional, the parameterization of a critical curve is determined up to proportionality, while for the length functional, the parameterization of a critical curve can be arbitrary. This leads to the strong singularity of the length functional, which is a hindrance in some situations.

Theorem 3.22 *For a curve parameterized by an interval $[a, b]$, the inequality $L(\gamma)^2 \leqslant 2(b - a)E(\gamma)$ holds, which becomes an equality only for a curve whose parameterization is proportional to arc length.*

Proof Let $f(t) = \sqrt{(\gamma'(t), \gamma'(t))}$. According to the Cauchy–Bunyakovsky–Schwarz inequality, we have

$$L(\gamma)^2 = \left(\int_a^b f(t) \cdot 1 \, dt \right)^2 \leqslant \int_a^b f^2(t)dt \cdot \int_a^b 1 \, dt = 2E(\gamma)(b-a).$$

The equality is attained only for $f(t) = \text{const.}$ $\square$

Let us determine what piecewise smooth curves are critical for the energy functional. Consider a variation $\gamma(s,t)$ of a curve $\gamma(t)$. We use the following notation:

$W(t) = \left. \frac{\partial \gamma(s,t)}{\partial s} \right|_{s=0}$ is the variation vector;

$V(t) = \frac{\partial \gamma(t)}{\partial t}$ is the velocity vector of the curve;

$A(t) = \nabla_V V$ is the acceleration vector of the curve;

$\Delta_t V = V(t+0) - V(t-0)$ is the jump of the velocity vector at a point t.

Theorem 3.23 (First Variation Formula) *The derivative* $\left. \frac{\partial E(\gamma(s,t))}{\partial s} \right|_{s=0}$ *equals*

$$- \sum_t (W(t), \Delta_t V) - \int_0^1 (W(t), A(t))dt.$$

Proof According to Theorem 3.13 on p. 96, we have

$$\frac{\partial}{\partial s} \left(\frac{\partial \gamma}{\partial t}, \frac{\partial \gamma}{\partial t} \right) = 2 \left(\nabla_{\partial \gamma / \partial s} \frac{\partial \gamma}{\partial t}, \frac{\partial \gamma}{\partial t} \right).$$

Therefore,

$$\frac{\partial E(\gamma(s,t))}{\partial s} = \frac{1}{2} \frac{\partial}{\partial s} \int_0^1 \left(\frac{\partial \gamma}{\partial t}, \frac{\partial \gamma}{\partial t} \right) dt = \int_0^1 \left(\nabla_{\partial \gamma / \partial s} \frac{\partial \gamma}{\partial t}, \frac{\partial \gamma}{\partial t} \right) dt.$$

Using the symmetry of the Christoffel symbols, we replace $\nabla_{\partial \gamma / \partial s} \frac{\partial \gamma}{\partial t}$ in the last integral by

$$\nabla_{\partial \gamma / \partial t} \frac{\partial \gamma}{\partial s} = \nabla_{\gamma'} \frac{\partial \gamma}{\partial s}.$$

The identity

$$\frac{d}{dt} \left(\frac{\partial \gamma}{\partial s}, \frac{\partial \gamma}{\partial t} \right) = \left(\nabla_{\gamma'} \frac{\partial \gamma}{\partial s}, \frac{\partial \gamma}{\partial t} \right) + \left(\frac{\partial \gamma}{\partial s}, \nabla_{\gamma'} \frac{\partial \gamma}{\partial t} \right)$$

makes it possible to perform integration by parts in the form

$$\int_{t_{i-1}}^{t_i} \left(\nabla_{\gamma'} \frac{\partial \gamma}{\partial s}, \frac{\partial \gamma}{\partial t} \right) dt = \left(\frac{\partial \gamma}{\partial s}, \frac{\partial \gamma}{\partial t} \right) \Big|_{t=t_{i-1}+0}^{t=t_i-0} - \int_{t_{i-1}}^{t_i} \left(\frac{\partial \gamma}{\partial s}, \nabla_{\gamma'} \frac{\partial \gamma}{\partial t} \right) dt.$$

Summing these expressions over $i = 1, \ldots, k$ and taking into account the vanishing of the vector $\frac{\partial \gamma}{\partial s}$ at $t = 0$ and $t = 1$, we obtain

$$\frac{\partial E(\gamma(s,t))}{\partial s} = -\sum_{i=1}^{k-1} \left(\frac{\partial \gamma}{\partial s}, \Delta_{t_i} \frac{\partial \gamma}{\partial t} \right) - \int_0^1 \left(\frac{\partial \gamma}{\partial s}, \nabla_{\gamma'} \frac{\partial \gamma}{\partial t} \right) dt. \tag{3.17}$$

Substituting $s = 0$, we arrive at the required expression

$$\frac{\partial E(\gamma(s,t))}{\partial s} \Big|_{s=0} = -\sum_t (W(t), \Delta_t V) - \int_0^1 (W(t), A(t)) dt.$$

$\square$

   Recall that a geodesic (with parameterization proportional to the natural one) is a smooth curve with zero acceleration $A(t)$.

Corollary *A curve is critical for the energy functional if and only if this curve is geodesic.*

Proof The second term in the expression for the variation of the energy functional shows that a variation in the direction of the acceleration vector, $A(t)$ reduces E. Therefore, a critical curve is geodesic on each interval of smoothness. The first term in the expression shows that at the endpoints of these intervals a variation W of the curve in the direction of $\Delta_t V$ reduces E, and hence a critical curve is smooth, that is, has no sharp corners. $\square$

3.19   The Second Variation Formula

For functions of many variables, a critical point is not necessarily a point of local minimum or maximum. The condition that a point is critical is necessary but not sufficient for local minimality. A sufficient condition for a function to have local minimum is the positive definiteness of its Hessian matrix.

   Not all geodesics are shortest curves joining given points. Through any two points of the sphere which are not diametrically opposite infinitely many geodesics of different lengths pass, and only one of these geodesics is shortest. To distinguish the shortest geodesic among the other ones, we need a sufficient minimality condition. It is this purpose that the second variation formula for the energy functional serves. This formula applies only to critical curves, that is, geodesics.

Consider a two-parameter variation $\gamma(s_1, s_2, t)$ of a geodesic $\gamma(t) = \gamma(0, 0, t)$. Let $\frac{\partial \gamma}{\partial s_1}(0, 0, t) = W_1(t)$ and $\frac{\partial \gamma}{\partial s_2}(0, 0, t) = W_2(t)$ be the variation vectors. The *Hessian determinant* $E_{**}(W_1, W_2)$ of the energy functional is the second derivative

$$\frac{\partial^2 E(\gamma(s_1, s_2))}{\partial s_1 \partial s_2}\bigg|_{(s_1, s_2) = (0,0)}.$$

For brevity, we denote this second derivative by $\frac{\partial^2 E}{\partial s_1 \partial s_2}(0, 0)$. We also use the notation $V = \frac{d\gamma(t)}{dt} = \gamma'$ for the velocity vector of the curve γ and set

$$\Delta_t \nabla_{\gamma'} W_1 = \nabla_{\gamma'} W_1(t + 0) - \nabla_{\gamma'} W_1(t - 0).$$

The second variation formula involves the Riemann curvature tensor $R(X, Y)Z$. We begin with a remark, which we will need shortly. The vector fields $W_1 = \frac{\partial \gamma}{\partial s_1}$, $W_2 = \frac{\partial \gamma}{\partial s_2}$, and $V = \frac{\partial \gamma}{\partial t}$ pairwise commute, because they are induced by commuting (according to Problem 3.27) coordinate vector fields via the map $(s_1, s_2, t) \to \gamma(s_1, s_2, t)$. Therefore,

$$\nabla_{W_1} V = \nabla_V W_1, \tag{3.18}$$

$$R(V, W_1)V = \nabla_V \nabla_{W_1} V - \nabla_{W_1} \nabla_V V = (\nabla_V)^2 W_1 - \nabla_{W_1} \nabla_V V. \tag{3.19}$$

Theorem 3.24 (Second Variation Formula) *For a two-parameter variation of a geodesic* $\gamma(t) = \gamma(0, 0, t)$*, the second derivative* $\frac{\partial^2 E}{\partial s_1 \partial s_2}(0, 0)$ *of the energy functional equals*

$$-\sum_t (W_2(t), \Delta_t \nabla_V W_1) - \int_0^1 \left(W_2, (\nabla_V)^2 W_1 - R(V, W_1)V \right) dt.$$

Proof Let us apply Eq. (3.17) obtained in deriving the first variation formula:

$$\frac{\partial E}{\partial s_2} = -\sum_t \left(\frac{\partial \gamma}{\partial s_2}, \Delta_t \frac{\partial \gamma}{\partial t} \right) - \int_0^1 \left(\frac{\partial \gamma}{\partial s_2}, \nabla_{\gamma'} \frac{\partial \gamma}{\partial t} \right) dt.$$

Differentiating with respect to s_1, we obtain

$$\frac{\partial^2 E}{\partial s_1 \partial s_2} = -\sum_t \left(\nabla_{\partial \gamma / \partial s_1} \frac{\partial \gamma}{\partial s_2}, \Delta_t \frac{\partial \gamma}{\partial t} \right) - \sum_t \left(\frac{\partial \gamma}{\partial s_2}, \nabla_{\partial \gamma / \partial s_1} \Delta_t \frac{\partial \gamma}{\partial t} \right)$$

$$- \int_0^1 \left(\nabla_{\partial \gamma / \partial s_1} \frac{\partial \gamma}{\partial s_2}, \nabla_{\gamma'} \frac{\partial \gamma}{\partial t} \right) dt - \int_0^1 \left(\frac{\partial \gamma}{\partial s_2}, \nabla_{\partial \gamma / \partial s_1} \nabla_{\gamma'} \frac{\partial \gamma}{\partial t} \right) dt.$$

We must evaluate this quantity at the point $(s_1, s_2) = (0, 0)$. By assumption the curve $\gamma(t) = \gamma(0, 0, t)$ is geodesic (has no sharp corners). Therefore, at the point $(0, 0)$,

$$\Delta_t \frac{\partial \gamma}{\partial t} = 0, \quad \nabla_{\gamma'} \frac{\partial \gamma}{\partial t} = 0.$$

Thus, the first and the third term vanish, and there remain only the second and the fourth one. Transforming them, we obtain the following equation, in which the notations $\frac{\partial \gamma}{\partial s_1}(0, 0, t) = W_1(t)$, $\frac{\partial \gamma}{\partial s_2}(0, 0, t) = W_2(t)$, and $V = \frac{d\gamma(t)}{dt} = \gamma'$ are used:

$$\frac{\partial^2 E}{\partial s_1 \partial s_2}(0, 0) = -\sum_t \left(W_2, \nabla_{\partial \gamma / \partial s_1} \Delta_t \frac{\partial \gamma}{\partial t} \right) - \int_0^1 \left(W_2, \nabla_{\partial \gamma / \partial s_1} \nabla_V V \right) dt.$$

Equation (3.18) implies

$$\nabla_{\partial \gamma / \partial s_1} \Delta_t \frac{\partial \gamma}{\partial t} = \Delta_t \nabla_V \frac{\partial \gamma}{\partial s_1} = \Delta_t \nabla_V W_1.$$

It remains to check that

$$\nabla_{\partial \gamma / \partial s_1} \nabla_V V = (\nabla_V)^2 W_1 - R(V, W_1)V.$$

This follows directly from (3.19).   □

The second variation formula shows that the Hessian determinant $E_{**}(W_1, W_2) = \frac{\partial^2 E}{\partial s_1 \partial s_2}(0, 0)$ is a symmetric bilinear function in W_1 and W_2 (its symmetry follows from the symmetry of the second derivative).

With each symmetric bilinear form a quadratic form is associated. In particular, the quadratic form associated with the bilinear form $E_{**}(W_1, W_2)$ is $E_{**}(W, W)$. This quadratic form can be expressed in terms of the one-parameter variation $\gamma(s, t)$ with variation vector W without resorting to a two-parameter variation as

$$E_{**}(W, W) = \frac{\partial^2 E(\gamma(s, t))}{\partial s^2} \bigg|_{s=0}.$$

Indeed, consider the two-parameter variation $\gamma(s_1, s_2, t) = \gamma(s_1 + s_2, t)$. We have

$$\frac{\partial \gamma(s_1, s_2, t)}{\partial s_i} = \frac{\partial \gamma(s, t)}{\partial s} = W(t) \ (i = 1, 2), \quad \frac{\partial^2 E(\gamma(s_1, s_2, t))}{\partial s_1 \partial s_2} = \frac{\partial^2 E(\gamma(s, t))}{\partial s^2}.$$

For a minimal geodesic γ, the quadratic form $E_{**}(W, W)$ is nonnegative definite, because the Hessian matrix is nonnegative definite at a point of minimum.

3.20 Jacobi Vector Fields and Conjugate Points

A vector field J given on a geodesic γ is called a *Jacobi vector field* if it satisfies the Jacobi differential equation

$$(\nabla_V)^2 J - R(V, J)V = 0,$$

where $V = \gamma'$ is the velocity vector. This is a second-order linear differential equation in two variables, and it has four linearly independent solutions. Each Jacobi vector field is completely determined by the initial conditions $J(0)$ and $\nabla_V J(0)$. Recall that the expression $(\nabla_V)^2 W_1 - R(V, W_1)V$ occurs in the second variation formula.

Let $p = \gamma(a)$ and $q = \gamma(b)$ be two points of a geodesic γ for $a \neq b$. These points are said to be *conjugate* along γ if there exists a nonzero Jacobi vector field J along γ which vanishes at $t = a$ and at $t = b$. The *multiplicity* of p and q as conjugate points equals the dimension of the vector space spanned by all such Jacobi fields.

HISTORICAL COMMENT In 1838 Carl Gustav Jacob Jacobi (1804–1851) introduced the notion of a point conjugate to a point of a geodesic and stated conditions for a geodesic to be shortest.

Consider the kernel of the form E_{**}, which consists of all vectors W_1 such that $E_{**}(W_1, W_2) = 0$ for all W_2.

Theorem 3.25 *A vector field W_1 belongs to the kernel of the form E_{**} if and only if W_1 is a Jacobi vector field.*

Proof Consider a Jacobi vector field J vanishing at points p and q. Let us apply the second variation formula:

$$E_{**}(J, W_2) = -\sum_t (W_2(t), \Delta_t \nabla_V J) - \int_0^1 \left(W_2, (\nabla_V)^2 J - R(V, J)V \right) dt.$$

In the case under consideration, $\Delta_t \nabla_V J = 0$ (J is smooth as a solution of a differential equation) and $(\nabla_V)^2 J - R(V, J)V = 0$ (J satisfies the Jacobi equation); therefore,

$$E_{**}(J, W_2) = -\sum_t (W_2(t), 0) - \int_0^1 (W_2, 0)\, dt = 0,$$

i.e., J belongs to the kernel of E_{**}.

Conversely, suppose that W_1 belongs to the kernel of E_{**}. Let us partition the interval $[0, 1]$ by points $0 = t_0 < t_1 < \cdots < t_k = 1$ so that the vector field W_1 is smooth on each of the intervals $[t_{i-1}, t_i]$. Consider a smooth function $f : [0, 1] \rightarrow$ $[0, 1]$ vanishing at the points $t_0, t_1, \ldots, t_k$ and positive at all other points. We set

$$W_2(t) = f(t) \left((\nabla_V)^2 W_1 - R(V, W_1)V \right);$$

of course, V and W_1 also depend on t. We have

$$-E_{**}(W_1, W_2) = \sum 0 + \int_0^1 f(t) \left\| (\nabla_V)^2 W_1 - R(V, W_1)V \right\|^2 dt.$$

This quantity equals 0; therefore, the restriction of W_1 to each of the intervals $[t_{i-1}, t_i]$ is a Jacobi field.

Now consider a vector field W_2' such that $W_2'(t_i) = \Delta_{t_i} \nabla_V W_1$ for $i = 1, 2, \ldots,$ $k-1$. We have

$$-E_{**}(W_1, W_2') = \sum_{i=1}^{k-1} \left\| \Delta_{t_i} \nabla_V W_1 \right\|^2 + \int_0^1 0 \, dt = 0;$$

this means that $\nabla_V W_1$ has no jumps. But the solution W_1 of the Jacobi equation is completely determined by the vectors $W_1(t_i)$ and $\nabla_V W_1(t_i)$. Hence the k Jacobi vector fields (restrictions to the intervals $[t_{i-1}, t_i]$) are compatible with one another and can be glued together so as to form a smooth vector field on the interval $[0, 1]$. □

Thus, the dimension of the kernel of E_{**} equals the multiplicity of p and q as conjugate points. The dimension of the space of Jacobi fields vanishing at $t = 0$ equals 2. Therefore, the multiplicity of conjugate points is not higher than 2. Let us show that, in fact, the multiplicity is at most 1 (and therefore the notion of the multiplicity of conjugate points makes sense only on many-dimensional Riemannian manifolds). To this end, it suffices to construct an example of a Jacobi field which vanishes at $t = 0$ and does not vanish at $t = 1$. We set $J(t) = tV(t)$, where $V(t) = \gamma'(t)$ is the velocity vector of a geodesic. We have

$$\nabla_V J = 1 \cdot V + t\nabla_V V = V,$$

because $\nabla_V V = 0$. Hence

$$(\nabla_V)^2 J = \nabla_V V = 0.$$

Moreover, $R(V, J)V = tR(V, V)V = 0$, because the Riemann curvature tensor is skew-symmetric with respect to the first two variables. Thus, $J(t) = tV(t)$ is a Jacobi field with $J(0) = 0$ and $J(1) \neq 0$.

Jacobi fields can be obtained by using variations with loose endpoints in the class of geodesics. Let $\gamma(s, t)$ be a family of geodesic, i.e., a family in which each curve $\gamma(s_0, t)$ is a geodesic; we do not fix the endpoints of these geodesics. Then the variation vector field $J(t) = \frac{\partial \gamma}{\partial s}(0, t)$ is a Jacobi field. Indeed, since each curve $\gamma(s_0, t)$ is a geodesic, it follows that $\nabla_{\partial\gamma/\partial t}\frac{\partial\gamma}{\partial t}(s, t) = 0$. Therefore,

$$0 = \nabla_{\partial\gamma/\partial s}\nabla_{\partial\gamma/\partial t}\frac{\partial\gamma}{\partial t} = \nabla_{\partial\gamma/\partial t}\nabla_{\partial\gamma/\partial s}\frac{\partial\gamma}{\partial t} - R\left(\frac{\partial\gamma}{\partial t}, \frac{\partial\gamma}{\partial s}\right)\frac{\partial\gamma}{\partial t}.$$

Moreover, $\nabla_{\partial\gamma/\partial s}\frac{\partial\gamma}{\partial t} = \nabla_{\partial\gamma/\partial t}\frac{\partial\gamma}{\partial s}$, whence

$$\nabla_{\partial\gamma/\partial t}\nabla_{\partial\gamma/\partial s}\frac{\partial\gamma}{\partial t} = (\nabla_{\partial\gamma/\partial t})^2\frac{\partial\gamma}{\partial s}.$$

Thus, at $s = 0$ we have

$$(\nabla_V)^2 J - R(V, J)V = 0,$$

as required.

The converse is also true: any Jacobi field can be obtained by a variation in the class of geodesics. But we will not prove this (for a proof, see, e.g., [Mi4], p. 81).

If e_1, e_2 is an orthonormal basis, then in this basis the matrix of the first quadratic form is the identity, and hence $R^2_{112} = -K$ and $R(e_1, e_2)e_1 = R^2_{112}e_2 = -Ke_2$. Suppose given a geodesic with natural parameterization; let $e_1 = V$ be its velocity vector, and let e_2 be the unit vector orthogonal to it. Finally, let J be a vector field orthogonal to the geodesic, i.e., $J = ye_2$. The vector field e_1 is parallel along the geodesic, and hence so is the vector field e_2. Therefore, $\nabla_V J = y'e_2$ and $(\nabla_V)^2 J = y''e_2$. Thus, for the vector field ye_2, which is orthogonal to the geodesic, the Jacobi differential equation has the form $(\nabla_V)^2 J + KJ = 0$, i.e.,

$$y'' + Ky = 0.$$

The Jacobi equation for $J = xe_1 + ye_2$ can be written in coordinates as $x'' = 0$ and $y'' + Ky = 0$. Conjugate points are associated only with Jacobi vector fields for which $x = 0$, i.e., with vector fields orthogonal to the geodesic. Clearly, if $y(t)$ is a solution of the equation $y'' + Ky = 0$, then, for any constant c, cy is a solution of this equation as well. Therefore, in studying conjugate points, it suffices to consider solutions of the equation $y'' + Ky = 0$ with initial conditions $y(0) = 0$ and $y'(0) = 1$ in the domain $t \geqslant 0$.

If $K \leqslant 0$ on the entire geodesic, then this geodesic contains no conjugate points. Indeed, in this case, $y''(t) \geqslant 0$, and hence $y'(t) = 1 + \int_0^t y''(t)dt \geqslant 1$, which means that the function $y(t)$ monotonically increases and cannot vanish at $t > 0$. Thus, conjugate points can exist only if the Gaussian curvature is positive at some points. If, in addition, the Gaussian curvature is bounded by a positive constant from below or above, then we can estimate the distance between neighboring conjugate points from above or below. For this purpose, we can use Sturm's theory, which makes it

possible to estimate distances between neighboring zeros of solutions to equations
of the form $y'' + Ky = 0$.

Theorem 3.26 (Sturm) *Suppose given two differential equations*

$$y'' + a(x)y = 0, \quad z'' + b(x)z = 0$$

*in the interval $[0, x_1]$, where $a(x) \leqslant b(x)$ in the entire interval; let $y(x)$ and $z(x)$
be solutions of these equations for which $y(0) = z(0) = 0$ and $y'(0) = z'(0) = 1$.
Then $y(x) \geqslant z(x)$ for $x > 0$ if both functions y and z are positive on the half-open
interval $(0, x]$.*

Proof Clearly, $y''z - yz'' = (b(x) - a(x))yz$ and $(y'z - yz')' = y''z - yz''$; therefore,

$$y'z - yz' = \int_0^x (b(x) - a(x))yz \, dx \geqslant 0.$$

This means that $(y/z)' \geqslant 0$.

According to L'Hôpital rule, we have $\lim_{x \to 0} \frac{y(x)}{z(x)} = \lim_{x \to 0} \frac{y'(x)}{z'(x)} = 1$; hence
$y(x) \geqslant z(x)$ for $x > 0$. $\qquad\square$

Corollary

(a) *If $z(x_1) = 0$ for some positive x_1 and the function z has no zeros between 0 and
x_1, then $y(x) \neq 0$ for $0 < x < x_1$.*

(b) *If $y(x_1) = 0$ for some positive x_1, then $z(x_2) = 0$ for some positive $x_2 \leqslant x_1$.*

HISTORICAL COMMENT Jacques Charles François Sturm (1803–1855) developed
comparison theory for solutions of differential equations in 1836.

Theorem 3.27 *Let $\gamma(t)$ be a geodesic with natural parameterization, and let $K(t)$
be the Gaussian curvature at $\gamma(t)$. Suppose given a positive number K.*

(a) *If $K(t) \leqslant K$ for all t, then no point $\gamma(t)$, $0 < t < \pi/\sqrt{K}$, of the geodesic is
conjugate to the point $\gamma(0)$.*

(b) *If $K \leqslant K(t)$ for all t, then the geodesic has a point $\gamma(t)$, $0 < t \leqslant \pi/\sqrt{K}$,
conjugate to $\gamma(0)$.*

Proof

(a) We set $a(t) = K(t)$ and $b(t) = K$ and apply assertion (a) of the corollary to
Sturm's theorem. The solution $z(t)$ equals $\frac{1}{\sqrt{K}} \sin(t\sqrt{K})$; therefore, $y(t)$ does
not vanish for $0 < t < \frac{\pi}{\sqrt{K}}$.

(b) We set $a(t) = K$ and $b(t) = K(t)$ and apply assertion (b) of the corollary to
Sturm's theorem. The solution $y(t)$ equals $\frac{1}{\sqrt{K}} \sin(t\sqrt{K})$; therefore, $z(t) = 0$
for some positive $t \leqslant \frac{\pi}{\sqrt{K}}$. $\qquad\square$

A small enough part of a geodesic is a shortest curve joining given points (see Theorem 3.19 on p. 104). But if a part of a geodesic contains conjugate points, then it does not have this property.

Theorem 3.28 *Let $\gamma(t)$, $t \in [a, b]$, be a geodesic, and let $\gamma(a)$ and $\gamma(\tau)$, $a < \tau < b$, be conjugate points. Then γ is not a shortest curve joining the points $\gamma(a)$ and $\gamma(b)$.*

Proof Let $J(t)$ be a nonzero Jacobi vector field on the geodesic $\gamma(t)$ for which $J(a) = J(\tau) = 0$. Consider the vector field $\tilde{J}(t)$ on $\gamma(t)$ which coincides with $J(t)$ for $a \leqslant t \leqslant \tau$ and vanishes for $\tau \leqslant t \leqslant b$. At the point $\gamma(\tau)$ the jump $\Delta_\tau \nabla_V \tilde{J} = \nabla_V J(\tau)$ of the velocity vector is nonzero. Indeed, otherwise, since $J(t) = y(t)e_2(t)$ and $\nabla_V J(\tau) = y'(\tau)e_2(\tau)$, the equations $y(\tau) = y'(\tau) = 0$ would imply that the vector field $J(t)$ must be zero.

Let us construct a vector field $X(t)$ on the curve $\gamma(t)$ so that $X(a) = X(b) = 0$ and $(X(\tau), \Delta_\tau \nabla_V \tilde{J}) = 1$ and consider the vector field $W = \frac{1}{c}\tilde{J} - cX$. We have

$$E_{**}(W, W) = \frac{1}{c^2}E_{**}(\tilde{J}, \tilde{J}) - 2E_{**}(\tilde{J}, X) + c^2 E_{**}(X, X).$$

It follows from the second variation formula that

$$E_{**}(W, W) = 0 - 2(X(\tau), \Delta_\tau \nabla_V \tilde{J}) + c^2 E_{**}(X, X)$$

$$= -2 + c^2 E_{**}(X, X).$$

If c is small enough, then $E_{**}(W, W) < 0$.

Take a vector field W for which $E_{**}(W, W) < 0$ and consider the variations

$$\alpha(s, t) = \exp_{\gamma(t)}(sW(t)),$$

$$\beta(s_1, s_2, t) = \alpha(s_1 + s_2, t) = \exp_{\gamma(t)}((s_1 + s_2)W(t)).$$

We have

$$\left.\frac{\partial E(\alpha(s, t))}{\partial s}\right|_{s=0} = \left.\frac{\partial^2 E(\beta(s_1, s_2, t))}{\partial s_1 \partial s_2}\right|_{(s_1, s_2)=(0,0)} = E_{**}(W, W) < 0.$$

Therefore, the map $s \mapsto E(\alpha(s, t))$ has a strict local maximum at $s = 0$. Thus, the curve γ cannot be a strict local minimum for the functional E. $\qquad \square$

It follows directly from Theorems 3.27 and 3.28 that if the Gaussian curvature of a surface is not smaller than a given positive number K at each point, then the shortest length of a geodesic on this surface is at most $\frac{\pi}{\sqrt{K}}$.

Recall that a metric space is complete if it contains all of its limit points. An example of a complete metric space is a surface without boundary embedded in $\mathbb{R}^3$ as a closed subset.

There are points on a surface with boundary or on a surface minus a point that cannot be joined by a geodesic. But on a surface without boundary, which is complete as a metric space, any two points can be joined by a geodesic. This follows from the Hopf–Rinow theorems, which we discuss in Sect. 5.4 (see Theorem 5.5 on p. 175 and Theorem 5.7 on p. 177). As a result, we obtain the following theorem.

Theorem 3.29 (Bonnet)   *If the Gaussian curvature of a complete surface without boundary is not smaller than a fixed positive number K at each point, then this surface is compact.*

Proof If the geodesic is longer than $\frac{\pi}{\sqrt{K}}$, then this geodesic is not minimal. Therefore, the distance between any two points of the surface does not exceed $\frac{\pi}{\sqrt{K}}$.
                                                                                              □

HISTORICAL COMMENT Bonnet proved the theorem on the maximum length of a geodesic on a surface whose Gaussian curvature is bounded below by a fixed positive number at each point in 1855.

To conclude this section, we give one more description of conjugate points: points p and $\exp_p(V)$ on a geodesic $\gamma(t) = \exp_p(tV)$ are conjugate if and only if V is a critical point of the map $\exp_p$. Before proving this assertion, we explain it, because it is concerned with such an intricate object as the tangent space to the tangent space. The vector V is a point of the tangent space at p. At the point V the differential of $\exp_p$ is defined, which is a linear map from the tangent space at V to the tangent space at $\exp_p(V)$. The point V is a critical point of $\exp_p$ if this linear map of tangent spaces is degenerate, i.e., some nonzero vector X in the tangent space at V is mapped to zero.

Theorem 3.30 *Points p and $\exp_p(V)$ on the geodesic $\gamma(t) = \exp_p(tV)$ are conjugate if and only if V is a critical point of the map $\exp_p$.*

Proof First, suppose that V is a critical point of the map $\exp_p$ and a nonzero vector X in the tangent space at V is mapped to zero. Let us represent X as the velocity vector of a curve $V(s)$ in the tangent space at p: $V(0) = V$ and $V'(0) = X$. Consider the family of curves $\exp_p(tV(s))$; for fixed s, each of these curves is geodesic, i.e., this family is a variation of the geodesic $\gamma(t) = \exp_p(tV)$ in the class of geodesics. Therefore, $W(t) = \frac{\partial}{\partial s}\big|_{s=0}\exp_p(tV(s))$ is a Jacobi field along the geodesic $\exp_p(tV)$. Clearly, $W(0) = 0$ and

$$W(1) = \frac{\partial}{\partial s}\bigg|_{s=0} \exp_p(V(s)) = (\exp_p)_*(V'(0)) = (\exp_p)_*(X) = 0.$$

It remains to check that the vector field $W(t)$ is not identically zero. To this end, we calculate its covariant derivative at $t = 0$:

$$\nabla_{\gamma'} W(0) = \nabla_{\gamma'}\big|_{t=0} \frac{\partial}{\partial s} \exp_p(V(s))$$

$$= \nabla_{\partial/\partial s}\big|_{s=0} \frac{\partial}{\partial t} \exp_p(V(s)) = \nabla_{\partial/\partial s}\big|_{s=0} V(s).$$

The last expression is the covariant derivative of the vector field $s \mapsto V(s)$ along the constant curve $s \mapsto p$. When the curve contracts to a point, the covariant derivative of the vector field transforms into the usual derivative with respect to s. Thus, $\nabla_{\gamma'} W(0) = V'(0) = X \neq 0$.

Now suppose that V is not a critical point of the map $\exp_p$. Take a basis X_1, X_2 of the tangent space at V; then $\exp_*(X_1)$, $\exp_*(X_2)$ is a basis of the tangent space at $\exp_p(V)$. Let us represent the vectors X_1 and X_2 as the velocity vectors of curves $V_1(s)$ and $V_1(s)$ in the tangent space at p such that $V_i(0) = V$ and $V_i'(0) = X_i$ and consider the variations $\exp_p(t V_i(s))$ and the corresponding variational fields $W_i(t)$. Clearly, $W_i(0) = 0$, and the vectors $W_i(1) = (\exp_p)_*(X_i)$ are linearly independent; hence their nontrivial linear combinations cannot vanish at $t = 1$. The dimension of the space of Jacobi fields vanishing at $t = 0$ equals 2, and therefore the Jacobi field on the geodesic $\exp_p(t V)$ cannot vanish both at $t = 0$ and at $t = 1$.   □

3.21   Jacobi's Theorem on a Normal Spherical Image

The *principal normal indicatrix* (also known as the *normal spherical image*) Γ of a space curve C is the set of heads of the principal normal vectors (with the same tail) to this curve. Recall that a principal normal vector is the vector e_2 in the Frenet–Serret formulas.

Theorem 3.31 (Jacobi) *If the principal normal indicatrix of a curve is a regular curve without self-intersections, then it divides the unit sphere into two parts of the same area.*

Proof Let s and $\bar{s}$ be the natural parameters of the curves C and Γ. It follows from the Frenet–Serret formula $e_2' = -ke_1 + \varkappa e_3$ that $\left(\frac{d\bar{s}}{ds}\right)^2 = \|e_2'\|^2 = k^2 + \varkappa^2$ and

$$\frac{de_2}{d\bar{s}} = \frac{de_2}{ds} \cdot \frac{ds}{d\bar{s}} = (-ke_1 + \varkappa e_3)\frac{ds}{d\bar{s}}. \qquad (3.20)$$

Differentiating this equation and taking into account the two other Frenet–Serret formulas, we obtain

$$\frac{d^2 e_2}{d\bar{s}^2} = (-k e_1 + \varkappa e_3)\frac{d^2 s}{d\bar{s}^2} + (-k' e_1 + \varkappa' e_3)\left(\frac{ds}{d\bar{s}}\right)^2 - (k^2 + \varkappa^2)e_2 \left(\frac{ds}{d\bar{s}}\right)^2;$$

$$(3.21)$$

here $k' = \frac{dk}{ds}$ and $\varkappa' = \frac{d\varkappa}{ds}$.

Let us calculate the geodesic curvature $\bar{k}_g$ of the curve Γ on the sphere. Let ε_1, ε_2, ε_3 be the Darboux frame of this curve. The curve Γ is traced by the head of the vector e_2, and therefore $\varepsilon_1 = \frac{de_2}{d\bar{s}}$. The vector e_2 is the normal vector to the sphere; hence $\varepsilon_3 = e_2$ and $\varepsilon_2 = e_2 \times \frac{de_2}{d\bar{s}}$. Thus,

$$\frac{d^2 e_2}{d\bar{s}^2} = \frac{d\varepsilon_1}{d\bar{s}} = \bar{k}_g \varepsilon_2 + \bar{k}_n \varepsilon_3.$$

Using (3.20), we obtain

$$\bar{k}_g = \left(\varepsilon_2, \frac{d^2 e_2}{d\bar{s}^2}\right) = \left(e_2 \times \frac{de_2}{d\bar{s}}, \frac{d^2 e_2}{d\bar{s}^2}\right)$$

$$= \left(-k e_2 \times e_1 + \varkappa e_2 \times e_3, \frac{d^2 e_2}{d\bar{s}^2}\right)\frac{ds}{d\bar{s}} = \left(k e_3 + \varkappa e_1, \frac{d^2 e_2}{d\bar{s}^2}\right)\frac{ds}{d\bar{s}}.$$

Now Eq. (3.21) gives

$$\bar{k}_g = (\varkappa' k - k' \varkappa)\left(\frac{ds}{d\bar{s}}\right)^3 = \frac{\varkappa' k - k' \varkappa}{k^2 + \varkappa^2} \cdot \frac{ds}{d\bar{s}} = \frac{d}{ds}\left(\arctan\frac{\varkappa}{k}\right) \cdot \frac{ds}{d\bar{s}},$$

whence $\bar{k}_g d\bar{s} = \frac{d}{ds}\left(\arctan\frac{\varkappa}{k}\right)ds$.

As a result, we obtain

$$\int_{\Gamma} \bar{k}_g d\bar{s} = \int_C \frac{d}{ds}\left(\arctan\frac{\varkappa}{k}\right)ds = 0,$$

because the curve C is closed and k does not vanish.

The rest of the proof is quite simple. Let R be one of the two domains of the sphere bounded by the curve Γ, and let A be its area. The Gaussian curvature K of a sphere of radius 1 equals 1, so that the Gauss–Bonnet formula yields

$$2\pi = \int_{\Gamma} \bar{k}_g d\bar{s} + \int_R K \, dA = \int_R dA = A.$$

$\square$

HISTORICAL COMMENT Jacobi proved the theorem on a normal spherical image in 1842.

3.22   Surfaces of Constant Gaussian Curvature

In polar geodesic coordinates (r, φ) the first quadratic form is $dr^2 + G(r, \varphi)d\varphi^2$, and the Gaussian curvature K is expressed in terms of G as

$$K = -\frac{1}{\sqrt{G}} \cdot \frac{\partial^2 \sqrt{G}}{\partial r^2}$$

(see formula (3.15) on p. 102). This expression can be written in the form

$$\frac{\partial^2 \sqrt{G}}{\partial r^2} + K\sqrt{G} = 0. \tag{3.22}$$

If the Gaussian curvature K is constant, then we obtain a second-order linear differential equation with constant coefficients.

Theorem 3.32 (Minding)   *Any two regular surfaces with the same constant Gaussian curvature are locally isometric. Any point of one of the surfaces and any orthonormal basis at this point can be mapped to any point of the other surface and any orthonormal basis at this point by a local isometry.*

Proof First, we solve Eq. (3.22), considering three cases: (1) $K = 0$, (2) $K > 0$, and (3) $K < 0$. Recall that $\lim_{r\to 0} G = 0$ and $\lim_{r\to 0} \frac{\partial \sqrt{G}}{\partial r} = 1$ (this was proved on p. 102).

(1) If $K = 0$, then $\frac{\partial^2 \sqrt{G}}{\partial r^2} = 0$. Therefore, $\frac{\partial \sqrt{G}}{\partial r} = f(\varphi)$ is a function of φ. The function $\frac{\partial \sqrt{G}}{\partial r}$ does not depend of r, and $\lim_{r\to 0} \frac{\partial \sqrt{G}}{\partial r} = 1$; hence $\sqrt{G} = r + g(\varphi)$. The function $g(\varphi)$ does not depend of r either; hence $g(\varphi) = \lim_{r\to 0} G = 0$. Thus, in case (1), we have $E = 1$, $F = 0$, and $G = r^2$.

(2) If $K > 0$, then the general solution of Eq. (3.22) has the form

$$\sqrt{G} = A(\varphi)\cos(\sqrt{K}r) + B(\varphi)\sin(\sqrt{K}r),$$

where $A(\varphi)$ and $B(\varphi)$ are some functions. Since $\lim_{r\to 0} G = 0$, we have $A(\varphi) = 0$. Therefore,

$$\frac{\partial \sqrt{G}}{\partial r} = B(\varphi)\sqrt{K}\cos(\sqrt{K}r).$$

From $\lim_{r\to 0} \frac{\partial \sqrt{G}}{\partial r} = 1$ it follows that $B(\varphi) = \frac{1}{\sqrt{K}}$. Thus, in case (2), $E = 1$, $F = 0$, and $G = \frac{1}{K}\sin^2(\sqrt{K}r)$.

(3) If $K < 0$, then the general solution of Eq. (3.22) has the form

$$\sqrt{G} = A(\varphi)\cosh(\sqrt{-K}r) + B(\varphi)\sinh(\sqrt{-K}r).$$

From the same considerations as in the previous case we obtain $E = 1$, $F = 0$, and $G = \frac{1}{-K} \sinh^2(\sqrt{-K}r)$.

To construct the required local isomorphism, we identify neighborhoods of the given points with neighborhoods of tangent spaces by means of the map $\exp^{-1}$. After that, we apply the isometry of the tangent spaces which maps one orthonormal basis to the other. $\square$

HISTORICAL COMMENT Minding proved the theorem on the local isometry of surfaces of constant Gaussian curvature in 1839. He also found equations for surfaces of revolution of constant curvature. In 1840 Minding derived trigonometric relations in a triangle formed by geodesics on a surface of constant curvature. He discovered that trigonometry on a surface of constant negative curvature is obtained from that on the sphere by replacing the usual trigonometric functions by the hyperbolic ones. But Minding overlooked the connection between this discovery and hyperbolic geometry. The first to notice this connection was Eugenio Beltrami (1835–1900); this had enabled him to construct the first model of hyperbolic geometry in 1868.

Now let us look at the structure of a surface of revolution of constant curvature. Recall that, according to Problem 3.11, if the parameterization of a meridian of a surface of revolution is natural, i.e., the parameterization of the surface has the form $(f(s) \cos v, f(s) \sin v, g(s))$, where $(f')^2 + (g')^2 = 1$, then the Gaussian curvature equals $-\frac{f''}{f}$.

First, consider the case of positive Gaussian curvature. In this case, we have $K = a^2$, where $a > 0$, and the function f is determined by the equation $f'' + a^2 f = 0$. The solutions of this equation have the form $f(s) = a_1 \cos(as) + a_2 \sin(as) = A \cos(as + b)$. Appropriately choosing the point from which the length of an arc is measured, we can assume that $b = 0$. We can also assume that $A > 0$. Since the function f must not vanish, it follows that $|s| < \frac{\pi}{2a}$. Moreover, $g' = \pm\sqrt{1 - (f')^2}$, whence we obtain

$$g(s) = \pm \int_0^s \sqrt{1 - a^2 A^2 \sin^2(at)}\, dt.$$

If $A > 1/a$, then the function $g(s)$ is defined only when $\sin|as| < \frac{1}{aA}$.

For $A = 1/a$, the surface is a sphere. Curves and axes of rotation for $A < 1/a$ and $A > 1/a$ are shown in Fig. 3.3. In both cases, the arc of the curve is open, i.e., it does not contain its endpoints. In the second case, the limit position of the tangent as an endpoint is approached is perpendicular to the axis of rotation.

In the case of zero Gaussian curvature, the equation for the function f has the form $f'' = 0$, and the solutions of this equation have the form $f(s) = as + b$. Moreover, $g' = \pm\sqrt{1 - (f')^2} = \pm\sqrt{1 - a^2}$, and hence $g = cs + d$, where $c = \pm\sqrt{1 - a^2}$. For $a = 0$, we obtain a right circular cylinder, for $c = 0$, we obtain a plane, and for $ac \neq 0$, we obtain a right circular cone.

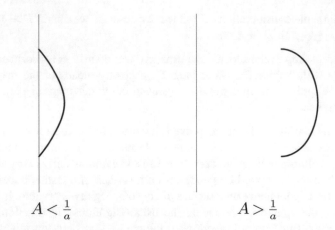

$$A < \frac{1}{a} \qquad\qquad A > \frac{1}{a}$$

Fig. 3.3 A surface of rotation of positive Gaussian curvature

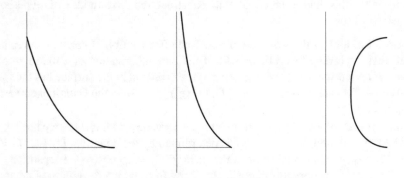

Fig. 3.4 Surfaces of revolution of negative Gaussian curvature

Finally, consider the case of negative Gaussian curvature. In this case, $K = -a^2$, where $a > 0$, the equation for f has the form $f'' - a^2 f = 0$, and the solutions of this equation have the form $f(s) = a_1 \cosh(as) + a_2 \sinh(as)$. Depending on how a_1 and a_2 are related, the solution can be written in one of the three forms (with real coefficients) $A \cosh(as + b)$ (if $a_1 > a_2$), Be^{as} (if $a_1 = a_2$), and $C \sinh(as + c)$ (if $a_1 < a_2$). In these cases, respectively, $g(s) = \pm \int_0^s \sqrt{1 - a^2 A^2 \sinh^2(at)}\,dt + D$, $g(s) = \pm \int_0^s \sqrt{1 - a^2 B^2 e^{2at}}\,dt + D$, and $g(s) = \pm \int_0^s \sqrt{1 - a^2 C^2 \cosh^2(at)}\,dt + D$. The corresponding curves and axes of rotation are shown in Fig. 3.4.

3.23   Rigidity (Unbendability) of the Sphere

Locally, the sphere is not rigid. Indeed, in the previous section we gave examples of surfaces of constant positive Gaussian curvature. They are locally isometric to the sphere but cannot be obtained from the sphere by a motion of space. Nevertheless, globally, the sphere is rigid.

Theorem 3.33 *Let S be a closed connected surface of constant Gaussian curvature K embedded in $\mathbb{R}^3$. Then S is the sphere $\mathbb{S}^2$.*

Proof A closed surface embedded in $\mathbb{R}^3$ is orientable. If we choose its orientation, then its principal curvatures will be defined (when no orientation is chosen, its principal curvatures are defined only up to sign). We assume that the principal curvatures k_1 and k_2 of the surface are chosen so that $k_1 \geqslant k_2$ at each point of S. These functions are continuous everywhere on the surface S and differentiable everywhere except possibly at umbilical points.

Lemma *Suppose that the Gaussian curvature of an oriented smooth surface is positive at a point p and that this point is a point of local maximum of the function k_1 and a point of local minimum of the function k_2. Then the point p is umbilical.*

Proof Suppose that the point p is not umbilical. Then in a neighborhood of p we can introduce coordinates (u, v) so that the coordinate lines are lines of curvature (see p. 107). Then the first and the second quadratic form are $E\,du^2 + G\,dv^2$ and $k_1 E\,du^2 + k_2 G\,dv^2$, and the Codazzi–Mainardi equations are written as

$$(k_1 E)_v = \frac{E_v}{2}(k_1 + k_2), \quad (k_2 G)_u = \frac{G_u}{2}(k_1 + k_2),$$

i.e.,

$$E(k_1)_v = \frac{E_v}{2}(-k_1 + k_2), \quad G(k_2)_u = \frac{G_u}{2}(k_1 - k_2). \tag{3.23}$$

According to Theorem 3.3 (see p. 76), in the case where $F = 0$, the Gaussian curvature K is expressed in terms of the coefficients of the first quadratic form as

$$K(EG)^2 = \begin{vmatrix} -\frac{1}{2}G_{uu} - \frac{1}{2}E_{vv} & \frac{1}{2}E_u & -\frac{1}{2}E_v \\ -\frac{1}{2}G_u & E & 0 \\ \frac{1}{2}G_v & 0 & G \end{vmatrix} - \begin{vmatrix} 0 & \frac{1}{2}E_v & \frac{1}{2}G_u \\ \frac{1}{2}E_v & E & 0 \\ \frac{1}{2}G_u & 0 & G \end{vmatrix};$$

therefore,

$$-2KEG = E_{vv} + G_{uu} + P_1 E_v + Q_1 G_u, \tag{3.24}$$

where P_1 and Q_1 are continuous functions of u and v.

Let us express E_v and G_u from Eq. (3.23), differentiate the obtained expressions with respect to v and u, and substitute the resulting expressions for E_{vv} and G_{uu} into Eq. (3.24):

$$-2KEG = -\frac{2E}{k_1 - k_2}(k_1)_{vv} + \frac{2G}{k_1 - k_2}(k_2)_{uu} + P_2(k_1)_v + Q_2(k_2)_u,$$

i.e.,

$$-(k_1 - k_2)KEG = -2E(k_1)_{vv} + 2G(k_2)_{uu} + P_3(k_1)_v + Q_3(k_2)_u.$$

At the point p the quantity $-(k_1 - k_2)KEG$ is negative. Since p is a point of local maximum for the function k_1 and a point of local minimum for k_2, it follows that $(k_1)_v = (k_2)_u = 0$, $(k_1)_{vv} \leqslant 0$, and $(k_2)_{uu} \geqslant 0$. Thus, at the point p the quantity $-2E(k_1)_{vv} + 2G(k_2)_{uu} + P_3(k_1)_v + Q_3(k_2)_u$ is nonnegative. This contradiction shows that p is a umbilical point. $\square$

Any closed surface embedded in $\mathbb{R}^3$ has at least one point of positive Gaussian curvature (this follows from Problem 3.12 on p. 74). Choose such a point on the surface S. Since the Gaussian curvature of S is constant, we have $K > 0$. The continuous function k_1 on S attains its maximum at some point p. The product $k_1 k_2 = K$ is a positive constant; therefore, the function k_2 attains its minimum at p. Thus, according to the lemma, the point p is umbilical, i.e., $k_1 = k_2$.

Now consider any point q of the surface S. By assumption $k_1(q) \geqslant k_2(q)$, whence

$$k_1(p) \geqslant k_1(q) \geqslant k_2(q) \geqslant k_2(p) = k_1(p).$$

Therefore, $k_1(q) = k_2(q)$, i.e., all points of S are umbilical. Moreover, the surface S is closed (by assumption), and according to Problem 3.19, this surface is a sphere.
                                                                                 $\square$

Corollary *The sphere $\mathbb{S}^2$ is rigid in the following sense: if there exists an isometry of $\mathbb{S}^2$ to a surface S embedded in $\mathbb{R}^3$, then S is the sphere $\mathbb{S}^2$.*

Proof Let $\varphi \colon \mathbb{S}^2 \to S$ be an isometry of the sphere to a surface S. Then S has constant curvature, because Gaussian curvature is invariant with respect to isometries. The surface S is compact and connected as a continuous image of a connected compact space. Therefore, according to Theorem 3.33, S is a sphere. $\square$

HISTORICAL COMMENT Theorem 3.33 was proved by Heinrich Liebmann (1874–1939) in 1899. It is known as Liebmann's theorem on the rigidity of the sphere. The lemma on which the proof given above is based was proposed by David Hilbert (1862–1943) in 1901. It is called Hilbert's lemma.

3.24   Convex Surfaces: Hadamard's Theorem

A *convex surface* is the boundary of a bounded convex body in $\mathbb{R}^3$.

The Gaussian curvature of a convex surface is nonnegative at each point. Indeed, a convex surface lies on one side of the tangent plane at each point, and therefore all normal curvatures are of the same sign (although some of them may vanish).

Theorem 3.34 (Hadamard)   *If the Gaussian curvature of a connected closed surface S embedded in $\mathbb{R}^3$ is positive at each point, then this surface is convex.*

Proof Let us choose the inward normal vector at each point of S. Then all normal curvatures are positive at each point of the surface.

Consider the open set D bounded by the surface S. Choose an arbitrary point A in D and let D_A denote the set consisting of all those points $B \in D$ for which the segment AB is contained in D. Obviously, the set D_A is open both in $\mathbb{R}^3$ and in the topological space D. Let us show that D_A is closed in D.

Let B_n be a sequence of points of D_A converging to a point $B_0 \in D$. Suppose that $B_0 \notin D_A$. This means that the surface S is tangent to the segment AB at some interior point P of this segment. Then the segment AB lies in the tangent plane to S at the point P, and the normal curvature corresponding to the direction of AB is nonpositive, because in a small neighborhood of P the normal vector and the section lie on different sides of the tangent plane. (The normal vector and the points B_n lie on one side of this plane and the section lies on the other side.)

Since D is connected, it follows that D_A coincides with D. Thus, the segment joining any two given points of D is entirely contained in D, which means that the set D is convex. $\square$

Remark Hadamard's theorem is valid not only for an embedded surface but also for an immersed one; thus, if the Gaussian curvature of an immersed closed surface is positive, then this surface is embedded. The proof of this theorem for hypersurfaces (Theorem 4.3) is given on p. 149.

HISTORICAL COMMENT Jacques Salomon Hadamard (1865–1963) proved the theorem about the convexity of a surface of positive Gaussian curvature in 1897.

3.25   The Laplace–Beltrami Operator

Recall that the Laplace operator in the plane with coordinates (x, y) assigns the function

$$\Delta f = \frac{\partial^2 f}{\partial x^2} + \frac{\partial^2 f}{\partial y^2}$$

to each function $f(x, y)$. The Laplace–Beltrami operator is a generalization to curved surfaces of the Laplace operator on the plane. The definition of the Laplace–Beltrami operator uses the metric of the surface in one way or another.

For the Laplace–Beltrami operator, the same notation Δ is used. However, by a well-established tradition, the Laplace–Beltrami operator Δ in the case of the plane (and, in general, of any Euclidean space) equals the negative of the classical Laplace operator rather than this operator itself. Such a notation is chosen to ensure the nonnegativity of the spectrum of the Laplace–Beltrami operator.

The simplest definition of the *Laplace–Beltrami operator* is in terms of the differential forms ω^1 and ω^2 dual to a moving orthonormal frame. Let $df = a_1\omega^1 + a_2\omega^2$. We set $*df = -a_2\omega^1 + a_1\omega^2$. Then $d*df = d(-a_2\omega^1 + a_1\omega^2) = (-\Delta f)\omega^1 \wedge \omega^2$, i.e., Δf is defined as the coefficient of the 2-form $\omega^1 \wedge \omega^2$ with the minus sign.

In the case of the plane, this approach leads to the usual Laplace operator (with the minus sign). Indeed, we have $df = \frac{\partial f}{\partial x}dx + \frac{\partial f}{\partial y}dy$ and $*df = -\frac{\partial f}{\partial y}dx + \frac{\partial f}{\partial x}dy$. Therefore,

$$d*df = d\left(-\frac{\partial f}{\partial y}dx + \frac{\partial f}{\partial x}dy\right) = -\frac{\partial^2 f}{\partial x^2}dy \wedge dx + \frac{\partial^2 f}{\partial x^2}dx \wedge dy$$

$$= \left(\frac{\partial^2 f}{\partial x^2} + \frac{\partial^2 f}{\partial y^2}\right)dx \wedge dy.$$

Another definition of the Laplace–Beltrami operator is in terms of the composition of divergence and gradient: $\Delta f = -\operatorname{div}(\operatorname{grad} f)$. The *gradient* of a function f is the vector field $\operatorname{grad}(f)$ characterized by the property

$$(\operatorname{grad} f, V) = \partial_V f$$

for any vector field V. Here $\partial_V f$ is the derivative of f in the direction of the vector field V.

The definition of the *divergence* of a vector field involves the differentiation of a vector field (i.e., covariant differentiation), rather than that of a function. The divergence $\operatorname{div}(V)$ of a vector field V is defined at each point p as the trace of the map $W \mapsto \nabla_W V$; here the vector W ranges over the tangent space at p.

The Laplace–Beltrami operator can be expressed in local coordinates in terms of the Riemannian metric g_{ij}. We will do this step by step: first, we will express the gradient, and then, the divergence.

Consider local coordinates u^i and the coordinate vector fields $r_i = \frac{\partial}{\partial u^i}$. We have $V = V^i r_i$ and $g_{ij} = (r_i, r_j)$; therefore,

$$\partial_V f = V^i \frac{\partial f}{\partial u^i} = V^i g_{ik} g^{kl} \frac{\partial f}{\partial u^l} = V^i (r_i, r_k) g^{kl} \frac{\partial f}{\partial u^l} = \left(V, g^{kl} \frac{\partial f}{\partial u^l} r_k\right).$$

Thus,

$$\operatorname{grad} f = g^{kl} \frac{\partial f}{\partial u^l} r_k.$$

To calculate the divergence, we need expression (3.7) on p. 93:

$$\nabla_W V = W^j \left(\frac{\partial V^k}{\partial u^j} + V^i \Gamma^k_{ij} \right) r_k.$$

From this expression we obtain

$$\operatorname{div} V = \frac{\partial V^j}{\partial u^j} + V^i \Gamma^j_{ij} = \frac{\partial V^j}{\partial u^j} + V^j \Gamma^i_{ij}.$$

Now we apply Theorem 3.7 (see p. 86) and express the Christoffel symbols in terms of the Riemannian metric:

$$\Gamma^i_{ij} = \frac{1}{2} g^{il} \left(\frac{\partial g_{il}}{\partial u^j} - \frac{\partial g_{ij}}{\partial u^l} + \frac{\partial g_{lj}}{\partial u^i} \right) = \frac{1}{2} g^{il} \frac{\partial g_{il}}{\partial u^j}.$$

Thus,

$$\operatorname{div} V = \frac{\partial V^j}{\partial u^j} + \frac{1}{2} V^j g^{il} \frac{\partial g_{il}}{\partial u^j} = \frac{\partial V^j}{\partial u^j} + \frac{1}{2} V^j \operatorname{tr} \left(g^{-1} \frac{\partial g}{\partial u^j} \right),$$

where g is the matrix of the first quadratic form. To perform further transformations, we need the matrix algebra formula $\frac{d}{dt} \ln(\det A) = \operatorname{tr} \left(A^{-1} \frac{dA}{dt} \right)$ (see Appendix, Theorem 8.2 on p. 258). Using it, we obtain

$$\operatorname{div} V = \frac{\partial V^j}{\partial u^j} + \frac{1}{2} V^j \frac{\partial}{\partial u^j} (\ln \det g) = \frac{1}{\sqrt{\det g}} \frac{\partial}{\partial u^j} (V^j \sqrt{\det g}).$$

As a result, we arrive at the following expression for the Laplace–Beltrami operator in local coordinates:

$$\Delta f = -\operatorname{div}(\operatorname{grad} f) = -\frac{1}{\sqrt{\det g}} \frac{\partial}{\partial u^i} \left(\sqrt{\det g} \, g^{ij} \frac{\partial f}{\partial u^j} \right). \tag{3.25}$$

Let us check that the two approaches to the definition of the Laplace–Beltrami operator are equivalent. To do this, it suffices to prove that, for any function φ defined on the given surface and vanishing outside a given small neighborhood U of the point p, both definitions of Δf give the same integral $\int_U \varphi \Delta f \, d\sigma$, where $d\sigma$ is the area form. We will use two systems of local coordinates, arbitrary local coordinates u^1, u^2 and local coordinates x^1, x^2 orthonormal at the point p. For the metric g_{ij} in the coordinates u^i, we obtain

$$g_{ij} = (r_i, r_j) = \left(\frac{\partial r}{\partial u^i}, \frac{\partial r}{\partial u^j} \right) = \left(\frac{\partial r}{\partial x^k} \frac{\partial x^k}{\partial u^i}, \frac{\partial r}{\partial x^l} \frac{\partial x^l}{\partial u^j} \right) = \frac{\partial x^k}{\partial u^i} \frac{\partial x^k}{\partial u^j};$$

in the last expression the summation over k is performed, although both k's are superscripts.

Let us calculate the integral $\int \varphi \Delta f \, d\sigma = \int \varphi \Delta f \sqrt{g} du^1 \wedge du^2$ for Δf defined in the first way. By definition, we have $\Delta f \, d\sigma = -d(*df)$, whence we obtain

$$\int \varphi \Delta f \, d\sigma = -\int \varphi d(*df) = \int d\varphi \wedge (*df); \qquad (3.26)$$

we have used the relations $\int_U \varphi d(*df) + \int_U d\varphi \wedge (*df) = \int_U d(\varphi(*df)) = \int_{\partial U} \varphi(*df) = 0$. Next,

$$\int d\varphi \wedge (*df) = \int \left(\frac{\partial \varphi}{\partial x^1} dx^1 + \frac{\partial \varphi}{\partial x^2} dx^2 \right) \wedge \left(-\frac{\partial f}{\partial x^2} dx^1 + \frac{\partial f}{\partial x^1} dx^2 \right)$$

$$= \int \left(\frac{\partial \varphi}{\partial x^1} \frac{\partial f}{\partial x^1} + \frac{\partial \varphi}{\partial x^2} \frac{\partial f}{\partial x^2} \right) d\sigma$$

$$= \int \left(\frac{\partial u^i}{\partial x^k} \frac{\partial u^j}{\partial x^k} \right) \frac{\partial f}{\partial u^j} \frac{\partial \varphi}{\partial u^i} d\sigma$$

$$= \int g^{ij} \frac{\partial f}{\partial u^j} \frac{\partial \varphi}{\partial u^i} \sqrt{\det g} \, du^1 \wedge du^2$$

$$= -\int \frac{1}{\sqrt{\det g}} \frac{\partial}{\partial u^i} \left(\sqrt{\det g} \, g^{ij} \frac{\partial f}{\partial u^j} \right) \varphi \sqrt{\det g} \, du^1 \wedge du^2.$$

The last equality follows from the equation

$$\frac{1}{\sqrt{\det g}} \frac{\partial}{\partial u^i} \left(\sqrt{\det g} \, g^{ij} \frac{\partial f}{\partial u^j} \varphi \right) = \frac{1}{\sqrt{\det g}} \frac{\partial}{\partial u^i} \left(\sqrt{\det g} \, g^{ij} \frac{\partial f}{\partial u^j} \right) \varphi + g^{ij} \frac{\partial f}{\partial u^j} \frac{\partial \varphi}{\partial u^i}.$$

Thus, the first approach leads to the expression (3.25) in local coordinates for the Laplace–Beltrami operator.

HISTORICAL COMMENT The Laplace–Beltrami operator for surfaces was introduced by Beltrami in 1864–1865.

Theorem 3.35 *If a function f on a closed orientable surface is such that $\Delta f = \lambda f$ for some constant λ, then $\lambda \geqslant 0$.*

Proof Let us apply formula (3.26) to $\varphi = f$:

$$\int f\Delta f\, d\sigma = \int df \wedge (*df) = \int \left(\left(\frac{\partial f}{\partial x^1} \right)^2 + \left(\frac{\partial f}{\partial x^2} \right)^2 \right) \geqslant 0.$$

Here the expression in local coordinates is written at one point. The integral is nonnegative because the expression has this form at each point.

Clearly, the sign of the integral

$$\int f\Delta f\, d\sigma = \lambda \int f^2\, d\sigma$$

coincides with that of λ. Therefore, $\lambda \geqslant 0$. □

Theorem 3.36 (Bochner)   *If a function f on a closed orientable surface is such that $\Delta f \geqslant 0$ at all points, then this function is constant.*

Proof Applying formula (3.26) to $\varphi = 1$, we obtain

$$\int \Delta f\, d\sigma = \int d(1)\varphi \wedge (*df) = 0.$$

Therefore, $\Delta f = 0$. Now let us apply the same formula to $\varphi = f$:

$$0 = \int f\Delta f\, d\sigma = \int df \wedge (*df) = \int \left(\left(\frac{\partial f}{\partial x^1} \right)^2 + \left(\frac{\partial f}{\partial x^2} \right)^2 \right).$$

Thus, $df = 0$ and $f = \text{const}$. □

HISTORICAL COMMENT   Salomon Bochner (1899–1982) proved in 1946 that if a function φ on a closed Riemannian manifold satisfies the condition $\Delta\varphi \geqslant 0$, then this function is constant. This assertion is often called Bochner's lemma. Bochner's lemma is closely related to a general theorem about elliptic operators proved by Eberhard Hopf (1902–1983) in 1927.

A function f for which $\Delta f = 0$ said to be *harmonic*.

Problem 3.36 Prove that the projection of a minimal surface onto any straight line is a harmonic function.

3.26   Solutions of Problems

3.1 A n s w e r: $\begin{pmatrix} 1 & 0 \\ 0 & r^2 \end{pmatrix}$ and $\begin{pmatrix} 1 & 0 \\ 0 & \frac{1}{r^2} \end{pmatrix}$. The Cartesian coordinates (x, y) are expressed

in terms of the polar ones as $x = r \cos\varphi$, $y = r \sin\varphi$. Therefore, $e_r = \left(\frac{\partial x}{\partial r}, \frac{\partial y}{\partial r} \right) =$

$(\cos\varphi, \sin\varphi)$ and $e_\varphi = \left(\frac{\partial x}{\partial \varphi}, \frac{\partial y}{\partial \varphi} \right) = (-r \sin\varphi, r \cos\varphi)$. Calculating the pairwise

inner products of these vectors, we find the matrix (g_{ij}).

3.2 A n s w e r: $\begin{pmatrix} 1 + (f')^2 & 0 \\ 0 & f^2 \end{pmatrix}$ and $\begin{pmatrix} \frac{1}{1+(f')^2} & 0 \\ 0 & \frac{1}{f^2} \end{pmatrix}$. Here $e_u = (f' \cos v, f' \sin v, 1)$

and $e_v = (-f \sin v, f \cos v, 0)$.

3.3 Let $\gamma(s)$ and $\bar{\gamma}(s)$ be two curves on S with common tangent at the point $\gamma(0) = \bar{\gamma}(0)$, where s is the natural parameter. Then, for small s, the distance $d(s)$ between $\gamma(s)$ and $\bar{\gamma}(s)$ is of order s^2; to be more precise, the ratio $d(s)/s^2$ tends to a finite limit as $s \to 0$. Therefore, if $\varepsilon_3(s)$ and $\bar{\varepsilon}_3(s)$ are the unit normals to the surface S at the points $\gamma(s)$ and $\bar{\gamma}(s)$, then $\varepsilon'_3(0) = \bar{\varepsilon}'_3(0)$. By assumption $\varepsilon_1(0) = \bar{\varepsilon}_1(0)$; it is also clear that $\varepsilon_3(0) = \bar{\varepsilon}_3(0)$. Therefore, $\varepsilon_2(0) = \bar{\varepsilon}_2(0)$, and the equation $\varepsilon'_3 = -k_n \varepsilon_1 - \varkappa_g \varepsilon_2$ implies $k_n(0) = \bar{k}_n(0)$ and $\varkappa_g(0) = \bar{\varkappa}_g(0)$.

3.4 The vectors e_2, e_3, ε_2, and ε_3 lie in the same plane (orthogonal to the vector $e_1 = \varepsilon_1$), and the bases e_2, e_3 and ε_2, ε_3 have the same orientation (see Fig. 3.5). Therefore, $e_2 = \varepsilon_2 \sin\theta + \varepsilon_3 \cos\theta$, $\varepsilon_2 = e_2 \sin\theta - e_3 \cos\theta$, and $\varepsilon_3 = e_2 \cos\theta + e_3 \sin\theta$. Hence $\varepsilon'_1 = e'_1 = k e_2 = (k \sin\theta)\varepsilon_2 + (k \cos\theta)\varepsilon_3$. On the other hand, $\varepsilon'_1 = k_g \varepsilon_2 + k_n \varepsilon_3$. Thus, $k_g = k \sin\theta$ and $k_n = k \cos\theta$.

Fig. 3.5 The two bases

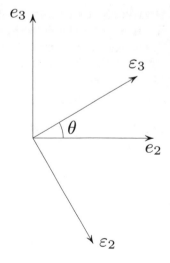

Let us calculate $\varkappa_g$. We have $\varepsilon_2' = -k_g \varepsilon_1 + \varkappa_g \varepsilon_3$, so that $\varkappa_g = (\varepsilon_3, \varepsilon_2')$. The Frenet–Serret formulas give

$$\varepsilon_2' = \cos\theta \frac{d\theta}{ds} e_2 - (k\sin\theta)e_1 + (\varkappa\sin\theta)e_3 + \sin\theta \frac{d\theta}{ds} e_3 + (\varkappa\cos\theta)e_2.$$

Therefore, $(\varepsilon_3, \varepsilon_2') = \varkappa + \frac{d\theta}{ds}$.

3.5 According to Problem 3.4, $k_g = k\sin\theta$ and $k_n = k\cos\theta$, whence $k_n^2 + k_g^2 = k^2(\sin^2\theta + \cos^2\theta) = k^2$.

3.6

(a) Let $k_n(\theta)$ be the normal curvature of the curve in the intersection of the surface and the plane making an angle of θ with the normal plane. Then $k_n(\theta) = k(\theta)\cos\theta$ in view of Problem 3.4. But all curves in the family have common tangent vector; therefore, according to Problem 3.3, all of them have the same normal curvature $k_n(\theta)$, which equals the curvature k of the normal section.

(b) Let $R(\theta) = 1/k(\theta)$ be the curvature radius of the curve in the intersection of the surface and the plane making an angle of θ with the normal plane, and let $R = R(0) = 1/k$. Since $k = k(\theta)\cos\theta$, we have $R(\theta) = R\cos\theta$. Therefore, the intersection of the family of osculating circles with the plane perpendicular to the surface and to the given tangent vector is a circle (see Fig. 3.6). This means that the osculating circles fill a sphere.

3.7 Let $\gamma(s)$ be a curve with natural parameterization lying on a sphere of radius R centered at the origin. Then $\varepsilon_1 = \gamma'$, $\varepsilon_3 = \pm\frac{1}{R}\gamma$, and $\varepsilon_2 = \varepsilon_3 \times \varepsilon_1$. Twice differentiating the identity $(\gamma, \gamma) = R^2 = $ const, we obtain $0 = \frac{d^2}{ds^2}(\gamma, \gamma) = 2(\gamma', \gamma') + 2(\gamma'', \gamma)$. Therefore,

$$k_n = (\varepsilon_1', \varepsilon_3) = \left(\gamma'', \pm\frac{1}{R}\gamma\right) = \mp\frac{1}{R}(\gamma', \gamma') = \mp\frac{1}{R}.$$

Fig. 3.6 Spheric sections

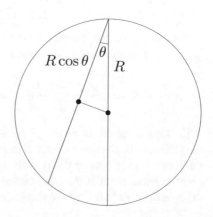

$R\cos\theta$ θ R

Next,

$$\frac{d\varepsilon_2}{ds} = \pm\frac{1}{R} \cdot \frac{d}{ds}(\gamma \times \gamma') = \pm\frac{1}{R}\gamma' \times \gamma' \pm \frac{1}{R}\gamma \times \gamma''$$

$$= \pm\frac{1}{R}\gamma \times \left(k_g\varepsilon_2 - \frac{1}{R^2}\gamma\right) = \pm\frac{k_g}{R}\gamma \times \varepsilon_2 = k_g\varepsilon_3 \times \varepsilon_2 = -k_g\varepsilon_1.$$

Hence it follows from $\varepsilon_2' = -k_g\varepsilon_1 + \varkappa_g\varepsilon_3$ that $\varkappa_g = 0$.

3.8 A n s w e r: $\pm\frac{\sqrt{R^2-r^2}}{Rr}$. The identity $k_n^2 + k_g^2 = k^2$ is to be used. In the case under consideration, $k^2 = 1/r^2$ and $k_n^2 = 1/R^2$. Therefore, $k_g^2 = \frac{R^2-r^2}{R^2 r^2}$.

3.9 For a straight line, the velocity vector ε_1 is constant, and hence $\varepsilon_1' = 0$. The zero vector is orthogonal to any plane.

3.10

(a) As a meridian we can take the initial plane curve. If ε_1 is the unit tangent vector to this curve, then the vector ε_1' is orthogonal to the curve. But then it is also orthogonal to the plane tangent to the surface, because the tangent plane passes through the tangent to the curve and is orthogonal to the plane of the curve.

(b) If ε_1 is the unit tangent vector to the circle, then the vector ε_1' is directed from the point of tangency to the center of the circle (or from the center of the circle). It is orthogonal to the tangent plane if and only if the tangent to the meridian is parallel to the axis of rotation.

3.11 Clearly, $r_u = (f'\cos v, f'\sin v, g')$, $r_v = (-f\sin v, f\cos v, 0)$, $r_{uu} = (f''\cos v, f''\sin v, g'')$, $r_{uv} = r_{vu} = (-f'\sin v, f'\cos v, 0)$, and $r_{vv} = (-f\cos v, -f\sin v, 0)$. Moreover, $n = \frac{1}{\sqrt{(f')^2+(g')^2}}(-g'\cos v, -g'\sin v, f')$.

Therefore, $E = (f')^2 + (g')^2$, $F = 0$, and $G = f^2$; we also have $L = \frac{g''f'-f''g'}{\sqrt{(f')^2+(g')^2}}$, $M = 0$, and $N = \frac{fg'}{\sqrt{(f')^2+(g')^2}}$. Hence the determinant of the first quadratic form equals $f^2((f')^2 + (g')^2)$, and the determinant of the second quadratic form equals $\frac{fg'(g''f'-f''g')}{(f')^2+(g')^2}$. Thus, $K = \frac{g'(g''f'-f''g')}{f((f')^2+(g')^2)^2}$.

Now suppose that the parameterization of the meridian is natural, i.e., $(f')^2 + (g')^2 = 1$. Differentiating, we obtain $f'f'' + g'g'' = 0$. Therefore,

$$K = \frac{g'g''f' - f''(g')^2}{f} = \frac{-f''(f')^2 - f''(g')^2}{f} = -\frac{f''}{f}.$$

3.12 Take a sphere of radius R enclosing the given surface, consider the family of spheres with the same center also enclosing this surface, and choose a sphere of minimum radius among them. This sphere and the surface have common tangent plane at some point p. At this point p we choose the normal vector directed inside the sphere for both the sphere and the surface. Then all normal sections of the sphere and the surface have positive curvature at p. Therefore, according to Problem 1.5, all normal curvatures of the surface at the point p are at least $1/R$, and the Gaussian curvature is at least $1/R^2$.

3.13

(a) Let (x^1, x^2, x^3) be coordinates in space $\mathbb{R}^3$. Clearly, $\frac{\partial F}{\partial x^i}(p) = 2(p^i - q^i)$. Therefore, the derivative of F in the direction of ξ equals $2(p - q, \xi)$. We have $2(p - q, \xi) = 0$ for all vectors ξ tangent to S at the point p if and only if the vector $\overrightarrow{pq}$ is orthogonal to S at p.

(b) It is sufficient to consider the case where the surface S is given as the graph of a function $z = f(x, y)$ with $\frac{\partial f}{\partial x}(0, 0) = 0$ and $\frac{\partial f}{\partial y}(0, 0) = 0$ and $f(0, 0) = 0$. In this case, the point $p = (0, 0, 0)$ is critical if and only if $q = (0, 0, t)$. Indeed, for a critical point q, the equation $(q, \xi) = 0$ must hold for all $\xi = (\xi^1, \xi^2, 0)$.

Clearly, $F(x, y) = x^2 + y^2 + (f(x, y) - t)^2$. A simple calculation shows that, at the point $p = (0, 0, 0)$,

$$\begin{pmatrix} F_{xx} & F_{xy} \\ F_{xy} & F_{yy} \end{pmatrix} = 2 \begin{pmatrix} 1 - tf_{xx} & tf_{xy} \\ tf_{xy} & 1 - tf_{yy} \end{pmatrix}.$$

Thus, the critical point $(0, 0, 0)$ is degenerate if and only if $\det(A - tB) = 0$, where $A = I$ is the matrix of the first quadratic form and B is the matrix of the second quadratic form. The equation $\det(A - tB) = 0$ is equivalent to the conditions $t \neq 0$ and $\det(B - t^{-1}A) = 0$, i.e., $t = 1/k_i$.

3.14 It can be assumed that the surface is given in the same form as in Example 3.4 and the line l is determined by the equations $y = 0$ and $z = 0$. Then $f(x, 0) = 0$ and $f_{xx}(0, 0) = 0$; therefore, at the point $(0, 0, 0)$ the Hessian matrix has the form $\begin{pmatrix} 0 & f_{xy} \\ f_{xy} & f_{yy} \end{pmatrix}$. Its determinant equals $-(f_{xy})^2 \leq 0$. The sign of the Hessian determinant coincides with that of the Gaussian curvature (moreover, in the situation under consideration, the matrix of the first quadratic form is the identity, and therefore the Hessian determinant equals the Gaussian curvature).

3.15 The equation $d\omega_i^j = \omega_i^k \wedge \omega_k^j$ yields

$$d\omega_1^2 = \omega_1^1 \wedge \omega_1^2 + \omega_1^2 \wedge \omega_2^2 + \omega_1^3 \wedge \omega_3^2 = \omega_1^3 \wedge \omega_3^2.$$

The proof of the remaining two relations is similar.

3.16 The equation $d\omega^j = \omega^i \wedge \omega_i^j$ yields

$$d\omega^3 = \omega^1 \wedge \omega_1^3 + \omega^2 \wedge \omega_2^3 + \omega^3 \wedge \omega_3^3 = \omega^1 \wedge \omega_1^3 + \omega^2 \wedge \omega_2^3.$$

We have $d\omega^3 = 0$ for the restrictions of forms to surfaces, which implies the required relation.

3.17 If the matrix of the first quadratic form is the identity, then $\omega_1^3 = -L\omega^1 - M\omega^2$ and $\omega_2^3 = -M\omega^1 - N\omega^2$. Therefore,

$$-\omega^1 \wedge \omega_2^3 + \omega^2 \wedge \omega_1^3 = -\omega^1 \wedge (-M\omega^1 - N\omega^2) + \omega^2 \wedge (-L\omega^1 - M\omega^2)$$

$$= (L + N)\omega^1 \wedge \omega^2 = 2H\omega^1 \wedge \omega^2.$$

3.18 The geodesic curvature of any geodesic is zero; therefore, for a digon G, the Gauss–Bonnet formula takes the form

$$\int_G K\,d\sigma = 2\pi - (\pi - \alpha_1) - (\pi - \alpha_2).$$

Moreover, $-\pi < \pi - \alpha_k < \pi$, and hence $2\pi - (\pi - \alpha_1) - (\pi - \alpha_2) > 0$. Thus, the Gaussian curvature must take a positive value at some point of the domain G.

3.19 First, we prove that the principal curvatures are constant on the whole surface. For this purpose, we differentiate the equation $n_1 + kr_1 = 0$ with respect to u^2 and the equation $n_2 + kr_2 = 0$ with respect to u^1. As a result, we obtain $n_{12} + k_2 r_1 + kr_{12} = 0$ and $n_{21} + k_1 r_2 + kr_{21} = 0$. But $n_{12} = n_{21}$ and $r_{12} = r_{21}$, whence $k_2 r_1 = k_1 r_2$. Since the vectors r_1 and r_2 are linearly independent, it follows that $k_1 = k_2 = 0$, i.e., both partial derivatives of the function k vanish.

The constancy of k implies the vanishing of the derivative of the vector $n + kr$ in any direction (on the surface) at each point of the surface; therefore, $n + kr = a$ is a constant vector. Since $r = \frac{a-n}{k}$, it follows that the given surface is the sphere of radius $1/k$ centered at a/k, because the vector n has length 1.

3.20 The eigenvalues of the operator L are the principal curvatures k_1 and k_2. This follows from Rodrigues' formulas and the geometric definition of L; although, this is also seen directly from the definition $L = BA^{-1}$, where A and B are the matrices of the first and second quadratic forms. Indeed, it suffices to note that $\det(B - dA) = \det(L - dI) \cdot \det A$.

Any operator satisfies its own characteristic equation; therefore, $L^2 - (k_1 + k_2)L + k_1 k_2 I = 0$, i.e., $L^2 - 2HL + KI = 0$.

3.21 First, we calculate $\Gamma_{ijk} = (r_{ij}, r_k)$. In the case under consideration, we have $r_1 = (\cos\varphi, \sin\varphi)$, $r_2 = (-r\sin\varphi, r\cos\varphi)$, $r_{11} = (0, 0)$, $r_{12} = r_{21} = (-\sin\varphi, \cos\varphi)$, and $r_{22} = (-r\cos\varphi, -r\sin\varphi)$. Hence $\Gamma_{122} = \Gamma_{212} = r$ and $\Gamma_{221} = -r$; the remaining symbols Γ_{ijk} are zero. Next, we have $\Gamma_{ij}^k = g^{kl}\Gamma_{ijl}$, where $g^{11} = 1/r^2$, $g^{22} = 1$, and $g^{12} = g^{21} = 0$ (see Problem 3.1). Therefore, $\Gamma_{12}^2 = \Gamma_{21}^2 = 1/r$ and $\Gamma_{22}^1 = -r$; the remaining symbols Γ_{ij}^k are zero.

3.22 First, we calculate $\Gamma_{ijk} = (r_{ij}, r_k)$. In our case, $r_1 = (f'\cos v, f'\sin v, 1)$, $r_2 = (-f\sin v, f\cos v, 0)$, $r_{11} = (f''\cos v, f''\sin v, 0)$, $r_{12} = r_{21} = (-f'\sin v, f'\cos v, 0)$, and $r_{22} = (-f\cos v, -f\sin v, 0)$. Hence $\Gamma_{111} = f'f''$, $\Gamma_{122} = \Gamma_{212} = ff'$, and $\Gamma_{221} = -ff'$; the remaining symbols Γ_{ijk} are zero. Next, we have $g^{11} = \frac{1}{1+(f')^2}$, $g^{22} = \frac{1}{f^2}$, and $g^{12} = g^{21} = 0$ (see Problem 3.2). Therefore, $\Gamma^1_{11} = \frac{f'f''}{1+(f')^2}$, $\Gamma^2_{12} = \Gamma^2_{21} = \frac{f'}{f}$, and $\Gamma^1_{22} = -\frac{ff'}{1+(f')^2}$; the remaining symbols Γ^k_{ij} are zero.

3.23 First, note that $r_i = \tilde{r}_p \frac{\partial y^p}{\partial x^i}$. Therefore,

$$r_{ij} = \frac{\partial \tilde{r}_p}{\partial x^j} \cdot \frac{\partial y^p}{\partial x^i} + \tilde{r}_p \frac{\partial^2 y^p}{\partial x^i \partial x^j}$$

$$= \tilde{r}_{pq} \frac{\partial y^q}{\partial x^j} \cdot \frac{\partial y^p}{\partial x^i} + \tilde{r}_p \frac{\partial^2 y^p}{\partial x^i \partial x^j}$$

$$= \left(\tilde{\Gamma}^s_{pq} \tilde{r}_s + \tilde{b}_{pq} n \right) \frac{\partial y^q}{\partial x^j} \cdot \frac{\partial y^p}{\partial x^i} + \tilde{r}_s \frac{\partial^2 y^s}{\partial x^i \partial x^j}.$$

On the other hand, $r_{ij} = \Gamma^k_{ij} r_k + b_{ij} n$ and $r_k = \tilde{r}_s \frac{\partial y^s}{\partial x^k}$. The vectors $\tilde{r}_1$ and $\tilde{r}_2$ are linearly independent and orthogonal to the vector n, which implies

$$\Gamma^k_{ij} \frac{\partial y^s}{\partial x^k} = \tilde{\Gamma}^s_{pq} \frac{\partial y^p}{\partial x^i} \cdot \frac{\partial y^q}{\partial x^j} + \frac{\partial^2 y^s}{\partial x^i \partial x^j},$$

whence

$$\Gamma^k_{ij} = \left(\tilde{\Gamma}^s_{pq} \frac{\partial y^p}{\partial x^i} \cdot \frac{\partial y^q}{\partial x^j} + \frac{\partial^2 y^s}{\partial x^i \partial x^j} \right) \frac{\partial x^k}{\partial y^s}.$$

3.24 Let Γ^k_{ij} and $\tilde{\Gamma}^s_{pq}$ be the Christoffel symbols in local coordinates (x^1, x^2) and (y^1, y^2). If $y^k = x^k + c^k_{ij} x^i x^j$, where the c^k_{ij} are constants, then $\frac{\partial y^j}{\partial x^i}(0) = \delta^j_i$ and $\frac{\partial^2 y^s}{\partial x^i \partial x^j}(0) = c^s_{ij} + c^s_{ji}$. Therefore, according to Problem 3.23, we have

$$\Gamma^k_{ij}(0) = \left(\tilde{\Gamma}^s_{pq}(0) \delta^p_i \delta^q_j + c^s_{ij} + c^s_{ji} \right) \delta^k_s$$

$$= \tilde{\Gamma}^k_{ij}(0) + c^k_{ij} + c^k_{ji}.$$

Now, setting $c^k_{ij} = -\frac{1}{2} \tilde{\Gamma}^k_{ij}(0)$, we obtain $\Gamma^k_{ij}(0) = 0$, because $\tilde{\Gamma}^k_{ij} = \tilde{\Gamma}^k_{ji}$.

Another method for constructing a coordinate system for which all Christoffel symbols vanish at a given point is described in the solution of Problem 3.28.

3.25 Choose a parameterization $r(u^1, u^2)$ of the manifold M^2 in a neighborhood of some point so that the vectors $r_1 = \frac{\partial r}{\partial u^1}$ and $r_2 = \frac{\partial r}{\partial u^2}$ form an orthonormal basis at this point. Let us show that $df(ar_1 + br_2) = an_1 + bn_2$, where n_1 and n_2 are the derivatives with respect to u^1 and u^2 of the normal vector $n(u^1, u^2)$ to the surface. Indeed, consider a curve $r(u^1(t), u^2(t))$ passing through the given point at $t = 0$. The velocity vector of this curve at this point equals $(u^1)' r_1 + (u^2)' r_2$. The

Fig. 3.7 The net of the cone

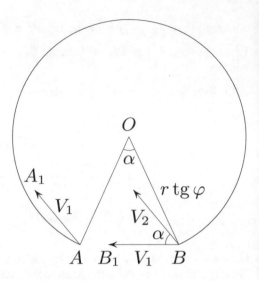

differential df maps this vector to the vector $(u^1)'n_1 + (u^2)'n_2$, as required. Thus, the entries of the matrix of the operator df in the basis r_1, r_2 are $(n_i, r_j) = -b_{ij}$.

3.26 Suppose that the radius of the circle in the intersection of the plane with the sphere is $r \sin\varphi$. Then the length of this circle equals $2\pi r \sin\varphi$ and $S = 2\pi r^2(1 - \cos\varphi)$. Consider the cone tangent to the sphere along the given circle (which also serves as the base of the cone). Its slant height equals $r \tan\varphi$. Consider the net of this cone (see Fig. 3.7).

The required angle α between the vectors V_1 and V_2 is equal to the angle AOB. Indeed, the parallel transport from A to B along the arc AB takes the vector V_1 to a parallel vector V_2. The vector at B corresponding to the vector V_1 is the vector $\overrightarrow{BB_1}$ for which $\angle OBB_1 = \angle OAA_1$; hence, at the point B, the angle between the vectors $V_1 = \overrightarrow{BB_1}$ and V_2 equals the angle AOB.

The length of the arc AB is equal to that of the circle, i.e., $2\pi r \sin\varphi$. Therefore,

$$2\pi - \alpha = \frac{2\pi r \sin\varphi}{r \tan\varphi} = 2\pi \cos\varphi,$$

whence $\alpha = 2\pi(1 - \cos\varphi) = S/r^2$.

Another solution of this problem is based on the corollary of Theorem 3.12.

3.27 Clearly, $\partial_{r_i}(f) = \frac{\partial f}{\partial u^i}$ and $\partial_{r_j}\partial_{r_i}(f) = \frac{\partial^2 f}{\partial u^j \partial u^i}$. Thus, the relation $[r_1, r_2] = 0$ follows from the equation $\frac{\partial^2 f}{\partial u^1 \partial u^2} = \frac{\partial^2 f}{\partial u^2 \partial u^1}$.

3.28 Let us write an equation for a geodesic $x^1 = tc^1$, $x^2 = tc^2$ at a point p:

$$\frac{d^2x^i}{dt^2} + \Gamma^i_{jk}\frac{dx^j}{dt} \cdot \frac{dx^k}{dt} = 0.$$

In the case under consideration, $\frac{dx^i}{dt} = c^i$ and $\frac{d^2x^i}{dt^2} = 0$. Therefore, $\Gamma^i_{jk}c^jc^k = 0$ for any numbers c^i. Thus, $\Gamma^i_{jk} = 0$.

Another way of constructing a coordinate system for which all Christoffel symbols at a given point vanish is described in the solution of Problem 3.24.

3.29 The parallels and meridians are orthogonal; therefore, it suffices to prove that the parallels are lines of curvature. Moving along a parallel, the unit normal vector n rotates about the axis of rotation of the surface; hence its derivative n' is proportional to a velocity vector in cicular motion.

3.30 Let $n_1(t)$ and $n_2(t)$ be the unit normal vectors to the surfaces at the points of the curve $\gamma(t)$. If γ is a line of curvature for the first surface, then $n'_1 = \lambda_1\gamma'$. If the vectors n_1 and n_2 are proportional (i.e., the surfaces are tangent along the curve γ), then $n'_2 = \pm\lambda_1\gamma'$, and therefore γ is a line of curvature for the second surface too. In what follows, we will assume that the vectors n_1 and n_2 are not proportional. Since these vectors are perpendicular to γ', we have

$$n'_2 = \lambda_2\gamma' + \mu n_1 + \nu n_2.$$

It is required to prove that $\mu = \nu = 0$. For the inner products of the expression for n'_2 with n_1 and n_2, we have

$$(n_1, n'_2) = \mu + \nu(n_1, n_2), \quad (n_2, n'_2) = \mu(n_2, n_1) + \nu.$$

Clearly, $(n_2, n'_2) = 0$ and $(n_1, n'_2) = (n_1, n_2)' - (n'_1, n_2) = -(n'_1, n_2) = -(\lambda_1\gamma', n_2) = 0$. Hence

$$\mu + \nu(n_1, n_2) = 0, \quad \mu(n_2, n_1) + \nu = 0.$$

The inequality $(n_1, n_2)^2 < 1$ implies $\mu = \nu = 0$.

3.31 For an asymptotic line, the first equation of motion for the Darboux frame is $\varepsilon'_1 = k_g\varepsilon_2$; therefore, $\varepsilon_2 = \pm e_2$, where e_2 is the second vector from the Frenet–Serret frame. Reversing, if necessary, the direction of the normal vector to the surface, we can achieve the equality $\varepsilon_2 = e_2$.

3.32 According to Euler's formula, asymptotic directions correspond to those angles φ for which $k_1\cos^2\varphi + k_2\sin^2\varphi = 0$, i.e., $\tan^2\varphi = -\frac{k_1}{k_2}$. Moreover, the angle between asymptotic directions equals 2φ. Thus, asymptotic directions are perpendicular if and only if $k_1 = -k_2$, i.e., $H = 0$.

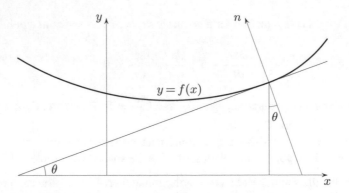

Fig. 3.8 A surface of revolution

3.33 Consider the surface obtained by rotating the graph of $y = f(x)$ about the Ox axis. According to Problem 3.29, the parallels and meridians of a surface of revolution are lines of curvature. Therefore, a surface of revolution is minimal if and only if the normal curvatures of the parallels and meridians have the same absolute value and opposite signs. It the normal vector n is chosen as in Fig. 3.8, then the normal curvature of the meridians is $\frac{y''}{(1+(y')^2)^{3/2}}$ and the normal curvature of the parallels is $-\frac{1}{y}\cos\theta$ (by virtue of Meusnier's formula). Moreover, $\tan\theta = y'$. As a result, we obtain the following equation for a minimal surface of revolution:

$$\frac{y''}{(1+(y')^2)^{3/2}} = \frac{1}{y} \cdot \frac{1}{(1+(y')^2)^{1/2}}.$$

There exists a point x for which $y'(x) \neq 0$. Let us find a solution in a neighborhood of this point. To this end, we multiply both sides of the equation by $2y'$:

$$\frac{2y'y''}{1+(y')^2} = \frac{2y'}{y}.$$

Setting $z = 1 + (y')^2$, we obtain $z' = 2y'y''$, whence

$$\frac{z'}{z} = \frac{2y'}{y}.$$

Therefore, $\ln z = \ln y^2 + \ln a = \ln ay^2$ and $1 + (y')^2 = z = ay^2$. Clearly, $a > 0$; let $a = k^2$. Then

$$1 + \left(\frac{dy}{dx}\right)^2 = (ky)^2,$$

i.e.,

$$\frac{k\,dy}{\sqrt{(ky)^2 - 1}} = k\,dx.$$

Thus, $\mathrm{arcosh}(ky) = kx + c$, i.e., $y = \frac{1}{k}\cosh(kx + c)$, so that the derivative y' can vanish only when $kx + c = 0$. For the graphs corresponding to $kx + c \geqslant 0$ and $kx + c \leqslant 0$ to be smoothly glued, it is required that k and c be the same for both graphs.

3.34 According to Problem 3.20, the Weingarten operator L satisfies the relation $L^2 = 2HL - KI$, where H is the mean curvature, K is the Gaussian curvature, and I is the identity operator. Thus, for a minimal surface, we have $L^2 = -KI$.

The differential of the spherical Gauss map equals $-L$. Therefore, the spherical Gauss map transforms the inner product (V, W) on a minimal surface into the inner product $(-LV, -LW) = (L^2V, W) = -K(V, W)$. If inner products are proportional, then they define equal angles between vectors.

3.35 The required variation is given by the map $(s, t) \mapsto \exp_{\gamma(t)}(sW(t))$.

3.36 It is sufficient to prove that each of the three coordinates in $\mathbb{R}^3$ is a harmonic function on a minimal surface. Let $x = (x^1, x^2, x^3)$ denote a point of a minimal surface. Consider a moving orthonormal frame ε_1, ε_2 and let ε_3 be the normal vector to the surface. Then $dx = \omega^1\varepsilon_1 + \omega^2\varepsilon_2$ and $dx \times \varepsilon_3 = -\omega^1\varepsilon_2 + \omega^2\varepsilon_1 = *dx$. Therefore,

$$-d(*dx) = dx \times d\varepsilon_3 = (\omega^1\varepsilon_1 + \omega^2\varepsilon_2) \times (\omega_3^1\varepsilon_1 + \omega_3^2\varepsilon_2) = (\omega^1 \wedge \omega_3^2 - \omega^2 \wedge \omega_3^1)\varepsilon_3.$$

Moreover, we have $\omega^1 \wedge \omega_3^2 - \omega^2 \wedge \omega_3^1 = 2H\omega^1 \wedge \omega^2$ according to Problem 3.17. Hence $(\Delta x)\omega^1 \wedge \omega^2 = -d(*dx) = (2H\varepsilon_3)\omega^1 \wedge \omega^2$, i.e., $\Delta x = 2H\varepsilon_3$. Thus, if $H = 0$, then the functions x^1, x^2, and x^3 are harmonic on a minimal surface.

Chapter 4
Hypersurfaces in $\mathbb{R}^{n+1}$: Connections

This is a transitional chapter from surfaces in three-dimensional space to Riemannian manifolds of any dimension. The intermediate object is a hypersurface, which is a direct generalization of a surface. If $M^n \subset \mathbb{R}^{n+1}$ is a submanifold of codimension 1, then M^n is called a *hypersurface* in $\mathbb{R}^{n+1}$. In Chaps. 1 and 3 we dealt with hypersurfaces in $\mathbb{R}^2$ and in $\mathbb{R}^3$. Now we will discuss (in Sccts. 4.1–4.6) hypersurfaces of an arbitrary dimension. (Here and in what follows, all manifolds, functions, vector fields, vector bundles, sections of vector bundles, and so on are assumed to be smooth.)

The rest of the chapter (Sects. 4.7–4.10) is concerned with the general theory of conncctions on vector bundles. The notion of a connection on a vector bundle generalizes those of covariant differentiation on a surface and of a connection on the tangent bundle of a hypersurface, which is introduced in Sect. 4.2.

4.1   The Weingarten Operator

If a hypersurface $M = M^n$ is orientable, then at each point $x \in M$ we can define a unit vector $N(x)$ normal to M so that it depends smoothly on x. (If the hypersurface M is nonorientable, then such a family of vectors can be defined only locally.) With each vector X in the tangent space $T_x M$ at a point x we associate the vector $L(X)$ defined as follows. Let $X = X^i e_i$, where $e_1, \ldots, e_{n+1}$ is the standard basis in $\mathbb{R}^{n+1}$. Then $-L(X) = \partial_X N = X^i \partial_i N$ is the derivative of N in the direction of X. It is easy to check that $L(X) \in T_x M$. Indeed, differentiating the equation $(N, N) = 1$, we obtain $0 = \partial_X (N, N) = 2(\partial_X N, N)$. Thus, L is a self-map of the space $T_x M$; clearly, it is linear, i.e., L is a linear operator. The operator L is called the *Weingarten operator*.

V. V. Prasolov, *Differential Geometry*, Moscow Lectures 8,
https://doi.org/10.1007/978-3-030-92249-8_4

The Weingarten operator is closely related to the *spherical Gauss map* $f: M \to \mathbb{S}^n$, which takes every point $x \in M$ to the vector $N(x)$ treated as a point of the unit sphere. Note that $-L(X) = df(X)$, because $dN(X) = X^i \partial_i N = \partial_X N$.

Example 4.1 For the standard sphere of radius R, the Weingarten operator equals $\pm \frac{1}{R} I$, where I is the identity operator (the minus sign corresponds to the outward normal N and the plus sign, to the inward one).

Proof For the sphere, the Gauss map f has the form $f(X) = \pm \frac{1}{R} X$.   □

A point of a hypersurface is said to be *umbilical* if the Weingarten operator L is proportional to the identity operator at this point. For example, all points of the sphere $\mathbb{S}^n$ are umbilical.

Theorem 4.1 *The Weingarten operator* $L: T_x M \to T_x M$ *is self-adjoint with respect to the inner product in* $\mathbb{R}^{n+1}$ *restricted to* $T_x M$; *i.e.,* $\big(L(X), Y\big) = \big(X, L(Y)\big)$ *for any* $X, Y \in T_x M$.

Proof Clearly, $(N, X) = 0$ and $(N, Y) = 0$. Therefore,

$$0 = \partial_X (N, Y) = (\partial_X N, Y) + (N, \partial_X Y),$$

$$0 = \partial_Y (N, X) = (\partial_Y N, X) + (N, \partial_Y X).$$

Thus,

$$\big(L(X), Y\big) - \big(X, L(Y)\big) = -(\partial_X N, Y) + (\partial_Y N, X)$$

$$= (N, \partial_X Y - \partial_Y X) = (N, [X, Y]) = 0,$$

because the vector $[X, Y]$ lies in the tangent space $T_x M$.   □

The Weingarten operator L makes it possible to define the *first, second, third,* etc. *bilinear forms* on $T_x M$ by setting the value of the kth bilinear form at a pair of vectors (X, Y) to equal $(L^{k-1} X, Y)$. All these forms are symmetric, because the operator L is self-adjoint. The first bilinear form is merely the inner product of vectors. In differential geometry these bilinear forms are not as important as the corresponding *first, second, third,* etc. *quadratic forms* $(L^{k-1} X, X)$.

The determinant of the operator L is called the *Gaussian curvature* of the surface at the given point and denoted by K. The trace of the operator L is called the *mean curvature* and denoted by H. The eigenvalues of the operator L are called the *principal curvatures*.

Remark For a hypersurface in $\mathbb{R}^3$, the definitions of the second quadratic form and the Gaussian curvature coincide with the corresponding definitions in Sect. 3.4.

Let k_i be one of the principal curvatures, and let $e_i = \frac{\partial r}{\partial u^i}$ be the corresponding eigenvector. For a hypersurface, Rodrigues' formula $\frac{\partial N}{\partial u^i} = -k_i \frac{\partial r}{\partial u^i}$ remains valid. Indeed,

$$\frac{\partial N}{\partial u^i} = \partial_i N = -L(e_i) = -k_i e_i = -k_i \frac{\partial r}{\partial u^i}.$$

On a hypersurface M^n consider an infinitesimal parallelotope spanned by vectors $a_1 e_1, \ldots, a_n e_n$. The spherical Gauss map takes it to the infinitesimal parallelotope with edges $-k_1 a_1 e_1, \ldots, -k_n a_n e_n$. The ratio of the volumes of these parallelotopes is equal to the absolute value of the Gaussian curvature. Therefore, the ratio of the volume of the image under the spherical Gauss map of an infinitesimal figure on M^n to the volume of this figure itself equals the absolute value of the Gaussian curvature of the hypersurface.

Example 4.2 Consider a hypersurface given as the graph of a function, i.e., as $x^{n+1} = f(x^1, \ldots, x^n)$, and suppose that the vector $(0, \ldots, 0, 1)$ is normal to it, i.e., $\frac{\partial f}{\partial x^i}(0) = 0$ for all i. Then, in the local coordinates $x^1, \ldots, x^n$, the second quadratic form at the origin is determined by the Hessian matrix $\frac{\partial^2 f}{\partial x^i \partial x^j}(0)$.

Proof Clearly, $\mathbf{II}(X, Y) = (L(X), Y) = (-\partial_X N, Y) = (N, \partial_X Y)$. The hypersurface is parametrically given by $r = (x^1, \ldots, x^n, f(x^1, \ldots, x^n))$. Therefore, the last coordinate of the vector $r_i(0) = \frac{\partial r}{\partial x^i}(0)$ equals $\frac{\partial f}{\partial x^i}(0)$, and the last coordinate of the vector $\partial_{r_i} r_j$ equals $\frac{\partial^2 f}{\partial x^i \partial x^j}(0)$. □

Example 4.3 For the hypersurface $x^{n+1} = \frac{1}{2}(k_1(x^1)^2 + \cdots + k_n(x^n)^2)$ with normal vector $(0, \ldots, 0, 1)$, the Weingarten operator at the origin is diagonal with diagonal elements k_i.

Any operator L satisfies the equation $p(L) = 0$, where p is the characteristic polynomial of L, i.e., $p(\lambda) = \det(L - \lambda I)$ (as usual, I denotes the identity operator). For $n = 2$ (i.e., for a surface in $\mathbb{R}^3$), the characteristic polynomial has the form $\lambda^2 - (\mathrm{tr} L)\lambda + \det L$ and, therefore, the Weingarten operator L satisfies the relation $K \cdot I - 2H \cdot L + L^2 = 0$. Thus, if $\mathbf{I}$, $\mathbf{II}$, and $\mathbf{III}$ are the first, second, and third quadratic forms, then

$$K \cdot \mathbf{I} - 2H \cdot \mathbf{II} + \mathbf{III} = 0.$$

In particular, for $n = 2$, the third quadratic form is expressed in terms of the first two; it is also clear that the remaining quadratic forms are expressed in terms of the first two as well. For a hypersurface of dimension n, all quadratic forms are expressed in terms of the first n quadratic forms.

4.2   Connections on Hypersurfaces

Given any two vector fields X and Y in $\mathbb{R}^{n+1}$, the vector field $\partial_X Y$, that is, the derivative of Y in the direction of X, is defined. If X and Y are vector fields on a hypersurface $M \subset \mathbb{R}^{n+1}$, then the vector field $\partial_X Y$ is not necessarily tangent to M. Consider the vector field on M defined as the orthogonal projection of the vector $\partial_X Y$ on $T_x M$ at each point $x \in M$. The vector field thus obtained is called the *covariant derivative* of Y in the direction of X and denoted by $\nabla_X Y$.

Theorem 4.2 (Gauss' Formula)   *Covariant derivative is given by the formula*

$$\nabla_X Y = \partial_X Y - (L(X), Y)N.$$

Proof It suffices to check that the vector $\partial_X Y - (L(X), Y)N$ lies in the tangent plane, i.e., is orthogonal to N. The inner product of the vectors $\partial_X Y - (L(X), Y)N$ and N equals

$$(\partial_X Y, N) - (Y, L(X)) = (\partial_X Y, N) + (Y, \partial_X N) = \partial_X(Y, N) = 0.$$

$\square$

It follows from Gauss' formula that the map ∇ is linear in X and Y. Moreover, it obeys Leibniz' rule

$$\nabla_X(fY) = f\nabla_X Y + (Xf)Y.$$

A map with these properties which takes any two vector fields on a hypersurface to a third vector field is called a *connection* on M.

Here we will consider not arbitrary but special connections, called the *Levi-Cività connections*, which have the following special properties:

- $\nabla_X Y - \nabla_Y X = [X, Y]$ *(symmetry)*;
- $\partial_X(Y, Z) = (\nabla_X Y, Z) + (Y, \nabla_X Z)$ *(compatibility with the metric)*.

Both these properties are easy to derive from Gauss' formula. Indeed, since the Weingarten operator is self-adjoint, we have

$$\nabla_X Y - \nabla_Y X = \partial_X Y - (L(X), Y)N - \partial_Y X + (L(Y), X)N$$
$$= \partial_X Y - \partial_Y X = [X, Y],$$

and since the vector N is orthogonal to any tangent to the hypersurface, we have

$$\partial_X(Y, Z) = (\partial_X Y, Z) + (Y, \partial_X Z)$$
$$= (\nabla_X Y, Z) + (Y, \nabla_X Z).$$

4.3   Geodesics on Hypersurfaces

Let $\gamma(t)$ be a parameterized curve on a hypersurface M, and let X be a vector field on M (for our purposes, it is sufficient that this vector field be defined on the curve γ). The vector field X is said to be *parallel along* γ if $\nabla_Y X = 0$ for $Y = \frac{d\gamma}{dt}$.

The geometric meaning of this notion is as follows. For a vector field X parallel along a curve, the vector $\partial_Y X = \frac{dX(t)}{dt}$ must be orthogonal to the tangent space. Therefore, we first perform the parallel translation of $X(t)$ from the point $\gamma(t)$ to the point $\gamma(t+dt)$ and then project the translate onto the tangent plane at this point. As a result, we obtain a vector $X(t + dt)$ for infinitesimal dt. Thus, the parallel transport along the curve changes a vector only in the normal direction; this change is necessary for the vector to remain tangent to the hypersurface.

Problem 4.1 Prove that parallel transport along a curve preserves (a) the length of vectors; (b) the angles between vectors.

A curve γ on a hypersurface is said to be *geodesic* if the vector field $\frac{d\gamma}{dt}$ is parallel along γ, i.e., $\nabla_{\frac{d\gamma}{dt}} \frac{d\gamma}{dt} = 0$.

Example 4.4 Any curve $\gamma(t)$ on the sphere $\mathbb{S}^n$ lying in a plane passing through the center of the sphere and having natural parameterization is geodesic.

Proof Let X be the velocity vector of such a curve $\gamma(t)$. By virtue of Gauss' formula, we have $\nabla_X X = \partial_X X - (L(X), X)N$. The section of the sphere $\mathbb{S}^n$ by a plane through its center is the circle $\mathbb{S}^1$. For a point uniformly moving along the circle $\mathbb{S}^1$, the acceleration vector $\frac{dX}{dt} = \partial_X X$ is directed from this point to the center of the circle. The center of $\mathbb{S}^1$ coincides with that of the sphere; therefore, the vector $\partial_X X$ is directed along the normal to the sphere. The vector $(L(X), X)N$ is directed along the normal as well, so that the tangent vector $\nabla_X X$ cannot have nonzero tangential component, and hence it must be zero.   □

4.4   Convex Hypersurfaces

A convex hypersurface is a hypersurface which is the boundary of a bounded convex body in $\mathbb{R}^{n+1}$. A convex hypersurface lies on one side of the tangent space at each point. All principal curvatures of such a surface are of the same sign (although some of them may vanish). This follows from Examples 4.2 and 4.3 (see p. 147).

Theorem 4.3 (Hadamard, 1897) *If all principal curvatures of a connected closed hypersurface immersed in $\mathbb{R}^{n+1}$ ($n \geqslant 2$) are positive (or negative), then this hypersurface is convex (in particular, it is oriented and embedded rather than only immersed).*

Proof We denote the given hypersurface by $h(M^n)$; here M^n is a manifold and $h\colon M^n \to \mathbb{R}^{n+1}$ is its immersion. Let $p \in M^n$ be a point, and let T_p be the tangent space to the hypersurface at the point $h(p)$, which consists of the velocity vectors of the images of curves passing through p. It follows from the assumptions of the theorem that the image of a small neighborhood of p under the map h lies on one side of T_p. Let us direct the normal vector $N(p)$ to the same side (then all principal curvatures are positive). This defines an orientation on the hypersurface, which makes it possible to define, in turn, the spherical Gauss map $f\colon M^n \to \mathbb{S}^n$. Since all principal curvatures are nonzero, it follows that the map df is nondegenerate at each point and hence f is a local homeomorphism. Any local homeomorphism from a connected compact manifold to a simply connected manifold is a (global) homeomorphism (see, e.g., [Pr2]). Thus, for different points p and q of the manifold M^n, the corresponding normal vectors $N(p)$ and $N(q)$ are different.

Consider the tangent space T_p. On the hypersurface there are points which lie on the same side of it as the vector $N(p)$. Let q be the farthest of them. Then $N(q) = -N(p)$. We claim that the hypersurface has no points lying on the other side. Indeed, suppose that such points exist and let r be the farthest of them. Then $N(p) = N(r)$, but this is not so.

Let us show that h is an embedding. Suppose that $h(p) = h(p')$ for a point p' different from p. Then p' is different from q, and hence the vector $N(p')$ is different from $N(p)$ and $N(q)$. Therefore, the tangent spaces T_p and $T_{p'}$ (passing through the point $h(p) = h(p')$) do not coincide. This implies that the hypersurface contains points on both sides of the tangent space T_p, which is false.

The convexity of the hypersurface follows from the fact that it lies on one side of every tangent space. $\square$

4.5 Minimal Hypersurfaces

A *minimal hypersurface* is a surface with zero mean curvature. As well as in the case of surfaces in $\mathbb{R}^3$, this name reflects the fact that any normal variation (that is, a variation in the normal direction) of a minimal hypersurface increases its volume.

To calculate the volume of a hypersurface (or of its part), we use the volume form. Let $\varepsilon_1, \ldots, \varepsilon_n$ be a moving orthonormal positively oriented frame on a given hypersurface (or on some domain of this hypersurface), and let $\omega^1, \ldots, \omega^n$ be the dual frame. Then the *volume form* on the hypersurface is $\omega^1 \wedge \cdots \wedge \omega^n$.

Let us find out how the volume form transforms under a change of basis. Consider another basis $e_i = a_i^j \varepsilon_j$. The matrix $A = (a_i^j)$ is the transition matrix from the basis ε to the basis e. Let $\omega^k = b_p^k \alpha^p$, where α is the dual basis for e. Then $b_p^k = b_p^k \alpha^p(e_i) = \omega^k(e_i) = a_i^j \omega^k(\varepsilon_j) = a_i^k$. Therefore, $\omega^1 \wedge \cdots \wedge \omega^n = (\det A)\alpha^1 \wedge \cdots \wedge \alpha^n$. The matrix g of the first quadratic form in the basis e equals AA^T, so that $\det A = \pm\sqrt{\det g}$. If both bases are oriented in the same way, then $\det A = \sqrt{\det g}$ and $\omega^1 \wedge \cdots \wedge \omega^n = \sqrt{\det g}\,\alpha^1 \wedge \cdots \wedge \alpha^n$.

We are interested in small normal variations of a bounded domain of a hypersurface. Therefore, we assume that M^n is a bounded hypersurface with boundary given in the parametric form $r(u^1, \ldots, u^n) \in \mathbb{R}^{n+1}$. A *normal variation* of this hypersurface is a family of hypersurfaces M_t^n of the form $r_t(u) = r(u) + th(u)N(u)$, where h is a function vanishing in a neighborhood of the boundary of the hypersurface and $N(u)$ is the unit normal to r. Consider the function $\mathrm{vol}(t) = \int_{M_t^n} dV_t$, where dV_t is the volume form on the hypersurface M_t^n. Let us show that

$$\mathrm{vol}'(0) = -\int_{M^n} nhH \, dV_0. \tag{4.1}$$

We will use the fact that if $g(t)$ is a family of matrices in which $g(0)$ is the identity matrix, then $\frac{d}{dt}\det g(0) = \frac{d}{dt}\mathrm{tr}g(0)$ (see Appendix, Theorem 8.1 on p. 257).

From (4.1) it follows that a hypersurface is minimal if and only if $\mathrm{vol}'(0) = 0$ for any normal variation of any of its domains. This is proved in precisely the same way as for surfaces (see p. 109). If a hypersurface is not minimal, then there exists a normal variation reducing its volume.

Consider the map $f_t : M_0^n \to M_t^n$ taking each point $r(u)$ to the point $r_t(u)$. Suppose that the vector ε_i is represented by a curve $\gamma_i(s)$, i.e., $\varepsilon_i = \left. \frac{dr(\gamma_i(s))}{ds} \right|_{s=0}$. Then the map f_t takes ε_i to the vector

$$\varepsilon_i(t) = \left. \frac{dr_t(\gamma_i(s))}{ds} \right|_{s=0} = \varepsilon_i + thN_i + th_iN,$$

where $N_i = \left. \frac{dN(\gamma_i(s))}{ds} \right|_{s=0}$ and $h_i = \left. \frac{dh(\gamma_i(s))}{ds} \right|_{s=0}$. Therefore, in the basis $\varepsilon_i(t)$ the matrix of the first quadratic form on the hypersurface M_t^n is

$$g_{ij}(t) = (\varepsilon_i(t), \varepsilon_j(t)) = g_{ij}(0) + th[(\varepsilon_i, N_j) + (\varepsilon_j, N_i)] = \delta_{ij} - 2th(\varepsilon_i, L\varepsilon_j) + o(t).$$

Clearly,

$$\int_{M_t^n = f_t(M^n)} dV_t = \int_{M^n} f_t^* dV_t = \int_{M^n} f_t^*(\sqrt{\det g(t)}\,\omega^1(t) \wedge \cdots \wedge \omega^n(t))$$

$$= \int_{M^n} \sqrt{\det g(t)}\,\omega^1 \wedge \cdots \wedge \omega^n = \int_{M^n} \sqrt{\det g(t)}\,dV_0.$$

Hence

$$\mathrm{vol}'(0) = \frac{d}{dt}\int_{M^n} \sqrt{\det g(t)}\,dV_0 \bigg|_{t=0} = \int_{M^n} \frac{d}{dt}\sqrt{\det g(t)} \bigg|_{t=0} dV_0$$

$$= \frac{1}{2}\int_{M^n} \sum_i g_{ii}'(0)\,dV_0.$$

We can assume that the vectors ε_i form an orthonormal eigenbasis of the self-adjoint operator L. Then

$$g'_{ii}(0) = -2h(\varepsilon_i, L\varepsilon_i) = -2k_i h,$$

and therefore

$$\text{vol}'(0) = -\int_{M^n} \sum_i k_i h \, dV_0 = -\int_{M^n} nHh \, dV_0.$$

4.6 Steiner's Formula

Consider a body bounded by an embedded oriented hypersurface $M^n \subset \mathbb{R}^{n+1}$. Let V be the volume of this body, and let X be its surface area (i.e., the volume of the manifold M^n). Consider also the set M^n_ε of those points that are obtained by translating a point $x \in M^n$ along the normal to the hypersurface at this point by a distance of ε. For sufficiently small ε, the set M^n_ε is an embedded hypersurface; in what follows, we assume that ε is small enough. Let $V(\varepsilon)$ be the volume of the body bounded by this hypersurface, and let $X(\varepsilon)$ be its surface area. Clearly, $V(r) = V \pm \int_0^r X(\varepsilon)d\varepsilon$ (the plus sign corresponds to the outward normal and the minus sign corresponds to the inward one). Therefore, if we manage to calculate the volume of the manifold M^n_ε, then we will be able to calculate $V(r)$ by integration with respect to ε.

To calculate the volume of the manifold M^n_ε, consider the spherical Gauss map[1]. The ratio of the volume of an infinitesimal figure on the hypersurface M^n_ε to the volume of the image of this figure under the spherical Gauss map equals $1/|K|$, where K is the Gaussian curvature (see p. 147). Therefore, $X(\varepsilon) = \int_{\mathbb{S}^n} \frac{dV}{|K(\varepsilon)|}$, where dV is the volume form and $K(\varepsilon)$ is the Gaussian curvature of the point of M^n_ε that is mapped to the corresponding point of the sphere.

The quantity $1/K$, where K is the Gaussian curvature, equals the product of the principal curvature radii. We also have $R_i(\varepsilon) = R_i + \varepsilon$, where $R_i = R_i(0)$ is the curvature radius of the initial hypersurface. Let us transport the integral along the unit sphere backward but to $M^n_0 = M^n$ rather than to M^n_ε. As a result, we will represent $X(\varepsilon)$ as the integral of $\prod \frac{R_i + \varepsilon}{R_i} = \prod(1 + \varepsilon k_i)$ over M^n. Multiplying out, we obtain a polynomial in ε whose coefficients are symmetric polynomials in the principal curvatures.

For example, consider a surface M^2 in $\mathbb{R}^3$. We have

$$X(\varepsilon) = \int_{M^2} (1 + \varepsilon(k_1 + k_2) + \varepsilon^2 k_1 k_2)d\sigma,$$

---

[1] This approach to calculating the volume of M^n_ε was communicated to me by A. G. Kulakov.

where $d\sigma$ is the area form on the surface. The product k_1k_2 is the Gaussian curvature K. Therefore, by the Gauss–Bonnet theorem, $\int_{M^2} k_1k_2\, d\sigma = 2\pi\chi(M^2)$. For example, if the surface is convex, then $\int_{M^2} k_1k_2\, d\sigma = 4\pi$. Moreover, $k_1 + k_2 = 2H$, where H is the mean curvature. Thus,

$$X(\varepsilon) = X(0) + 2\varepsilon\int_{M^2} H\, d\sigma + 2\varepsilon^2\pi\chi(M^2).$$

Integrating this formula with respect to ε, we obtain

$$V(\varepsilon) = V(0) + \varepsilon X(0) + \varepsilon^2\int_{M^2} H\, d\sigma + \varepsilon^3\frac{2\pi}{3}\chi(M^2).$$

HISTORICAL COMMENT In 1840 Steiner obtained formulas for the volume and surface area of the body B_ε consisting of all points at a distance of at most ε from a convex body B in the plane or three-dimensional space. To many-dimensional bodies this result was generalized by Hermann Weyl (1885–1955) in 1939.

4.7 Connections on Vector Bundles

An attempt to calculate the derivative of a vector field (e.g., the acceleration of a parameterized curve) on a manifold right away runs into the difficulty involved in the necessity to consider differences of vectors lying in different tangent spaces. Such a differentiation requires a method for identifying different tangent spaces.

A vector field on a manifold is a special case of a section of a vector bundle over a manifold (a vector field is a section of the tangent bundle). A connection on a vector bundle associates each section with its covariant derivative in the direction of a vector field. Covariant differentiation makes it possible to define the parallel transport of a fiber along a curve (see p. 156). Using parallel transport, we can identify different fibers, transporting them along a curve. This identification connects different fibers—that is where the name *connection* comes from. A connection is completely determined by this identification of fibers: using the parallel transport of a fiber along a curve, we can define the differentiation of a fiber in the direction of a vector field, and this differentiation coincides with the ordinary one (see p. 157).

Let M^n be a manifold. A *connection* ∇ on M^n is a map taking each pair of vector fields X and Y on M^n to a vector field $\nabla_X Y$ on M^n so that this map is linear in X and Y and, for any function f on M^n, $\nabla_{fX}Y = f\nabla_X Y$ and

$$\nabla_X(fY) = f\nabla_X Y + (\partial_X f)Y \quad \text{(\textit{Leibniz' rule}).}$$

The vector field $\nabla_X Y$ is called the *covariant derivative* of Y in the direction of X.

In a similar way we can define a connection on any vector bundle ξ over a manifold M^n. Namely, a *connection* ∇ on a bundle ξ associates a vector field X on the manifold M^n and a section s of the bundle ξ with a section $\nabla_X(s)$ of the bundle ξ so that this correspondence is linear in X and s and, for any function f on M^n, the relation $\nabla_{fX}(s) = f\nabla_X(s)$ and Leibniz' rule

$$\nabla_X(fs) = f\nabla_X(s) + (\partial_X f)s$$

hold. Thus, a connection on a manifold is a connection on its tangent bundle.

Problem 4.2 Let ∇^1 and ∇^2 be connections on a bundle ξ, and let λ and μ be constants. Prove that $\lambda\nabla^1 + \mu\nabla^2$ is a connection if and only if $\lambda + \mu = 1$.

Example 4.5 Consider a trivial bundle with fiber $\mathbb{R}^n$. Each section s is determined by a set of n functions $s^1, \ldots, s^n$ for which $s(x) = \sum s^i(x)e_i(x)$. The formula $\nabla_X(s) = \sum(\partial_X s^i)e_i$ defines a connection.

Proof The relation $\nabla_{fX}(s) = f\nabla_X(s)$ holds because $\partial_{fX} = f\partial_X$. Since $\partial_X(fs^i) = f(\partial_X s^i) + (\partial_X f)s^i$, it follows that $\nabla_X(fs) = f\nabla_X(s) + (\partial_X f)s$. $\square$

At every point $p \in M^n$ the section $\nabla_X(s)$ is completely determined by the restrictions of X and s to any neighborhood U of p. Indeed, take a neighborhood V of p whose closure is contained in U. Let us construct a function f on M^n so that $f(x) = 1$ for $x \in V$ and $f(x) = 0$ for $x \notin U$. The equality $\nabla_{fX}(s) = f\nabla_X(s)$ shows that $\nabla_X(s)$ at p is completely determined by the restriction of X to U. Since the function f is constant in a neighborhood of p, it follows that $(\partial_X f)(p) = 0$. Therefore, according to Leibniz' rule, we have $\nabla_X(fs) = \nabla_X(s)$ at the point p. The section fs is completely determined by the restriction of s to U.

Thus, it makes sense to speak of the restriction of a connection to an open set $U \subset M^n$. In particular, it makes sense to speak about a connection in a given coordinate neighborhood over which the bundle is trivial.

Problem 4.3 Prove that on any vector bundle over a manifold M^n there exists at least one connection.

Consider a trivial d-dimensional vector bundle ξ over a neighborhood U in $\mathbb{R}^n$. Let $e_1, \ldots, e_d$ be a basis in the fiber of ξ, and let $\varepsilon_1, \ldots, \varepsilon_n$ be a basis in $\mathbb{R}^n$. Then a section s is determined by functions $s^1(x), \ldots, s^d(x)$ for which $s = s^j e_j$, and a vector field X is determined by functions $X^1(x), \ldots, X^n(x)$ for which $X = X^i \varepsilon_i$. Let ∇ be any connection on the bundle ξ. Consider the functions $\Gamma_{ij}^k(x)$, where $i = 1, \ldots, n$ and $j, k = 1, \ldots, d$, defined as the coefficients in the expansion $\nabla_{\varepsilon_i} e_j = \Gamma_{ij}^k e_k$ of the section $\nabla_{\varepsilon_i} e_j$ in the basis $e_1, \ldots, e_d$.

The functions Γ_{ij}^k are completely determined by the connection ∇. Indeed,

$$\nabla_X(s) = X^i \nabla_{\varepsilon_i}(s^j e_j) \text{ and } \nabla_{\varepsilon_i}(s^j e_j) = s^j \nabla_{\varepsilon_i} e_j + \frac{\partial s^j}{\partial x^i} e_j.$$

The functions Γ_{ij}^k are called the *Christoffel symbols*.

We have already encountered the Christoffel symbols in the special case of a surface in space. Those Christoffel symbols were symmetric: $\Gamma_{ij}^k = \Gamma_{ji}^k$. Now we cannot hope for symmetry, because the indices i and j range over sets of different cardinalities. But even in the case where $d = n$, the Christoffel symbols may be asymmetric; for this reason, a special torsion tensor is introduced, which is designed just for measuring their asymmetry.

In what follows, in studying Riemannian manifolds, we will deal with the Levi-Cività connection, which is constructed from the Riemannian metric. For the Levi-Cività connection, the Christoffel symbols are symmetric. Therefore, we do have to monitor the order of indices i and j in the symbols Γ_{ij}^k, but only in this chapter. Note that there is no convention concerning the order of writing the subscripts i and j. The formula $\nabla_{\varepsilon_j} e_i = \Gamma_{ij}^k e_k$ for asymmetric symbols Christoffel is also frequent in the literature.

Theorem 4.4 *Let Γ_{ij}^k and $\tilde{\Gamma}_{rs}^t$ be the Christoffel symbols of a connection on the tangent bundle of a manifold M^n with respect to local coordinates x and y. Then*

$$\tilde{\Gamma}_{rs}^t = \Gamma_{ij}^k \frac{\partial y^t}{\partial x^k} \cdot \frac{\partial x^i}{\partial y^r} \cdot \frac{\partial x^j}{\partial y^s} + \frac{\partial y^t}{\partial x^k} \cdot \frac{\partial^2 x^k}{\partial y^r \partial y^s}.$$

Proof Let $e_k = \frac{\partial}{\partial x^k}$ and $\tilde{e}_t = \frac{\partial}{\partial y^t}$ be the basis vectors for the local coordinates x and y. Then $\tilde{e}_t = \frac{\partial x^k}{\partial y^t} e_k$. Therefore,

$$\tilde{\Gamma}_{rs}^t \frac{\partial x^k}{\partial y^t} e_k = \nabla_{\tilde{e}_r} \tilde{e}_s = \nabla_{\frac{\partial x^i}{\partial y^r} e_i} \left(\frac{\partial x^j}{\partial y^s} e_j \right) = \frac{\partial x^i}{\partial y^r} \nabla_{e_i} \left(\frac{\partial x^j}{\partial y^s} e_j \right)$$

$$= \frac{\partial x^i}{\partial y^r} \cdot \frac{\partial x^j}{\partial y^s} \nabla_{e_i} e_j + \frac{\partial x^i}{\partial y^r} \cdot \frac{\partial}{\partial x^i} \left(\frac{\partial x^j}{\partial y^s} \right) e_j$$

$$= \frac{\partial x^i}{\partial y^r} \cdot \frac{\partial x^j}{\partial y^s} \Gamma_{ij}^k e_k + \frac{\partial x^i}{\partial y^r} \cdot \frac{\partial y^p}{\partial x^i} \cdot \frac{\partial^2 x^j}{\partial y^p \partial y^s} e_j$$

$$= \left(\frac{\partial x^i}{\partial y^r} \cdot \frac{\partial x^j}{\partial y^s} \Gamma_{ij}^k + \frac{\partial^2 x^k}{\partial y^r \partial y^s} \right) e_k;$$

we have used the identity $\frac{\partial x^i}{\partial y^r} \cdot \frac{\partial y^p}{\partial x^i} = \delta_p^r$. Thus,

$$\tilde{\Gamma}_{rs}^t \frac{\partial x^k}{\partial y^t} = \frac{\partial x^i}{\partial y^r} \cdot \frac{\partial x^j}{\partial y^s} \Gamma_{ij}^k + \frac{\partial^2 x^k}{\partial y^r \partial y^s}.$$

We fix r, s, and t, multiply this equation by $\frac{\partial y^t}{\partial x^k}$ for $k = 1, \ldots, n$, and sum the resulting relations. Again applying the same identity, we obtain the required relation. $\square$

Covariant differentiation in the dual bundle, i.e., the covariant differentiation of 1-forms, is defined so that

$$\partial_X\big(\omega(Y)\big) = (\nabla_X\omega)(Y) + \omega\big(\nabla_X Y\big).$$

Thus, the definition of the covariant derivative $\nabla_X\omega$ is as follows:

$$(\nabla_X\omega)(Y) = \partial_X\big(\omega(Y)\big) - \omega\big(\nabla_X Y\big).$$

HISTORICAL COMMENT The theory of connections on vector bundles was developed by Jean-Louis Koszul (1921–2018) in 1950.

4.8 Geodesics

The equation $\nabla_{X^k \varepsilon_k}(s) = X^k \nabla_{\varepsilon_k}(s)$ shows that if we are interested in $\nabla_X(s)$ at a point p, then it suffices to know the vector field X only at the point p; the behavior of X in a neighborhood of p is of no interest. Moreover, if $X = \gamma'$ and we are interested in $\nabla_X(s)$ on some parameterized curve $\gamma: (a, b) \to M^n$, then it suffices to know only the restriction of s to the curve γ; the behavior of the section s in a neighborhood of γ is of no interest. Indeed, let $\gamma(t) = x^i(t)\varepsilon_i$. Then $\gamma'(t) = \frac{dx^i}{dt}\varepsilon_i$ and

$$\nabla_{\gamma'}(s) = \nabla_{\frac{dx^i}{dt}\varepsilon_i}(s^j e_j) = \frac{dx^i}{dt}\cdot\frac{\partial s^j}{\partial x^i}e_j + \frac{dx^i}{dt}s^j\Gamma_{ij}^k e_k$$

$$= \left(\frac{dx^i}{dt}\cdot\frac{\partial s^j}{\partial x^i} + \frac{dx^i}{dt}s^k\Gamma_{ik}^j\right)e_j = \left(\frac{ds^j}{dt} + \frac{dx^i}{dt}s^k\Gamma_{ik}^j\right)e_j.$$

Thus, to the section s we can assign the section $\nabla_{\gamma'}(s)$, which is defined on the curve γ and depends only on the restriction of s to γ.

We say that a section s is *parallel* along a curve γ if $\nabla_{\gamma'}(s) = 0$ at each point of γ. In local coordinates this condition is written as the system of linear differential equations

$$\frac{ds^j}{dt} + \Gamma_{ik}^j\frac{dx^i}{dt}s^k = 0, \quad j = 1, \ldots, \dim\xi. \tag{4.2}$$

In an interval (a, b) this system has precisely one solution satisfying a given initial condition $s(t_0)$, $t_0 \in (a, b)$. Therefore, if a bundle ξ is equipped with a connection and points p and q are joined by a curve on the manifold, then the fibers p and q can be identified. This identification is an isomorphism of vector spaces. Indeed, if sections s_1 and s_2 satisfy Eq. (4.2), then so does any section of the form $\lambda s_1 + \mu s_2$, where λ and μ are constants.

Given any vector bundle $p\colon E \to M^n$ over a manifold M^n, the tangent space at each point e of the total space E contains a distinguished *vertical space*, that is, the tangent space to the fiber over the point $x = p(e) \in M^n$. In the general case, there is no distinguished horizontal space complementary to the vertical one. But a connection can be used to single out a horizontal space by means of parallel transport. For every vector $X \in T_x M^n$, consider a curve $\gamma(t)$ determining it, for which $\gamma(0) = x$ and $\gamma'(0) = X$. The section passing through a point $e \in E$ and parallel along the curve γ is a curve $e_X(t)$ in the space E. The tangent vectors at $t = 0$ to such curves generate a *horizontal space* at the point e.

Using a connection, we can identify the fibers of a bundle over the endpoints p and q of some curve (the identification depends on the curve). In turn, the covariant derivative $\nabla_X s$ can be expressed in terms of the parallel transport of a section along a curve as follows. First, we choose a curve $\gamma(t)$ for which $\gamma'(0) = X$ is the given vector. Let $s(t)$ be the section s over the point $\gamma(t)$, and let s_ε be a section which is transported along the curve γ and coincides with the section s over the point $\gamma(\varepsilon)$. Then

$$\nabla_X s = \lim_{\varepsilon \to 0} \frac{s_\varepsilon(0) - s(0)}{\varepsilon}. \tag{4.3}$$

Here $s_\varepsilon(0)$ is the parallel transport to the point $\gamma(0)$ of the section s over $\gamma(\varepsilon)$.

To prove (4.3), we choose a basis $e_1, \ldots, e_d$ in the fiber of the bundle over some point of γ and transport it along the entire curve. We denote the obtained sections over the points $\gamma(t)$ by $e_1(t), \ldots, e_d(t)$. These sections form a basis over each point, and the section s can be expanded in this basis: $s(t) = s^i(t)e_i(t)$. Over the point $\gamma(\varepsilon)$ this expansion has the form $s(\varepsilon) = s^i(\varepsilon)e_i(\varepsilon)$. Since the section s_ε is obtained by the transport along the curve of the section over $\gamma(\varepsilon)$, it follows that the expansion coefficients are constant and the expansion has the form $s_\varepsilon(t) = s^i(\varepsilon)e_i(t)$. Thus, $s_\varepsilon(0) = s^i(\varepsilon)e_i(0)$ and

$$\lim_{\varepsilon \to 0} \frac{s_\varepsilon(0) - s(0)}{\varepsilon} = \lim_{\varepsilon \to 0} \frac{s^i(\varepsilon) - s^i(0)}{\varepsilon} e_i(0) = \frac{ds^i}{dt}(0)e_i(0).$$

Since the vector fields $e_i(t)$ are parallel along the curve γ, we have $\nabla_{\gamma'} e_i(t) = 0$. Hence

$$\nabla_X s = \nabla_{\gamma'} s^i(t)e_i(t)\Big|_{t=0} = \frac{ds^i}{dt}(0)e_i(0).$$

This completes the proof of formula (4.3).

Now consider the case where the bundle under consideration is a tangent bundle. Suppose given a connection ∇ on a manifold M^n. A curve γ on M^n is called a *geodesic* (with respect to the connection ∇) if $\nabla_{\gamma'}(\gamma') = 0$ for all points of γ. In other words, a geodesic is determined by the condition that the velocity vector is parallel along the curve γ. In local coordinates a geodesic is determined by the system of differential equations

$$\frac{d^2 x^j}{dt^2} + \Gamma^j_{ik} \frac{dx^i}{dt} \cdot \frac{dx^k}{dt} = 0, \quad j = 1, \ldots, n. \tag{4.4}$$

A geodesic is determined by a system of n second-order differential equations. This system is equivalent to the following system of $2n$ first-order equations:

$$\frac{dx^j}{dt} = v^j, \quad j = 1, \ldots, n, \tag{4.5}$$

$$\frac{dv^j}{dt} = -\Gamma^j_{ik} \frac{dx^i}{dt} \cdot \frac{dx^k}{dt}, \quad j = 1, \ldots, n.$$

Therefore, for each point $(x_0, v_0) \in \mathbb{R}^{2n}$, we can choose an $\varepsilon > 0$ so that if a point (x_1, v_1) lies in the ε-neighborhood of (x_0, v_0), then, for $|t| < \varepsilon$, system (4.5) has a unique solution $t \mapsto x(t)$ satisfying the initial conditions $x(0) = x_1$ and $\frac{dx}{dt}(0) = v_1$. This solution smoothly depends on x_1, v_1, and t.

The geodesics have the following homogeneity property: if $x(t)$ is a geodesic, then so is $x(ct)$. Indeed, let $y(t) = x(ct)$. Then $\frac{dy}{dt} = c\frac{dx}{dt}$ and $\frac{d^2 y}{dt^2} = c^2 \frac{d^2 x}{dt^2}$. Therefore, if $x(t)$ satisfies system (4.4), then $y(t)$ satisfies this system as well.

Let us draw a geodesic $\gamma_V(t)$ for which $\frac{d\gamma_V}{dt}(0) = V$ and $\gamma_V(0) = x$ through a point $x \in M^n$. The geodesic $\gamma_V(t)$ is uniquely determined for small t, but it may happen that it admits no extension to large t values (if such an extension exists, then it is unique). If $\gamma_V(t)$ is defined for $t = 1$, then we set $\exp_x(V) = \gamma_V(1)$.

Theorem 4.5 *For any point $x \in M^n$, there exists a neighborhood U of zero in $T_x M^n$ such that the map $\exp_x$ is defined on U and is a diffeomorphism from a neighborhood of zero in the tangent space at x to a neighborhood of the point x in the manifold M^n.*

Proof We have already proved that there exists an $\varepsilon > 0$ and a neighborhood U_1 of zero in $T_x M^n$ such that $\gamma_V(t)$ is defined for $|t| \leqslant \varepsilon$ and $V \in U_1$. Let U_2 be the image of U_1 under the contraction map $V \mapsto \varepsilon V$. Then, for each vector $W = \varepsilon V \in U_2$, the homogeneity property implies that $\gamma_W(1) = \gamma_{\varepsilon V}(1) = \gamma_V(\varepsilon)$ is defined. Thus, on U_2 a map $V \mapsto \exp_x(V)$ is defined; this map is smooth.

In the space $T_x M^n$ consider the curve $\alpha(t) = tV$, where V is a fixed vector. For $t = 0$, the velocity vector of this curve equals V. The map $\exp_x$ takes this curve to the curve $\gamma_{tV}(1) = \gamma_V(t)$. For $t = 0$, the velocity vector of $\gamma_V(t)$ on the manifold M^n equals V. Thus, the differential of $\exp_x$ at the point 0 is the identity operator on the space $T_x M^n$. Therefore, we can apply the inverse function theorem and find a neighborhood U of 0 in $T_x M^n$ such that the restriction of $\exp_x$ to U is a diffeomorphism. $\square$

4.9 The Curvature Tensor and the Torsion Tensor

Before proceeding to discuss the curvature and torsion tensors, we give a general definition of a tensor. We begin with tensors on a vector space V. Under a change of basis $\hat{e}_i = A_i^j e_j$ the coordinates of each vector change according to the contravariant law $\hat{v}^i = (A^{-1})^i_j v^j$, and the coordinates of each covector change according to the covariant law $\hat{e}_i = A_i^j e_j$ ("covariant" means "varying in the same way as under direct change of basis," and "contravariant" means "varying in the same way as under inverse change of basis"). The components of a tensor of type (p, q), where p is the number contravariant indices and q is the number of covariant ones, change according to the law

$$\hat{T}^{i'_1 \ldots i'_p}_{j'_1 \ldots j'_q} = (A^{-1})^{i'_1}_{i_1} \ldots (A^{-1})^{i'_p}_{i_p} A^{j_1}_{j'_1} \ldots A^{j_q}_{j'_q} T^{i_1 \ldots i_p}_{j_1 \ldots j_q}.$$

A *tensor* of type (p, q) is an element of the tensor product of p copies of the space V and q copies of the dual space V^*.

A tensor bundle of type (p, q) over a manifold M^n is a bundle whose fiber over a point $x \in M^n$ is the tensor product of p copies of the space $V = T_x M^n$ and q copies of the dual space V^*. For each coordinate neighborhood $U \subset M^n$ with coordinate functions $x^1, \ldots, x^n$, the bases $\frac{\partial}{\partial x^i}$ of the space $T_x M^n$ and dx^i of the dual space determine bases in the fibers over the points of the neighborhood U. A tensor field is a section of some tensor bundle. The tensor field in a coordinate neighborhood U has the form

$$T^{i_1 \ldots i_p}_{j_1 \ldots j_q} \frac{\partial}{\partial x^{i_1}} \otimes \cdots \otimes \frac{\partial}{\partial x^{i_p}} \otimes dx^{j_1} \otimes \cdots \otimes dx^{j_q}.$$

Under a change of coordinates the components of a tensor change as follows. Let $T(x)^{i_1 \ldots i_p}_{j_1 \ldots j_q}$ be the components of a tensor in local coordinates $x^1, \ldots, x^n$, and let $T(y)^{\alpha_1 \ldots \alpha_p}_{\beta_1 \ldots \beta_q}$ be the components of the same tensor in local coordinates $y^1, \ldots, y^n$. Then

$$T(y)^{\alpha_1 \ldots \alpha_p}_{\beta_1 \ldots \beta_q} = T(x)^{i_1 \ldots i_p}_{j_1 \ldots j_q} \frac{\partial y^{\alpha_1}}{\partial x^{i_1}} \cdots \frac{\partial y^{\alpha_p}}{\partial x^{i_p}} \frac{\partial x^{j_1}}{\partial y^{\beta_1}} \cdots \frac{\partial x^{j_q}}{\partial y^{\beta_q}}.$$

A tensor field is often briefly referred to simply as a tensor.

Example 4.6 The Kronecker symbol δ_i^j is a tensor of type $(1, 1)$.

Proof We must check that

$$\delta_\beta^\alpha(y) = \delta_i^j(x) \frac{\partial y^\alpha}{\partial x^i} \frac{\partial x^j}{\partial y^\beta},$$

where $\delta_{\beta}^{\alpha}(y) = \delta_{\beta}^{\alpha}$ and $\delta_i^j(x) = \delta_i^j$. Clearly,

$$\delta_i^j \frac{\partial y^{\alpha}}{\partial x^i} \frac{\partial x^j}{\partial y^{\beta}} = \frac{\partial y^{\alpha}}{\partial x^i} \frac{\partial x^i}{\partial y^{\beta}} = \delta_{\beta}^{\alpha}.$$

□

A tensor field of type $(1, 1)$ can be regarded as a linear operator on the tangent space at each point. Then the Kronecker symbol is the identity self-map of every tangent space.

Example 4.7 The Riemannian metric $g_{ij} = \left(\frac{\partial r}{\partial x^i}, \frac{\partial r}{\partial x^j} \right)$ on a hypersurface $r(x^1, \ldots, x^n)$ is a tensor of type $(0, 2)$.

Proof The required equation

$$g_{\beta_1, \beta_2}(y) = g_{j_1 j_2}(x) \frac{\partial x^{j_1}}{\partial y^{\beta_1}} \frac{\partial x^{j_2}}{\partial y^{\beta_2}}$$

follows from

$$\frac{\partial r}{\partial y^{\beta}} = \frac{\partial r}{\partial x^j} \frac{\partial x^j}{\partial y^{\beta}}.$$

□

We also need the notion of the commutator of vector fields on a manifold. In local coordinates the commutator $[X, Y]$ of vector fields X and Y is defined in the same way as the commutator of vector fields in Euclidean space:

$$[X, Y]^i = X^j \frac{\partial Y^i}{\partial x^j} - Y^j \frac{\partial X^i}{\partial x^j}.$$

An invariant definition of a commutator can be obtained by considering a vector field X as an operator ∂_X which takes each function f to its derivative $\partial_X f$ in the direction of a vector field X. Under this approach the commutator $[X, Y]$ acts as the operator $\partial_X \partial_Y - \partial_Y \partial_X$.

Now we can define the curvature tensor and the torsion tensor for a connection ∇ on a tangent bundle. The *torsion tensor* T of type $(1, 2)$ and the *curvature tensor* R of type $(1, 3)$ are defined at given vector fields X, Y, and Z by

$$T(X, Y) = \nabla_X Y - \nabla_Y X - [X, Y],$$

$$R(X, Y)Z = \nabla_X \nabla_Y Z - \nabla_Y \nabla_X Z - \nabla_{[X,Y]} Z.$$

Let us check that T and R are indeed tensors of the specified types. We begin with the torsion tensor. For $X = x^i \partial_i$ and $Y = y^j \partial_j$, we have

$$\nabla_X Y = x^i y^j \nabla_{\partial_i} \partial_j + x^i (\partial_i y^j) \partial_j = x^i y^j \Gamma_{ij}^k \partial_k + x^i (\partial_i y^j) \partial_j,$$

$$\nabla_Y X = x^i y^j \Gamma_{ji}^k \partial_k + y^i (\partial_i x^j) \partial_j.$$

Therefore,

$$\nabla_X Y - \nabla_Y X - [X, Y] = x^i y^j (\Gamma_{ij}^k - \Gamma_{ji}^k) \partial_k. \tag{4.6}$$

Let $T(\partial_i, \partial_j) = T_{ij}^k \partial_k$. Then $T_{ij}^k = \Gamma_{ij}^k - \Gamma_{ji}^k$. By virtue of Theorem 4.4 on p. 155, the Christoffel symbols $\Gamma_{ij}^k(x)$ and $\Gamma_{rs}^t(y)$ in local coordinates x and y are related by

$$\Gamma_{rs}^t(y) = \Gamma_{ij}^k(x) \frac{\partial y^t}{\partial x^k} \cdot \frac{\partial x^i}{\partial y^r} \cdot \frac{\partial x^j}{\partial y^s} + \frac{\partial y^t}{\partial x^k} \cdot \frac{\partial^2 x^k}{\partial y^r \partial y^s}.$$

Therefore, the Christoffel symbols themselves are not tensor components, but their differences $T_{ij}^k = \Gamma_{ij}^k - \Gamma_{ji}^k$ are.

A connection whose torsion identically vanishes is said to be *symmetric*.

Problem 4.4 Prove that the torsion at a point p is zero if and only local coordinates in a neighborhood of p can be chosen so that $\Gamma_{ij}^k(p) = 0$ for all i, j, k.

The fact that the torsion is a tensor can be proved without using an explicit expression for the transformation of the Christoffel symbols under a change of coordinates. It follows directly from formula (4.6) that the torsion is bilinear, that is, $T(fX, gY) = fgT(X, Y)$ for any functions f and g. Let us show that the bilinearity of the torsion implies that the torsion is a tensor. Under a change of local coordinates the coordinates of vectors and the coordinate vectors themselves vary according to the laws $u^i(x) = u^\alpha(y) \frac{\partial x^i}{\partial y^\alpha}$ and $\partial_j(x) = \partial_\beta(y) \frac{\partial y^\beta}{\partial x^j}$. Therefore,

$$u^\alpha(y) v^\beta(y) T_{\alpha\beta}^\gamma(y) \partial_\gamma(y) = u^i(x) v^j(x) T_{ij}^k \partial_k(x)$$

$$= u^\alpha(y) \frac{\partial x^i}{\partial y^\alpha} v^\beta(y) \frac{\partial x^j}{\partial y^\beta} T_{ij}^k(x) \partial_\gamma(y) \frac{\partial y^\gamma}{\partial x^k},$$

and hence

$$T_{\alpha\beta}^\gamma(y) = T_{ij}^k(x) \frac{\partial x^i}{\partial y^\alpha} \frac{\partial x^j}{\partial y^\beta} \frac{\partial y^\gamma}{\partial x^k},$$

as required.

We have already performed all needed calculations for the curvature tensor in the proof of Theorem 3.15 on p. 99; in that proof the trilinearity of the curvature was shown, which implies that the curvature components R^l_{ijk} can be determined from the equation

$$R(X^j \partial_j, Y^k \partial_k)(Z^i \partial_i) = X^j Y^k Z^i R^l_{ijk} \partial_l.$$

The trilinearity of the curvature implies that the curvature is a tensor (the argument is precisely the same as in the proof that the bilinearity of the torsion implies that the torsion is a tensor).

A simple calculation shows that

$$R^l_{ijk} = \frac{\partial \Gamma^l_{ki}}{\partial x^j} - \frac{\partial \Gamma^l_{ji}}{\partial x^k} + \Gamma^p_{ki} \Gamma^l_{jp} - \Gamma^p_{ji} \Gamma^l_{pk}. \tag{4.7}$$

The only difference between this calculation and that in the previous section is that previously the Christoffel symbols were symmetric and the order of subscripts did not matter, while now this order must be taken into account.

It is seen directly from the definition of the tensors T and R that $T(X, Y) = -T(Y, X)$ and $R(X, Y)Z = -R(Y, X)Z$. For the components of the tensors T and R, this means that $T^k_{ij} = -T^k_{ji}$ and $R^l_{ijk} = -R^l_{ikj}$. In the case of a connection with zero torsion, there also arise some other relations between the components of the curvature tensor.

Theorem 4.6 (First Bianchi Identity) *If $T = 0$, then*

$$R(X, Y)Z + R(Z, X)Y + R(Y, Z)X = 0, \ i.e., \ R^l_{ijk} + R^l_{kij} + R^l_{jki} = 0.$$

First Proof Choose local coordinates in a neighborhood of a given point so that all Christoffel symbols vanish at this point (for a torsion-free connection, Problem 4.4 can be used). Then relation (4.7) takes the form

$$R^l_{ijk} = \frac{\partial \Gamma^l_{ki}}{\partial x^j} - \frac{\partial \Gamma^l_{ji}}{\partial x^k}.$$

Therefore, $R^l_{kij} = \frac{\partial \Gamma^l_{jk}}{\partial x^i} - \frac{\partial \Gamma^l_{ik}}{\partial x^j}$ and $R^l_{jki} = \frac{\partial \Gamma^l_{ij}}{\partial x^k} - \frac{\partial \Gamma^l_{kj}}{\partial x^i}$. Summing these three equations and again using the symmetry of the Christoffel symbols, we obtain the required identity. □

Second Proof We must prove that if $\nabla_U V - \nabla_V U = [U, V]$ for any vector fields U and V, then the sum of the three expressions

$$\nabla_Y \nabla_X Z - \nabla_X \nabla_Y Z - \nabla_{[Y,X]} Z,$$

$$\nabla_Z \nabla_Y X - \nabla_Y \nabla_Z X - \nabla_{[Z,Y]} X,$$

$$\nabla_X \nabla_Z Y - \nabla_Z \nabla_X Y - \nabla_{[X,Z]} Y$$

vanishes. First, note that

$$\nabla_X(\nabla_Z Y - \nabla_Y Z) + \nabla_Y(\nabla_X Z - \nabla_Z X) + \nabla_Z(\nabla_Y X - \nabla_X Y)$$
$$= \nabla_X[Z, Y] + \nabla_Y[X, Z] + \nabla_Z[Y, X].$$

Next,

$$\nabla_X[Z, Y] - \nabla_{[Z,Y]}X + \nabla_Y[X, Z] - \nabla_{[X,Z]}Y + \nabla_Z[Y, X] - \nabla_{[Y,X]}Z$$
$$= [X, [Z, Y]] + [Y, [X, Z]] + [Z, [Y, X]] = 0$$

by virtue of the Jacobi identity (see Appendix, Theorem 8.3).   □

HISTORICAL COMMENT In 1902 Luigi Bianchi (1856–1928) proved two identities for the Riemann tensor. The second Bianchi identity is discussed on p. 166.

4.10   The Curvature Matrix of a Connection

Recall that a connection ∇ on a bundle ξ is a map which takes every pair consisting of a vector field X on a manifold M^n and a section s of a bundle ξ to a section $\nabla_X(s)$ of the bundle ξ so that this map is linear in X and s and, for any function f on M^n, the equation $\nabla_{fX}(s) = f\nabla_X(s)$ and the Leibniz' rule

$$\nabla_X(fs) = f\nabla_X(s) + (\partial_X f)s$$

hold.

A connection ∇ can be regarded as a map $\bar{\nabla}: \Gamma(\xi) \to \Gamma(T^* \otimes \xi)$, where $\Gamma(\xi)$ is the space of sections of the bundle ξ and T^* is the cotangent bundle of the manifold M^n, i.e., the space of 1-forms. Indeed, note that, given a vector field X and a tensor $\bar{\nabla}s = \omega^i \otimes s_i$, we can define the section $\omega^i(X)s_i$. Therefore, given a map $\bar{\nabla}: \Gamma(\xi) \to \Gamma(T^* \otimes \xi)$, we can assign the section $\omega^i(X)s_i$ to each vector field X and any section s. Clearly, this correspondence is linear in X, and for any function f on M^n, we have $\omega^i(fX) = f\omega^i(X)$. For the map $(X, s) \mapsto \omega^i(X)s_i$ to be a connection, it is required that the map $\bar{\nabla}$ have the properties

$$\bar{\nabla}(s_1 + s_2) = \bar{\nabla}s_1 + \bar{\nabla}s_2,$$
$$\bar{\nabla}(fs) = f\bar{\nabla}s + df \otimes s.$$

The first property means that the correspondence is linear in s, and the second one ensures the fulfillment of Leibniz' rule: it suffices to note that $\partial_X f = (df)(X)$.

Now suppose given a connection ∇ assigning a section $\nabla_X s$ to any vector field X and section s. Choose local coordinates $x^1, \ldots, x^n$ on the manifold M^n and consider the vector fields $e_i = \frac{\partial}{\partial x^i}$ in these local coordinates. We set $\bar{\nabla} s = \omega^i \otimes s_i$, where $\omega^i(e_j) = \delta^i_j$ and $s_i = \nabla_{e_i} s$. Let us check that $\bar{\nabla} s$ does not depend on the choice of local coordinates. Consider other local coordinates $y^1, \ldots, y^n$ in the same neighborhood. Let $\varepsilon_k = \frac{\partial}{\partial y^k} = A^i_k e_i$, and let $\theta^i(\varepsilon_j) = \delta^i_j$. Then $\omega^i = A^i_k \theta^k$. Thus, we must show that

$$A^i_k \theta^k \otimes \nabla_{e_i} = \theta^k \otimes \nabla_{A^i_k e_i}.$$

This follows from the linearity of $\nabla_X s$ in X (with respect to multiplication by functions rather than constants).

We introduced the notation $\bar{\nabla}$ just not to get lost in the proof. In what follows, we will use the same symbol for the assignment of a section to a vector field X and a section s and for the corresponding map $\Gamma(\xi) \to \Gamma(T^* \otimes \xi)$.

A connection on a trivial bundle over a neighborhood U in $\mathbb{R}^n$ is determined by a matrix $\omega = (\omega^j_i)$, whose entries serve as differential 1-forms on U. Indeed, let $e_1, \ldots, e_d$ be the basis sections of a bundle ξ, and let e be the column with components $e_1, \ldots, e_d$. Then $\nabla e_i = \omega^j_i \otimes e_j$, so that $\nabla e = \omega \otimes e$. The matrix ω is called the *connection matrix*. This matrix is related to the Christoffel symbols as $\Gamma^k_{ij} = \omega^k_j(\varepsilon_i)$, where $\varepsilon_1, \ldots, \varepsilon_n$ is the basis of $\mathbb{R}^n$. Indeed,

$$\Gamma^k_{ij} e_k = \nabla_{\varepsilon_i} e_j = \nabla e_j(\varepsilon_i) = (\omega^k_j \otimes e_k)(\varepsilon_i) = \omega^k_j(\varepsilon_i) e_k.$$

Thus, a connection can be defined by a connection matrix. Let $e' = Ae$ be another set of sections, where A is a square matrix whose entries are functions A^i_k, and let ω' be the corresponding connection matrix. Then

$$\omega' = dA \cdot A^{-1} + A\omega A^{-1}. \tag{4.8}$$

Indeed, in view of the properties of a connection, we have $\nabla(Ae) = A\nabla e + dA \otimes e$. On the other hand, $\nabla(Ae) = \nabla e' = \omega' \otimes e'$. Therefore,

$$\omega' \otimes e' = A\nabla e + dA \otimes e = A\omega \otimes A^{-1}e' + dA \otimes A^{-1}e' = (A\omega A^{-1} + dA \cdot A^{-1}) \otimes e'.$$

Now let us express the curvature tensor in terms of the connection matrix. This expression contains the differentials of the 1-forms ω^i_j. We use the following expression for the differential of a 1-form ω:

$$d\omega(X, Y) = \partial_X(\omega(Y)) - \partial_Y(\omega(X)) - \omega([X, Y])$$

(see Appendix, Theorem 8.4). We will also use the relation $\nabla_X e_i = \omega^j_i(X) e_j$, which follows directly from the equation $\nabla e_i = \omega^j_i \otimes e_j$.

Clearly,

$$\nabla_X \nabla_Y e_i = \nabla_X \omega_i^j(Y)e_j = \partial_X(\omega_i^j(Y))e_j + \omega_i^j(Y)\nabla_X e_j$$

$$= \partial_X(\omega_i^j(Y))e_j + \omega_i^j(Y)\omega_j^k(X)e_k$$

$$= \partial_X(\omega_i^j(Y))e_j + \omega_i^k(Y)\omega_k^j(X)e_j.$$

Therefore,

$$R(X, Y)e_i = (\nabla_X \nabla_Y - \nabla_Y \nabla_X - \nabla_{[X,Y]})e_i$$

$$= \big(\partial_X(\omega_i^j(Y)) - \partial_Y(\omega_i^j(X))$$

$$- \omega_i^j([X, Y]) - \omega_i^k(X)\omega_k^j(Y) + \omega_i^k(Y)\omega_k^j(X)\big)e_j.$$

Here

$$-\omega_i^k(X)\omega_k^j(Y) + \omega_i^k(Y)\omega_k^j(X) = -(\omega(X)\omega(Y) - \omega(Y)\omega(X))_i^j.$$

To show that matrices whose entries have super- and subscripts are indeed multiplied by this rule, consider two basis changes, $e' = Ae$ ($e_k' = A_k^j e_j$) and $e'' = Be'$ ($e_i'' = B_i^k e_k'$). The transformation $e'' = Be' = BAe$ for the matrix entries is written as $e_i'' = B_i^k e_k' = B_i^k A_k^j e_i$, i.e., $(BA)_i^j = B_i^k A_k^j$.

The expression $\omega(X)\omega(Y) - \omega(Y)\omega(X)$ can be denoted by $\omega \wedge \omega(X, Y)$. Then $R(X, Y)e_i = \Omega_i^j(X, Y)e_j$, where $\Omega_i^j = d\omega_i^j - (\omega \wedge \omega)_i^j$.

Remark For $p > 1$, a pairing between the spaces $\Lambda^p V$ and $\Lambda^p V^*$ can be defined in different ways. If the pairing is assumed to originate from the pairing between the dual spaces $V \otimes \cdots \otimes V$ and $V^* \otimes \cdots \otimes V^*$, then it has the form

$$(\omega_1 \wedge \cdots \wedge \omega_p)(x_1 \wedge \cdots \wedge x_p) = \frac{1}{p!} \begin{vmatrix} \omega_1(x_1) & \dots & \omega_p(x_1) \\ & \dots\dots\dots & \\ \omega_1(x_p) & \dots & \omega_p(x_p) \end{vmatrix}.$$

But usually the pairing is defined by

$$(\omega_1 \wedge \cdots \wedge \omega_p)(x_1 \wedge \cdots \wedge x_p) = \begin{vmatrix} \omega_1(x_1) & \dots & \omega_p(x_1) \\ & \dots\dots\dots & \\ \omega_1(x_p) & \dots & \omega_p(x_p) \end{vmatrix}.$$

This corresponds to the pairing between the spaces of tensors multiplied by $p!$.

The matrix $\Omega = d\omega - \omega \wedge \omega$ of 2-forms is called the *curvature matrix* of the connection. Let us check that, choosing another set of sections $e' = Ae$, we obtain the matrix $\Omega' = A\Omega A^{-1}$. First, note that $dA^{-1} = -A^{-1}dA \cdot A^{-1}$, because $d(A^{-1}A) = 0$. Equation (4.8) can be written in the form

$$\omega' A = dA + A\omega.$$

Let us differentiate this equation, having in mind that ω' is a 1-form and A is a 0-form:

$$d\omega' \cdot A - \omega' \wedge dA = dA \wedge \omega + A d\omega.$$

Therefore,

$$d\omega' = (\omega' \wedge dA + dA \wedge \omega + A d\omega)A^{-1}$$
$$= dA \cdot A^{-1} \wedge dA \cdot A^{-1} + A\omega A^{-1} \wedge dA \cdot A^{-1} + dA \wedge \omega A^{-1} + A d\omega A^{-1}.$$

It is also clear that

$$\omega' \wedge \omega' = (dA \cdot A^{-1} + A\omega A^{-1}) \wedge (dA \cdot A^{-1} + A\omega A^{-1})$$
$$= dA \cdot A^{-1} \wedge dA \cdot A^{-1} + A\omega A^{-1} \wedge dA \cdot A^{-1}$$
$$+ dA \cdot A^{-1} \wedge A\omega A^{-1} + A\omega A^{-1} \wedge A\omega A^{-1}.$$

Thus,

$$\Omega' = d\omega' - \omega' \wedge \omega' = A d\omega A^{-1} - A\omega \wedge \omega A^{-1} = A\Omega A^{-1}.$$

Theorem 4.7 (Second Bianchi Identity) $d\Omega = \omega \wedge \Omega - \Omega \wedge \omega.$

Proof Clearly,

$$d\Omega = d(d\omega - \omega \wedge \omega)$$
$$= -d\omega \wedge \omega + \omega \wedge d\omega$$
$$= (-\Omega - \omega \wedge \omega) \wedge \omega + \omega \wedge (\Omega + \omega \wedge \omega)$$
$$= \omega \wedge \Omega - \Omega \wedge \omega.$$

$\square$

Now consider a connection ∇ on the tangent bundle of a manifold M^n. Let e_1, $\ldots$, e_n be vector fields on an open set $U \subset M^n$ linearly independent at each point. For these vector fields, we can consider the dual 1-forms ω^1, $\ldots$, ω^n, for which $\omega^i(e_j) = \delta^i_j$. Let us calculate the differentials of these 1-forms. For this purpose, we need the 1-forms ω^j_i being the entries of the connection matrix. Recall that they are determined from the relation $\nabla_X e_i = \omega^j_i(X)e_j$.

For a connection on the tangent bundle, we defined the covariant derivative of a 1-form as (see p. 156)

$$(\nabla_X \omega)(Y) = \partial_X(\omega(Y)) - \omega(\nabla_X Y).$$

For covariant derivatives thus defined, we have

$$d\omega(X, Y) = \partial_X(\omega(Y)) - \partial_Y(\omega(X)) - \omega([X, Y])$$
$$= (\nabla_X \omega)(Y) - (\nabla_Y \omega)(X) + \omega(\nabla_X Y - \nabla_Y X - [X, Y])$$
$$= (\nabla_X \omega)(Y) - (\nabla_Y \omega)(X) - \omega(T(X, Y)),$$

where T is the torsion tensor of the connection ∇.

The definition of the covariant derivative of a 1-form implies $0 = \partial_X(\omega^i(e_j)) = (\nabla_X \omega^i)(e_j) + \omega^i(\nabla_X e_j)$. We also have $\omega^i(\nabla_X e_j) = \omega^i(\omega_j^k(X)e_k) = \omega_j^k(X)\delta_j^k = \omega_j^i(X) = \omega_k^i(X)\omega^k(e_j)$. Therefore,

$$0 = (\nabla_X \omega^i)(e_j) + \omega^i(\nabla_X e_j) = (\nabla_X \omega^i + \omega_k^i(X)\omega^k)e_j,$$

and hence $\nabla_X \omega^i = -\omega_k^i(X)\omega^k$. This implies

$$d\omega^j(X, Y) - (\omega^k \wedge \omega_k^j)(X, Y)$$

$$= \partial_X(\omega^j(Y)) - \partial_Y(\omega^j(X)) - \omega^j([X, Y]) - \left(\omega^k(X)\omega_k^j(Y) - \omega^k(Y)\omega_k^j(X)\right)$$

$$= \partial_X(\omega^j(Y)) + \omega^k(Y)\omega_k^j(X) - \partial_Y(\omega^j(X)) - \omega^k(X)\omega_k^j(Y) - \omega^j([X, Y])$$

$$= \partial_X(\omega^j(Y)) - (\nabla_X \omega^j)(Y) - \partial_Y(\omega^j(X)) + (\nabla_Y \omega^j)(X) - \omega^j([X, Y])$$

$$= \omega^j(\nabla_X Y - \nabla_Y X - [X, Y]) = \omega^j(T(X, Y)).$$

Thus, $d\omega^j = \omega^k \wedge \omega_k^j + \omega^j(T)$. In particular, for a torsion-free connection, we have $d\omega^j = \omega^k \wedge \omega_k^j$.

4.11 Solutions of Problems

4.1 Let us prove that parallel transport along a curve preserves inner product. Let $V(t)$ and $W(t)$ be vector fields parallel along a given curve. Then

$$\frac{d}{dt}(V(t), W(t)) = \left(\frac{dV(t)}{dt}, W(t)\right) + \left(V(t), \frac{dW(t)}{dt}\right) = 0,$$

because the vectors $\frac{dV(t)}{dt}$ and $\frac{dW(t)}{dt}$ are orthogonal to all tangent vectors.

4.2 Summing the equations $\nabla^1_X(fs) = f\nabla^1_X(s) + (\partial_X f)s$ and $\nabla^2_X(fs) = f\nabla^2_X(s) + (\partial_X f)s$ multiplied by λ and μ, we obtain $\nabla_X(fs) = f\nabla_X(s) + (\lambda + \mu)(\partial_X f)s$, where $\nabla = \lambda\nabla^1 + \mu\nabla^2$. Therefore, ∇ obeys Leibniz' rule if and only if $\lambda + \mu = 1$. The equation $\nabla_{fX}(s) = f\nabla_X(s)$ holds for any λ and μ.

4.3 Let us cover M^n by charts over which the bundle is trivial. On a trivial bundle over each of these charts we construct a connection as in Example 4.5. We take a partition of unity subordinate to the cover by charts and apply it to glue the constructed connections together into a single connection. This can be done by using Problem 4.2, or, to be more precise, its obvious generalization.

4.4 If in some system of local coordinates we have $\Gamma^k_{ij}(p) = 0$ for all i, j, k, then all components of the torsion tensor vanish at p in these coordinates, and therefore all components of the torsion tensor vanish at the point p in any coordinate system.

Now suppose that the torsion vanishes at a point p, i.e., $\Gamma^k_{ij}(p) = \Gamma^k_{ji}(p)$ for all i, j, k in any coordinate system. In a neighborhood of p we introduce two coordinate systems x and y and denote the Christoffel symbols in these coordinates by Γ^k_{ij} and $\tilde{\Gamma}^k_{ij}$. These Christoffel symbols are related by

$$\tilde{\Gamma}^t_{rs} = \Gamma^k_{ij}\frac{\partial y^t}{\partial x^k} \cdot \frac{\partial x^i}{\partial y^r} \cdot \frac{\partial x^j}{\partial y^s} + \frac{\partial y^t}{\partial x^k} \cdot \frac{\partial^2 x^k}{\partial y^r \partial y^s}.$$

Let us show that in a sufficiently small neighborhood of p the change of coordinates $x \to y$ can be performed in such a way that the coordinates of p are zero in both coordinate systems and the coordinates of any other point q are related by

$$x^k(q) = y^k(q) - \frac{1}{2}\Gamma^k_{ij}(p)y^i(q)y^j(q).$$

Indeed, it follows from the symmetry of the Christoffel symbols that $\frac{\partial x^k}{\partial y^i}(q) = \delta^k_i - \Gamma^k_{ij}(p)y^j(q)$; therefore, $\frac{\partial x^k}{\partial y^i}(p) = \delta^k_i$. Thus, at the point p the Jacobi matrix of the map $(y^1(q), \ldots, y^n(q)) \mapsto (x^1(q), \ldots, x^n(q))$ is the identity matrix, and any such map is locally invertible. It is also clear that

$$\frac{\partial^2 x^k}{\partial y^r \partial y^s}(p) = -\Gamma^k_{rs}(p),$$

whence it follows that

$$\tilde{\Gamma}^t_{rs}(p) = \Gamma^k_{ij}(p)\delta^t_k\delta^i_r\delta^j_s - \delta^t_k\Gamma^k_{rs}(p) = \Gamma^t_{rs}(p) - \Gamma^t_{rs}(p) = 0$$

for all $r, s,$ and t.

Chapter 5
Riemannian Manifolds

A *Riemannian manifold* M^n is a smooth manifold equipped with an inner product (*Riemannian metric*) $g(X, Y)$ in the tangent space at each point; moreover, this inner product is smooth in the sense that $g(X, Y)$ is a smooth function on M^n for any smooth vector fields X and Y on M^n.

In topology manifolds are considered up to a diffeomorphism. In differential geometry Riemannian manifolds are considered up to an *isometry*, that is, a diffeomorphism which transforms the Riemannian metric of one manifold into the Riemannian metric of the other.

5.1 Levi-Cività Connection

A connection ∇ on the tangent bundle TM^n is said to be *compatible with the metric* if

$$\partial_X g(Y, Z) = g(\nabla_X Y, Z) + g(Y, \nabla_X Z).$$

If a connection ∇ is compatible with a Riemannian metric g, then it preserves the inner products of vectors parallel along any curve. Indeed, suppose that vector fields Y and Z are parallel along a curve and X is a vector tangent to this curve at some point. Then $\nabla_X Y = 0$ and $\nabla_X Z = 0$, whence $\partial_X g(Y, Z) = 0$.

Recall that a connection ∇ is torsion-free (has zero torsion) if

$$\nabla_X Y - \nabla_Y X = [X, Y].$$

Theorem 5.1 (Levi-Cività) *On a Riemannian manifold there exists a unique torsion-free connection compatible with the metric.*

© The Author(s), under exclusive license to Springer Nature Switzerland AG 2022
V. V. Prasolov, *Differential Geometry*, Moscow Lectures 8,
https://doi.org/10.1007/978-3-030-92249-8_5

Proof For brevity, we introduce the notation $(X, Y) = g(X, Y)$. If a connection is compatible with the metric and has zero torsion, then

$$(\nabla_X Y, Z) = \partial_X(Y, Z) - (Y, \nabla_X Z)$$
$$= \partial_X(Y, Z) - (Y, \nabla_Z X) - (Y, [X, Z])$$
$$= \partial_X(Y, Z) - \partial_Z(Y, X) + (\nabla_Z Y, X) - (Y, [X, Z])$$
$$= \partial_X(Y, Z) - \partial_Z(Y, X) + (\nabla_Y Z, X) + ([Z, Y], X) - (Y, [X, Z])$$
$$= \partial_X(Y, Z) - \partial_Z(Y, X) + \partial_Y(Z, X) - (Z, \nabla_Y X)$$
$$+ ([Z, Y], X) - (Y, [X, Z])$$
$$= \partial_X(Y, Z) - \partial_Z(Y, X) + \partial_Y(Z, X) - (Z, \nabla_X Y)$$
$$- (Z, [Y, X]) + ([Z, Y], X) - (Y, [X, Z]).$$

As a result, reducing this expression to a more symmetric form, we obtain

$$2(\nabla_X Y, Z) = \partial_X(Y, Z) + \partial_Y(X, Z) - \partial_Z(X, Y) + \qquad (5.1)$$
$$+ ([X, Y], Z) - ([X, Z], Y) - ([Y, Z], X).$$

Since inner product is nondegenerate, knowing the inner products $(\nabla_X Y, Z)$ for all Z, we can calculate $\nabla_X Y$. Therefore, using (5.1), we can express $\nabla_X Y$ in terms of the Riemannian metric. Conversely, a straightforward but fairly long calculation (see the solution of Problem 5.1) shows that this formula indeed determines the vector $\nabla_X Y$, and the operation ∇ defined by this formula is a torsion-free connection compatible with the Riemannian metric.   □

HISTORICAL COMMENT Theorem 5.1 was proved in 1929 by Levi-Cività. Formula (5.1) was obtained in 1950 by Koszul. It is known as Koszul's formula.

Problem 5.1 Prove that formula (5.1) determines a torsion-free connection compatible with the Riemannian metric.

The connection in Theorem 5.1 is called the *Levi-Cività connection*.

Formula (5.1) makes it possible to express the Christoffel symbols of the Levi-Cività connection in terms of the metric. For this purpose, we choose a local chart and identify it with a domain in $\mathbb{R}^n$. As the vector fields X, Y, and Z we take the coordinate vector fields e_i, e_j, and e_k in $\mathbb{R}^n$. Taking into account the equation $\nabla_{e_i} e_j = \Gamma_{ij}^l e_l$, we obtain

$$g_{kl}\Gamma_{ij}^l = (\Gamma_{ij}^l e_l, e_k) = (\nabla_{e_i} e_j, e_k) = \frac{1}{2}\left(\frac{\partial g_{jk}}{\partial x_i} + \frac{\partial g_{ik}}{\partial x_j} - \frac{\partial g_{ij}}{\partial x_k}\right), \qquad (5.2)$$

where $g_{ij} = g(e_i, e_j)$.

5.2 Symmetries of the Riemann Tensor

The curvature tensor of the Levi-Cività connection is often called the *Riemann tensor*. Applying the superscript lowering operation to the Riemann tensor, we can pass from the tensor R^l_{ijk} to the tensor $R_{lijk} = g_{lp} R^p_{ijk}$, for which the symmetry properties have simpler form.

Theorem 5.2 *The tensor R_{lijk} satisfies the following identities:*

(a) $R_{lijk} = -R_{likj}$;
(b) $R_{lijk} + R_{lkij} + R_{ljki} = 0$;
(c) $R_{lijk} = -R_{iljk}$;
(d) $R_{lijk} = R_{jkli}$.

Proof Recall that it follows directly from the definition of the curvature tensor that $R^l_{ijk} = -R^l_{ikj}$. Lowering the superscript, we obtain identity (a). Identity (b) follows from the first Bianchi identity (see Theorem 4.6 on p. 162). Indeed, multiplying the identity $R^p_{ijk} + R^p_{kij} + R^p_{jki} = 0$ by g_{lp} for each p and summing the resulting identities, we obtain (b).

Let us prove identity (c). First, we show that it is equivalent to the identity

$$(R(X, Y)Z, W) + (R(X, Y)W, Z) = 0. \tag{5.3}$$

For $X = X^j \partial_j$, $Y = Y^k \partial_k$, $Z = Z^i \partial_i$, and $W = W^l \partial_l$, we have

$$(R(X, Y)Z, W) = (R(X^j \partial_j, Y^k \partial_k)(Z^i \partial_i), W^l \partial_l)$$
$$= (X^j Y^k Z^i R^p_{ijk} \partial_p, W^l \partial_l)$$
$$= X^j Y^k Z^i W^l g_{lp} R^l_{ijk} = X^j Y^k Z^i W^l R_{lijk},$$
$$(R(X, Y)W, Z) = X^j Y^k W^i Z^l R_{lijk} = X^j Y^k Z^i W^l R_{iljk}.$$

In turn, identity (5.3) is equivalent to

$$(R(X, Y)Z, Z) = 0. \tag{5.4}$$

Indeed, (5.4) implies

$$0 = (R(X, Y)(Z + W), Z + W)$$
$$= (R(X, Y)Z, Z) + (R(X, Y)W, W) + (R(X, Y)Z, W) + (R(X, Y)W, Z)$$
$$= (R(X, Y)Z, W) + (R(X, Y)W, Z).$$

The vector $R(X, Y)Z$ at a point $x \in M^n$ depends only on the vectors $X(x)$, $Y(x)$, and $Z(x)$; therefore, it suffices to prove (5.4) in the case where $[X, Y] = 0$. In this case, we have

$$R(X, Y)Z = \nabla_X \nabla_Y Z - \nabla_Y \nabla_X Z.$$

The connection ∇ is compatible with the metric; therefore, $\partial_X (Z, Z) = 2(\nabla_X Z, Z)$, whence

$$\partial_X \partial_Y (Z, Z) = 2(\nabla_X \nabla_Y Z, Z) + 2(\nabla_Y Z, \nabla_X Z).$$

Similarly,

$$\partial_Y \partial_X (Z, Z) = 2(\nabla_Y \nabla_X Z, Z) + 2(\nabla_X Z, \nabla_Y Z).$$

Since the vector fields X and Y commute, it follows that $\partial_X \partial_Y (Z, Z) = \partial_Y \partial_X (Z, Z)$, and hence

$$(\nabla_X \nabla_Y Z, Z) = (\nabla_Y \nabla_X Z, Z),$$

as required.

Identity (d) follows directly from identities (a), (b), and (c). Indeed, let us write identity (b) in four ways:

$$R_{lijk} + R_{lkij} + R_{ljki} = 0,$$

$$-R_{ijkl} - R_{iljk} - R_{iklj} = 0,$$

$$-R_{jkli} - R_{jikl} - R_{jlik} = 0,$$

$$R_{klij} + R_{kjli} + R_{kijl} = 0.$$

Summing these four identities and using (a) and (c), we obtain

$$2R_{lijk} - 2R_{jkli} = 0, \quad \text{i.e.,} \quad R_{lijk} = R_{jkli},$$

as required.   □

Corollary *In dimension 2 all nonzero components of the tensor R_{lijk} are equal to* $\pm R_{1212}$.

5.3   Geodesics on Riemannian Manifolds

For the Levi-Cività connection on a Riemannian manifold M^n, geodesics have various special properties caused by the presence of the Riemannian metric and the symmetry of the connection. For example, for a geodesic on a Riemannian manifold, the length of the velocity vector is constant. Indeed, we have

$$\frac{d}{dt}(\gamma', \gamma') = 2(\nabla_{\gamma'}\gamma', \gamma') = 0.$$

Theorem 5.3 (Gauss' Lemma)   *For any point* $x \in M^n$ *and any nonzero vector* $V_0 \in T_x M^n$, *the curve* $\exp_x(tV_0)$ *is orthogonal to the hypersurface* $\{\exp_x(V) \mid \|V\| = \text{const}\}$.

Proof Let $V(s)$ be a curve on the sphere $\|V\| = \text{const}$. We set $f(s,t) = \exp_x\left(tV(s)\right)$. It is required to prove that $\left(\frac{\partial f}{\partial t}, \frac{\partial f}{\partial s}\right) = 0$ for $t = 1$. Clearly,

$$\frac{\partial}{\partial t}\left(\frac{\partial f}{\partial t}, \frac{\partial f}{\partial s}\right) = \left(\nabla_{\frac{\partial}{\partial t}}\frac{\partial f}{\partial t}, \frac{\partial f}{\partial s}\right) + \left(\frac{\partial f}{\partial t}, \nabla_{\frac{\partial}{\partial t}}\frac{\partial f}{\partial s}\right).$$

The curve $\exp_x\left(tV(s_0)\right)$ is geodesic, whence $\nabla_{\frac{\partial}{\partial t}}\frac{\partial f}{\partial t} = 0$. Since the coordinate vector fields $\frac{\partial}{\partial t}$ and $\frac{\partial}{\partial s}$ commute, it follows from the symmetry of the connection that

$$\left(\frac{\partial f}{\partial t}, \nabla_{\frac{\partial}{\partial t}}\frac{\partial f}{\partial s}\right) = \left(\frac{\partial f}{\partial t}, \nabla_{\frac{\partial}{\partial s}}\frac{\partial f}{\partial t}\right) = \frac{1}{2}\frac{\partial}{\partial s}\left(\frac{\partial f}{\partial t}, \frac{\partial f}{\partial t}\right) = 0,$$

because $\frac{\partial f}{\partial t}$ is the velocity vector of the geodesic. Thus, $\frac{\partial}{\partial t}\left(\frac{\partial f}{\partial t}, \frac{\partial f}{\partial s}\right) = 0$, and hence $\left(\frac{\partial f}{\partial t}, \frac{\partial f}{\partial s}\right)$ does not depend on t. It remains to check that $\left(\frac{\partial f}{\partial t}, \frac{\partial f}{\partial s}\right) = 0$ for $t = 0$. But $f(s, 0) = \exp_x(0) = x$ is a constant point. Therefore, $\frac{\partial f}{\partial s}(s, 0) = 0$.   □

A connected Riemannian manifold M^n can be endowed with the structure of a metric space as follows. Given a smooth curve $\gamma : [a, b] \to M^n$, we define its *length* $l(\gamma)$ by

$$l(\gamma) = \int_a^b \|\gamma'(t)\| \, dt,$$

where $\|\gamma'(t)\|$ is the length of the velocity vector of this curve (this vector lies in the tangent space, and the length of a tangent vector is determined by the Riemannian metric). The length of a piecewise smooth curve equals the sum of lengths of the smooth curves from which it consists. For points x, $y \in M^n$, the *distance $d(x, y)$* is defined as

$$d(x, y) = \inf_\gamma l(\gamma),$$

where the greatest lower bound is taken over all piecewise smooth curves joining x and y.

Problem 5.2 Check that that $d(x, y)$ has all properties of a metric.

Problem 5.3 Check that the topology induced by the metric $d(x, y)$ coincides with the standard topology of the manifold.

Problem 5.4 Is it true that any two points x and y of a connected Riemannian manifold can be joined by a smooth curve γ for which $d(x, y) = l(\gamma)$?

Theorem 5.4 *Suppose that points x and y are joined by a smooth curve γ for which $d(x, y) = l(\gamma)$. Then γ is a geodesic (to be more precise, γ becomes a geodesic after a change of parameter).*

Proof It suffices to prove that any point z of the curve γ has a neighborhood inside which the arc of γ that contains z is a geodesic. Let D be an open ball in $T_z M^n$ centered at the origin and such that the restriction of $\exp_z$ to D is a diffeomorphism, and let $U = \exp_z D$. □

Lemma 5.1 *Let $\omega \colon [a, b] \to U$ be a smooth curve represented in the form $\omega(t) = \exp_z(r(t)V(t))$, where $r(t)$ is a number and $V(t)$ is a vector of length 1. Then $l(\omega) \geq |r(a) - r(b)|$, and the equality is attained if and only if the function r is monotone and the function V is constant.*

Proof We set $f(r, t) = \exp_z(rV(t))$. Then $\omega(t) = f(r(t), t)$, whence

$$\frac{d\omega(t)}{dt} = \frac{\partial f}{\partial r} r'(t) + \frac{\partial f}{\partial t}.$$

According to Gauss' lemma, the curve $f(r, t_0)$ is orthogonal to $f(r_0, t)$; therefore, the vectors $\frac{\partial f}{\partial r}$ and $\frac{\partial f}{\partial t}$ are orthogonal. The curve $\exp_z(rV(t_0))$ is geodesic and has velocity vector $V(t_0)$ at the point z; hence $\left\| \frac{\partial f}{\partial r} \right\| = \|V(t_0)\| = 1$. Thus,

$$\left\| \frac{d\omega(t)}{dt} \right\|^2 = |r'(t)|^2 + \left\| \frac{\partial f}{\partial t} \right\|^2 \geq |r'(t)|^2,$$

which means that

$$l(\omega) = \int_a^b \left\| \frac{d\omega(t)}{dt} \right\| dt \geq \int_a^b |r'(t)| dt \geq |r(a) - r(b)|,$$

and the equality is attained only if $\frac{\partial f}{\partial t} = 0$, i.e., $V(t) = \text{const}$ and the function r is monotone. $\qquad\qquad\square$

In the case where r is smaller than the radius of the ball D, we say that the hypersurface $\{\exp_z(V)| \ \|V\| = r\}$ is a *geodesic sphere* of radius r centered at z.

Lemma 5.1 can be formulated as follows: the length of any curve in U joining geodesic spheres of radii r_1 and r_2 centered at z is at least $|r_1 - r_2|$. Let γ_1 be a part of the curve γ which is entirely contained in U and joins the point z with some point $z' = \exp_z(rV)$, where $\|V\| = 1$. For each $\varepsilon \in (0, r)$, the curve γ_1 contains a curve joining concentric geodesic spheres of radii ε and r; therefore, $l(\gamma_1) \geq r - \varepsilon$. For $\varepsilon \to 0$, we have $l(\gamma_1) \geq r$. The equality is attained only in the case where γ_1 is a geodesic.

Corollary *Suppose that points x and y are joined by a piecewise smooth curve γ (consisting of finitely many smooth links) for which $d(x, y) = l(\gamma)$. Then the curve γ is geodesic (in particular, it is smooth).*

Proof It is sufficient to check that Lemma 5.1 remains valid for a piecewise smooth curve ω. Such a curve ω consists of smooth arcs joining concentric geodesic spheres of radii $r_1, r_2, \ldots, r_k$. Its length equals $|r_1 - r_k|$ if and only if the sequence $r_1, r_2, \ldots, r_k$ is monotone and the function $V(t)$ is constant. $\qquad\square$

Problem 5.5 Prove that any geodesic on the sphere $\mathbb{S}^n$ is a great circle or its part. (A *great circle* is the intersection of the sphere with a plane through its center.)

5.4 The Hopf–Rinow Theorem

A Riemannian manifold M^n is said to be *geodesically complete* if, for any point $x \in M^n$ and any vector $V \in T_x M^n$, the point $\exp_x(V)$ is defined, or, in other words, if any geodesic can be extended without bound. The Riemannian metric on a geodesically complete Riemannian manifold is said to be *geodesically complete* as well.

Example 5.1 The Riemannian manifold $\mathbb{R}^2 \setminus \{0\}$ is not geodesically complete (the geodesic going to the point 0 cannot be extended beyond this point).

Example 5.2 The Riemannian manifold $\mathbb{R}^1 \times \mathbb{S}^1$ is geodesically complete (any geodesic on a cylinder can be extended without bound).

Theorem 5.5 (Hopf–Rinow [Ho]) *On a connected geodesically complete Riemannian manifold M^n any two points x and y can be joined by a shortest geodesic.*

Proof [Rh] We begin with producing a geodesic which is a likely candidate, and then we prove that this is indeed the required geodesic.

Let $d = d(x, y)$. Consider a geodesic sphere S of sufficiently small radius r centered at x. The set S is compact; hence there exists a point $z \in S$ for which

Fig. 5.1 The choice of the
point z'

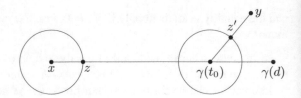

$d(z, y) = \min_{w \in S} d(w, y)$. Let $z = \exp_x(\varepsilon V)$, $\|V\| = 1$. We want to prove that the curve $\gamma(t) = \exp_x(tV)$, $t \in [0, d]$, is the required geodesic joining points x and y (obviously, it has length d).

We will prove that the distance between points $\gamma(t)$ and y equals $d - t$ for each $t \in [r, d]$; then, for $t = d$, we will have $\gamma(d) = y$. First, we prove this for $t = r$. Any curve joining the points x and y intersects the geodesic sphere S in some point w. Therefore,

$$d = d(x, y) = \min_{w \in S} \big(d(x, w) + d(w, y)\big) = r + \min_{w \in S} d(w, y) = r + d(z, y).$$

Thus, $d(z, y) = d - r$, whence $z = \gamma(r)$. Let t_0 be the least upper bound for those $t \in [r, d]$ at which the required equality holds. Then, by continuity, it also holds at t_0. Suppose that $t_0 < d$. Let S' be a geodesic sphere of sufficiently small radius r' centered at $\gamma(t_0)$. Choose a point z' of S' for which $d(z', y) = \min_{w' \in S'} d(w', y)$ (see Fig. 5.1). We have

$$d - t_0 = d\big(\gamma(t_0), y\big) = \min_{w' \in S'} \Big(d\big(\gamma(t_0), w'\big) + d(w', y)\Big) = r' + d(z', y).$$

Therefore,

$$d(z', y) = d - (t_0 + r'). \tag{5.5}$$

We claim that $z' = \gamma(t_0 + r')$. Indeed, on the one hand, by the triangle inequality, we have

$$d(x, z') \geq d(x, y) - d(y, z') = t_0 + r'.$$

On the other hand, if we go first from x to $\gamma(t_0)$ along the geodesic γ and then from $\gamma(t_0)$ to z' along a shortest geodesic, then we will obtain precisely a path of length $t_0 + r'$. This path has minimum length; therefore, according to the corollary of Theorem 5.4, it is geodesic. Thus, $z' = \gamma(t_0 + r')$, and (5.5) implies

$$d\big(\gamma(t_0 + r'), y\big) = d - (t_0 + r'),$$

i.e., the required equality also holds for $t_0 + r' > t_0$. This contradiction proves that $t_0 = d$. $\square$

The manifolds in Examples 5.1 and 5.2 are homeomorphic. Thus, these examples show that the notion of geodesic completeness is not topological: it depends on the choice of a Riemannian metric. But the geodesic completeness of a connected Riemannian manifold is equivalent with the completeness of this Riemannian manifold treated as a metric space. Before proving this, we recall the definition of a complete metric space.

Let X be a metric space with metric d. A sequence of points $x_1, x_2, \ldots$ of X is called a *Cauchy sequence* if, for any $\varepsilon > 0$, we can choose a number N so that $d(x_n, x_m) < \varepsilon$ for all $m, n > N$. A metric space is said to be *complete* if any Cauchy sequence has a limit point in this space.

Theorem 5.6 *Suppose that any closed bounded subset of a metric space X is compact. Then X is a complete metric space.*

Proof Let $\{x_i\}$ be a Cauchy sequence in X. Then, for each $\varepsilon > 0$, we can choose an N so that $d(x_n, x_m) < \varepsilon$ for $n, m > N$. Consider the closed subsets $C_i = \overline{\{x_i, x_{i+1}, \ldots\}}$ of X. The closed set $C = C_1$ is bounded, because $d(x_1, x_i) \leq d(x_1, x_{N+1}) + d(x_{N+1}, x_i) < d(x_1, x_{N+1}) + \varepsilon$ for $i > N$. Therefore, C is compact.

Suppose that the open sets $U_i = C \setminus C_i$ cover C, i.e., $\cap C_i = \varnothing$. Then the cover of the compact set C by these open sets contains a finite subcover, and hence some finite intersection of C_i is empty, which is not true. Therefore, $\cap C_i$ contains a point x. We have $d(x_n, x) \leq \varepsilon$ for $n > N$, because $x \in C_n$. Hence x is a limit point of the Cauchy sequence $\{x_i\}$. □

The Hopf–Rinow theorem readily implies the following completeness criterion for a Riemannian manifold (this criterion was also proved by Hopf and Rinow [Ho]).

Theorem 5.7 *A connected Riemannian manifold M^n is complete as a metric space if and only if it is geodesically complete.*

Proof First, suppose that M^n is geodesically complete. Let us show that any closed bounded set $C \subset M^n$ is compact. Suppose that the distance from a point $x \in C$ to any other point of the set C is at most d. Take a closed ball of radius d in $T_x M^n$ and consider its image B under the map $\exp_x$. The set B is compact. Moreover, according to the Hopf–Rinow theorem, it contains C. Thus, C is a closed subset of the compact space B, and hence it is compact. Theorem 5.6 implies that M^n is complete as a metric space.

Now suppose that M^n is complete as a metric space. We must show that, given any point $x \in M^n$ and any vector $V \in T_x M^n$, the curve $\gamma(t) = \exp_x(tV)$ is defined for all $t > 0$. Suppose that the least upper bound for those t at which $\gamma(t)$ is defined equals $\alpha < \infty$. Choose a sequence $t_k \in (0, \alpha)$ converging to α as $k \to \infty$. Clearly, t_k is a Cauchy sequence. Therefore, so is the sequence $\gamma(t_k)$, because

$$d\big(\gamma(t_k), \gamma(t_m)\big) \leq l(\gamma|_{[t_k, t_m]}) = \|V\| \cdot |t_k - t_m|.$$

Thus, the sequence $\gamma(t_k)$ converges to some point $y \in M^n$. Let U be a bounded neighborhood of y, and let $\bar{U}$ be its closure. Since the metric space M^n is complete, if follows that $\bar{U}$ is compact. In the restriction of the tangent bundle to $\bar{U}$ consider the set of vectors of length $\|V\|$. If the neighborhood U is small enough, then this set is the direct product of two compact subsets $\bar{U}$ and S^{n-1}, and hence it is compact. Thus, the sequence t_k has a subsequence t_{k_i} for which the sequence of vectors $\gamma'(t_{k_i})$ converges to a vector $W \in T_y M^n$. The velocity vector of the curve $\exp_y(\tau W)$ at $\tau = 0$ is W. The curve $\gamma(\tau + \alpha)$ contains a sequence of points such that this sequence itself converges to y and the velocity vectors at the points of the sequence converge to W. Both curves are geodesic, so that it follows from the uniqueness of a geodesic that $\gamma(\tau + \alpha) = \exp_y(\tau W)$ for small negative τ. But $\exp_y(\tau W)$ is also defined for small positive τ; therefore, the geodesic $\gamma(t)$ is defined for $t = \tau + \alpha$, provided that τ is a small positive number. We have arrived at a contradiction. $\square$

Remark Looking carefully at the proofs of the Hopf–Rinow theorems, we see that, to ensure the geodesic completeness of a connected manifold M^n, it is not necessary to require that any geodesic can be extended without bound; it is sufficient that any geodesic through some fixed point can be extended without bound. Indeed, according to Problem 5.6, any closed bounded subset of such a manifold M^n is compact, so that M^n is a complete metric space (by Theorem 5.6), which means that this manifold is geodesically complete.

Problem 5.6 Suppose that any geodesic from some fixed point x in a connected Riemannian manifold M^n can be extended without bound. Prove that
   (a) the point x can be joined to any other point $y \in M^n$ by a minimal geodesic;
   (b) any closed bounded subset of M^n is compact.

HISTORICAL COMMENT Heinz Hopf (1894–1971) and his student Willi Rinow (1907–1979) proved Theorems 5.5 and 5.7 in 1931. The constructive proof of Theorem 5.5 given above was proposed by Georges de Rham (1903–1990) in 1952.

5.5   The Existence of Complete Riemannian Metrics

Two Riemannian metrics on a manifold M^n are said to be *conformally equivalent* if, for each point $x \in M^n$, the inner products in the tangent space $T_x M^n$ corresponding to these Riemannian metrics are proportional (the proportionality coefficient depends only on the point x).

Theorem 5.8 (Nomizu–Ozeki [No]) *For any Riemannian metric g on a connected manifold M^n, there exists a geodesically complete Riemannian metric g' conformally equivalent to g.*

Proof For each point $x \in M^n$, we define $r(x)$ as the least upper bound for those numbers r for which the closure of the set $D_{x,r} = \{y \in M^n \mid d(x, y) < r\}$ is compact. If $r(x_0) = \infty$ for some point x_0, then M^n is compact and hence the metric g is complete. In what follows, we assume that $r(x) < \infty$ for all x.

Let us show that the function $r(x)$ is continuous. To this end, it suffices to check that $|r(x) - r(y)| \leq d(x, y)$. If $z \in D_{x,r}$, then $d(x, z) < r$ and $d(y, z) \leq d(y, x) + d(x, z) < d(y, x) + r$; hence $z \in D_{y, r+d(x,y)}$. Thus, $D_{x,r} \subset D_{y, r+d(x,y)}$. Therefore, if $r + d(x, y) < r(y)$, then the closure of $D_{y, r+d(x,y)}$ is compact, so that the closure of $D_{x,r}$ is compact as well, i.e., $r \leq r(x)$. It follows that $r(x) \geq r(y) - d(x, y)$. The proof of the inequality $r(y) \geq r(x) - d(x, y)$ is similar.

Let $\omega(x)$ be a smooth function such that $\omega(x) > 1/r(x)$ for all $x \in M^n$. Then the Riemannian manifold M^n with metric $g' = (\omega(x))^2 g$ is geodesically complete. To prove this, it suffices to show that $D'_{x, 1/3} = \{y \in M^n | d'(x, y) < 1/3\}$ is contained in $D_{x, r(x)/2}$. Indeed, then the closure of $D'_{x, 1/3}$ is compact, and hence any geodesic can be extended by a distance of $1/3$, which means that any geodesic can be extend without bound.

Suppose that $d(x, y) \geq r(x)/2$. Let $\gamma(t)$, $a \leq t \leq b$, be a piecewise smooth curve joining the points x and y. Its length in the metric g equals $L = \int_a^b \left\| \frac{dx}{dt} \right\| dt$. Clearly, $L \geq d(x, y) \geq r(x)/2$. The length of the same curve in the metric g' equals

$$L' = \int_a^b \omega(\gamma(t)) \left\| \frac{dx}{dt} \right\| dt = \omega(\gamma(t_0)) \int_a^b \left\| \frac{dx}{dt} \right\| dt = \omega(\gamma(t_0)) L$$

for some $t_0 \in [a, b]$. Therefore, $L' = \omega(\gamma(t_0)) L > \frac{1}{r(\gamma(t_0))} L$. Next, $|r(\gamma(t_0)) - r(x)| \leq d(x, x_0) \leq L$, whence $r(\gamma(t_0)) \leq r(x) + L$. It follows that $L' > \frac{L}{r(x) + L}$. But $L \geq r(x)/2$, so that $L' > 1/3$. Thus, $d'(x, y) \geq 1/3$, and hence $D'_{x, 1/3}$ is contained in $D_{x, r(x)/2}$.   □

We say that a Riemannian metric on a manifold M^n is *bounded* if the distance between any two points of M^n in this metric is smaller than some constant.

Theorem 5.9 (Nomizu–Ozeki [No])   *Any Riemannian metric g on a manifold M^n is conformally equivalent to a bounded Riemannian metric g'.*

Proof In view of Theorem 5.8, we can assume that the metric g is geodesically complete. Let us fix a point $x_0 \in M^n$ and consider the function $d(x, x_0)$. This function is continuous. Take a smooth function $\omega(x)$ for which $\omega(x) > d(x, x_0)$. Let us show that the Riemannian metric $g' = e^{-2\omega(x)} g$ is bounded.

The metric g is geodesically complete, and hence, for every point $x \in M^n$, there exists a geodesic γ which joins the points x and x_0 and has length $d(x, x_0) = L$. We can assume that $\gamma(0) = x_0$ and the parameterization s of the curve γ is natural. Any part of γ is a shortest geodesic, so that $d(\gamma(s), x_0) = s$ for all $s \in [0, L]$.

The length of the velocity vector $\frac{d\gamma}{ds}$ in the metric g' equals $e^{-\omega(\gamma(s))}$. Therefore, the length of γ in the metric g' equals $L' = \int_0^L e^{-\omega(\gamma(s))} ds$. Since $\omega(\gamma(s)) > d(\gamma(s), x_0) = s$, it follows that

$$L' < \int_0^L e^{-s} ds < \int_0^\infty e^{-s} ds = 1.$$

Thus, $d'(x, x_0) < 1$ for any point $x \in M^n$.   □

Problem 5.7 Prove that if any Riemannian metric on a manifold M^n is geodesically complete, then M^n is compact.

5.6 Covariant Differentiation of Tensors

Specifying a connection on a manifold is equivalent to specifying a covariant derivative $\nabla_X Y$ for each pair of vector fields X and Y. In considering arbitrary connections, we already mentioned that, using the covariant differentiation of vector fields, we can define the covariant differentiation of covector fields (differential 1-forms). It is defined so that

$$\partial_X\big(\omega(Y)\big) = (\nabla_X\omega)(Y) + \omega\big(\nabla_X Y\big).$$

Thus, the definition of the covariant derivative $\nabla_X\omega$ is as follows:

$$(\nabla_X\omega)(Y) = \partial_X\big(\omega(Y)\big) - \omega\big(\nabla_X Y\big).$$

The covariant derivative of a function is the derivative of this function in the direction of a vector field X: $\nabla_X f = \partial_X f$.

In local coordinates covariant derivatives in the directions of the vector fields $\frac{\partial}{\partial x^i}$ are denoted by ∇_i. The covariant derivative of a vector field with coordinates V^j equals

$$(\nabla_i V)^j = \frac{\partial V^j}{\partial x^i} + V^k\Gamma^j_{ik}.$$

Let us calculate the covariant derivative of a covector field in local coordinates:

$$(\nabla_i\omega)(e_j) = \partial_i(\omega(e_j)) - \omega(\nabla_i e_j)$$

$$= \frac{\partial\omega_j}{\partial x^i} - \omega(\Gamma^k_{ij}e_k) = \frac{\partial\omega_j}{\partial x^i} - \omega_k\Gamma^k_{ij}.$$

Thus, the covariant derivative of a covector field with coordinates ω_j equals

$$(\nabla_i\omega)_j = \frac{\partial\omega_j}{\partial x^i} - \omega_k\Gamma^k_{ij}.$$

In particular, $\nabla_i dx^k = -\Gamma^k_{ij}dx^j$.

The covariant derivative of a tensor field of any type is defined by Leibniz' rule:

$$\nabla_X(T\otimes S) = (\nabla_X T)\otimes S + T\otimes(\nabla_X S),$$

where T and S are any tensor fields. By definition, the covariant derivative is also linear: if T and S are tensor fields of the same type, then

$$\nabla_X(T + S) = \nabla_X T + \nabla_X S.$$

For example, let us calculate the covariant derivative of a tensor field A_k^i of type $(1, 1)$:

$$\nabla_l(A_k^i e_i \otimes dx^k) = \frac{\partial A_k^i}{\partial x^l} e_i \otimes dx^k + A_k^i \nabla_l(e_i) \otimes dx^k + A_k^i e_i \otimes \nabla_l(dx^k)$$

$$= \frac{\partial A_k^i}{\partial x^l} e_i \otimes dx^k + A_k^i \Gamma_{li}^m e_m \otimes dx^k - A_k^i \Gamma_{lm}^k e_i \otimes dx^m$$

$$= \frac{\partial A_k^i}{\partial x^l} e_i \otimes dx^k + A_k^m \Gamma_{lm}^i e_i \otimes dx^k - A_m^i \Gamma_{lk}^m e_i \otimes dx^k.$$

Thus,

$$\nabla_l A_k^i = \frac{\partial A_k^i}{\partial x^l} + A_k^m \Gamma_{lm}^i - A_m^i \Gamma_{lk}^m.$$

A similar calculation shows that the covariant derivatives of tensor fields of types $(2, 0)$ and $(0, 2)$ are

$$\nabla_l A^{ik} = \frac{\partial A^{ik}}{\partial x^l} + A^{mk} \Gamma_{lm}^i + A^{im} \Gamma_{lm}^k,$$

$$\nabla_l A_{ik} = \frac{\partial A_{ik}}{\partial x^l} - A_{mk} \Gamma_{li}^m - A_{im} \Gamma_{lk}^m.$$

The covariant derivative of a tensor field of type (p, q) with components $T_{j_1 j_2 \cdots j_q}^{i_1 i_2 \cdots i_p}$ is a tensor of type $(p, q + 1)$ whose components are calculated by the formula

$$\nabla_l T_{j_1 j_2 \cdots j_q}^{i_1 i_2 \cdots i_p} = \frac{\partial T_{j_1 j_2 \cdots j_q}^{i_1 i_2 \cdots i_p}}{\partial x^l} + \sum_{k=1}^{p} T_{j_1 j_2 \cdots j_q}^{i_1 \cdots k \cdots i_p} \Gamma_{lk}^{i_k} - \sum_{m=1}^{q} T_{j_1 \cdots m \cdots j_q}^{i_1 i_2 \cdots i_p} \Gamma_{lj_m}^m, \tag{5.6}$$

where the Γ_{ij}^k are the Christoffel symbols.

Theorem 5.10 (Ricci's Lemma) *The covariant derivatives of the tensors g_{ij} and g^{ij} vanish, i.e., these tensor fields are parallel.*

Proof First, let us check that $\nabla_l g_{ik} = 0$ for all i, k, l. By definition,

$$\nabla_l g_{ik} = \frac{\partial g_{ik}}{\partial x^l} - g_{mk} \Gamma_{li}^m - g_{im} \Gamma_{lk}^m.$$

According to formula (5.2) on p. 170, we have

$$g_{mk}\Gamma^m_{li} = \frac{1}{2}\left(\frac{\partial g_{ik}}{\partial x_l} + \frac{\partial g_{kl}}{\partial x_i} - \frac{\partial g_{il}}{\partial x_k}\right), \quad g_{im}\Gamma^m_{lk} = \frac{1}{2}\left(\frac{\partial g_{ik}}{\partial x_l} + \frac{\partial g_{il}}{\partial x_k} - \frac{\partial g_{kl}}{\partial x_i}\right).$$

Therefore,

$$g_{mk}\Gamma^m_{li} + g_{im}\Gamma^m_{lk} = \frac{\partial g_{ik}}{\partial x_l},$$

as required.

Now the relation $\nabla_l g^{ik} = 0$ follows from

$$0 = \nabla_l(g_{ik}g^{ik}) = (\nabla_l g_{ik})g^{ik} + g_{ik}\nabla_l g^{ik} = g_{ik}\nabla_l g^{ik}.$$

$\square$

Theorem 5.11 (second Bianchi Identity)   *The covariant derivatives of the Riemann curvature tensor satisfy the identity*

$$\nabla_l R^h_{ijk} + \nabla_j R^h_{ikl} + \nabla_k R^h_{ilj} = 0.$$

Proof The proof of the second Bianchi identity is similar to that of the first Bianchi identity for any symmetric connection (see Theorem 4.6 on p. 162). Choose local coordinates in a neighborhood of a given point x so that all Christoffel symbols at this point vanish. Then the covariant derivative at this point is merely the partial derivative. For the curvature tensor we have obtained the expression (see p. 162)

$$R^l_{ijk} = \frac{\partial \Gamma^l_{ki}}{\partial x^j} - \frac{\partial \Gamma^l_{ji}}{\partial x^k} + \Gamma^p_{ki}\Gamma^l_{jp} - \Gamma^p_{ji}\Gamma^l_{pk}.$$

The calculation of the covariant derivative at the point x yields

$$\nabla_l R^h_{ijk} = \frac{\partial^2 \Gamma^h_{ki}}{\partial x^l \partial x^j} - \frac{\partial^2 \Gamma^h_{ji}}{\partial x^l \partial x^k}.$$

Indeed, calculating the derivative of $\Gamma^p_{ki}\Gamma^l_{jp}$, we obtain the sum of two terms each of which contains, as a factor, a Christoffel symbol, which is zero.

A similar calculation shows that

$$\nabla_j R^h_{ikl} = \frac{\partial^2 \Gamma^h_{li}}{\partial x^j \partial x^k} - \frac{\partial^2 \Gamma^h_{ki}}{\partial x^j \partial x^l}, \quad \nabla_k R^h_{ilj} = \frac{\partial^2 \Gamma^h_{ji}}{\partial x^k \partial x^l} - \frac{\partial^2 \Gamma^h_{li}}{\partial x^k \partial x^j}.$$

Summing the three expressions for covariant derivatives, we obtain the required identity.

$\square$

In coordinate-free form, the second Bianchi identity is written as

$$(\nabla_W R)(X, Y)Z + (\nabla_X R)(Y, W)Z + (\nabla_Y R)(W, X)Z = 0.$$

Indeed, if $X = X^j e_j$, $Y = Y^k e_k$, $Z = Z^i e_i$, and $W = W^l e_l$, then

$$(\nabla_W R)(X, Y)Z = X^j Y^k Z^i W^l (\nabla_l R_{ijk}^h)e_h.$$

Remark A different formulation of the second Bianchi identity is given in Theorem 4.7 on p. 166. We will not use or prove the equivalence of these formulations for the Levi-Cività connection.

5.7 Sectional Curvature

The Riemann tensor is a complicated object with four indices. But it can be described in terms of a simpler object, sectional curvature. The sectional curvature assigns a number defined by using the Riemann tensor to each two-dimensional plane in the tangent space. The Riemann tensor is completely determined by the sectional curvature. Sectional curvature is closely related to Gaussian curvature. In particular, for a surface, the tangent space is two-dimensional, and the sectional curvature assigned to this two-dimensional space is equal to the Gaussian curvature. Recall that, for a surface, the nonzero components of the Riemann tensor equal, up to sign, the Gaussian curvature.

We set $R(X, Y, Z, W) = (R(X, Y)Z, W)$. Then $R(X^j e_j, Y^k e_k, Z^i e_i, W^l e_l) = (R(X^j e_j, Y^k e_k)Z^i e_i), W^l e_l) = (R_{ijk}^p X^j Y^k Z^i e_p, W^l e_l) = g_{pl} R_{ijk}^p X^j Y^k Z^i W^l = R_{lijk} X^j Y^k Z^i W^l$. Consider the plane Π generated by vectors X and Y in the space $T_x M^n$. Let us check that the number

$$K_\Pi = \frac{R(X, Y, Y, X)}{(X, X)(Y, Y) - (X, Y)^2}.$$

depends only on the plane Π.

Let $P = \begin{pmatrix} X^1 & X^2 \\ Y^1 & Y^2 \end{pmatrix}$ be the matrix of the coordinates of the vectors X and Y in some orthonormal basis of the plane Π. Then $k(X, Y) = (X, X)(Y, Y) - (X, Y)^2$ is the determinant of the matrix $P P^T$. Therefore, $k(X', Y') = (\det A)^2 k(X, Y)$, where A is the transition matrix from the basis X, Y to the basis X', Y'.

Consider the restriction of the polylinear function $R(X, Y, Z, W)$ to the plane Π. From the symmetry properties of the curvature tensor it follows, in particular, that this function is skew-symmetric with respect to the first two and the last two arguments. Any bilinear skew-symmetric function $B(X, Y) = b_{ij} X^i Y^j$ in the plane has the form $c(X^1 Y^2 - X^2 Y^1)$, where c is a constant. Indeed, since $B(X, X) = 0$, we have $b_{11} = b_{22} = 0$ and $b_{12} + b_{21} = 0$. Let us first fix Z and W. Then we obtain $R(X, Y, Z, W) = B(Z, W)(X^1Y^2 - X^2Y^1)$, where $B(Z, W)$ is a skew-symmetric bilinear function, i.e., $B(Z, W) = c(Z^1W^2 - Z^2W^1)$. Thus, $R(X, Y, Y, X) = -c(X^1Y^2 - X^2Y^1)^2$. Therefore, $R(X', Y', Y', X') = (\det A)^2 R(X, Y, Y, X)$, where A is the transition matrix from the basis X, Y to the basis X', Y'. As a result, we see that, for any pair of linearly independent vectors X, Y in the plane Π, the ratio of $R(X, Y, Y, X)$ to $(X, X)(Y, Y) - (X, Y)^2$ is the same, as required.

The number K_Π is called the *sectional curvature* corresponding to the plane Π. If the basis X, Y is orthonormal, then $K_\Pi = R(X, Y, Y, X)$. In particular, for $n = 2$, the sectional curvature at a point p is the Gaussian curvature at p. Indeed, if $X = e_1$ and $Y = e_2$, then $j = l = 1$ and $k = i = 2$, whence $R(X, Y, Y, X) = R_{1212} = K$. In the general case, the sectional curvature at a point p is the Gaussian curvature at p of the image of the plane Π under the map $\exp_p$ (see Theorem 5.18).

Theorem 5.12 *The sectional curvature completely determines the curvature tensor.*

Proof First, we show that $R(X, Y)Z$ can be represented in terms of expressions of the form $R(X, Y)Y$. To this end, we sum the three equations

$$R(X, Y + Z)(Y + Z) = R(X, Y)Y + R(X, Y)Z$$
$$+ R(X, Z)Y + R(X, Z)Z,$$
$$-R(Y, X + Z)(X + Z) = -R(Y, X)X + R(X, Y)Z$$
$$+ R(Z, Y)X - R(Y, Z)Z,$$
$$0 = R(X, Y)Z + R(Y, X)Z.$$

The sum of the next-to-last terms on the right-hand sides vanishes by virtue of the first Bianchi identity: $R(X, Z)Y + R(Z, Y)X + R(X, Y)Z = 0$. As a result, we obtain

$$3R(X, Y)Z = R(X, Y + Z)(Y + Z) - R(Y, X + Z)(X + Z)$$
$$- R(X, Y)Y - R(X, Z)Z + R(Y, X)X + R(Y, Z)Z.$$

Now let us show that $(R(X, Y)Y, Z)$ can be represented in terms of expressions of the form $(R(X, Y)Y, X)$. It follows from the symmetry properties of the Riemann tensor that

$$(R(X, Y)Y, Z) = (R(Y, Z)X, Y) = (R(Z, Y)Y, X).$$

Therefore, for fixed Y, the bilinear form $B(X, Z) = (R(X, Y)Y, Z)$ is symmetric. Thus, we can use the formula

$$2B(X, Z) = B(X + Z, X + Z) - B(X, X) - B(Z, Z),$$

which leads to the required representation. For example, the doubled first term in the expression for $3R(X, Y)Z$ is represented as

$$2(R(X, Y + Z)(Y + Z), W) = (R(X + W, Y + Z)(Y + Z), X + W)$$
$$- (R(X, Y + Z)(Y + Z), X) - (R(W, Y + Z)(Y + Z), W).$$

$\square$

Problem 5.8 Show that the second derivative with respect to α and β of

$$R(X + \alpha Z, Y + \beta W, X + \alpha Z, Y + \beta W) - R(X + \alpha W, Y + \beta Z, X + \alpha W, Y + \beta Z)$$

at the point $\alpha = 0$, $\beta = 0$ equals $6R(X, Y, Z, W)$. Prove Theorem 5.12 by using this fact.

A Riemannian manifold M^n is called a *space of constant curvature* if its sectional curvature is constant, i.e., depends neither on a point $x \in M^n$ nor on a plane in $T_x M^n$. The following theorem shows that, for $n \geq 3$, the former condition is redundant: the independence of the sectional curvature of a point follows from its independence of a plane at each point.

Theorem 5.13 (F. Schur) *A connected Riemannian manifold of dimension $n \geq 3$ whose sectional curvature is constant (does not depend on the choice of a plane Π) at each point is a space of constant curvature.*

Proof Suppose that the sectional curvature at a point p equals K for any plane Π. First, we prove that, in this case, we have $R(X, Y, Z, W) = K \cdot R_1(X, Y, Z, W)$ at this point, where $R_1(X, Y, Z, W) = (Y, Z)(X, W) - (X, Z)(Y, W)$. We set $k(X, Y) = R(X, Y, Y, X)$ and $k_1(X, Y) = R_1(X, Y, Y, X) = (X, X)(Y, Y) - (X, Y)^2$. By assumption, $k(X, Y) = K \cdot k_1(X, Y)$ at the point p. In the proof of Theorem 5.12 we represented $R(X, Y, Z, W)$ as the sum of terms of the form $R(X, Y, Y, X) = k(X, Y)$:

$$6R(X, Y, Z, W) = 2R(X, Y + Z, Y + Z, W) + \ldots$$
$$= R(X + W, Y + Z, Y + Z, X + W) + \ldots$$
$$= k(X + W, Y + Z) + \ldots$$

In view of this representation, the equation $k(X, Y) = K \cdot k_1(X, Y)$ implies $R(X, Y, Z, W) = K \cdot R_1(X, Y, Z, W)$.

According to Ricci's lemma, the tensor field g_{ij} is parallel; therefore, the tensor field R_1 with components $g_{ik}g_{jl} - g_{ij}g_{kl}$ is parallel as well: $\nabla_U R_1 = 0$ for any U. Thus,

$$(\nabla_U R)(X, Y, Z, W) = \partial_U K \cdot R_1(X, Y, Z, W).$$

This means that, for any X, Y, and Z, we have

$$(\nabla_U R)(X, Y)Z = \partial_U K \cdot ((Y, Z)X - (X, Z)Y).$$

We write similar equations for $(\nabla_X R)(Y, U)Z$ and $(\nabla_Y R)(U, X)Z$ and sum the three equations, using the second Bianchi identity

$$(\nabla_W R)(X, Y)Z + (\nabla_X R)(Y, W)Z + (\nabla_Y R)(W, X)Z = 0$$

(see p. 183). As a result, we obtain

$$0 = \partial_U K \cdot ((Y, Z)X - (X, Z)Y) + \partial_X K \cdot ((U, Z)Y - (Y, Z)U)$$
$$+ \partial_Y K \cdot ((X, Z)U - (U, Z)X).$$

Let X be an arbitrary vector, and let vectors Y, Z, and U be chosen so that the vectors X, Y, and Z are pairwise orthogonal (this can be done because $n \geq 3$), $U = Z$, and $\|Z\| = 1$. Then $(Y, Z) = (X, Z) = 0$ and $(U, Z) = 1$, which implies

$$\partial_X K \cdot Y - \partial_Y K \cdot X = 0.$$

Since the vectors X and Y are linearly independent, it follows that $\partial_X K = 0$ for any vector X. Therefore, K is a constant.   □

HISTORICAL COMMENT The theorem that the constancy of the sectional curvature in all directions implies its constancy at all points was proved by Friedrich Schur (1856–1932) in 1886.

5.8   Ricci Tensor

The Riemann tensor is too complicated. It is easier to study it by considering some of its components, such as the sectional curvature. The Ricci tensor is, as well as the sectional curvature, a component of the Riemann tensor. As opposed to the sectional curvature, it does not completely determine the Riemann tensor. Nevertheless, the Ricci tensor reflects some important properties of Riemannian manifolds.

The Ricci tensor assigns the trace of the map $Z \mapsto R(Z, Y)X$ to each pair of vector fields X and Y:

$$\mathrm{Ric}(X, Y) = \mathrm{tr}(Z \mapsto R(Z, Y)X).$$

In local coordinates, we obtain the map

$$Z^j e_j \mapsto R(Z^j e_j, Y^k e_k)X^i e_i = (R^p_{ijk} Y^k X^i)Z^j e_p.$$

The trace of the map $Z^j e_j \mapsto A^p_j Z^j e_p$ equals A^j_j (the summation is with respect to j); therefore, $\text{Ric}(X, Y) = R^j_{ijk} X^i Y^k = R^k_{ikj} X^i Y^j$. Thus, in local coordinates $\text{Ric} = R_{ij} dx^i \otimes dx^j$, where $R_{ij} = R^k_{ikj}$.

If the basis $e_1, \ldots, e_n$ is orthonormal, then $R^l_{ijk} = R_{lijk}$; hence $R_{ij} = \sum_{k=1}^n R_{kikj}$ in an orthonormal basis. The symmetry properties of the Riemann tensor imply $R_{ij} = R_{ji}$, i.e., the Ricci tensor is symmetric. Therefore, the Ricci tensor is completely determined by the values $R(X, X)$, where X ranges over all unit vectors.

Problem 5.9 Let $e_1, \ldots, e_n$ be an orthonormal basis. Prove that $\text{Ric}(e_n, e_n) = \sum_{j=1}^{n-1} K(e_j, e_n)$, where $K(e_j, e_n)$ is the sectional curvature corresponding to the plane spanned by the vectors e_j and e_n.

The *scalar curvature* S is the trace of the Ricci tensor with respect to the Riemannian metric: $S = g^{ij} R_{ij}$.

Problem 5.10 Let $e_1, \ldots, e_n$ be an orthonormal basis. Prove that $S = \sum_{j \neq k,\ j,k=1}^n K(e_j, e_k)$, where $K(e_j, e_k)$ is the sectional curvature corresponding to the plane spanned by the vectors e_j and e_k.

HISTORICAL COMMENT Ricci-Curbastro introduced the Ricci tensor in 1903.

5.9 Riemannian Submanifolds

Let M be a submanifold of a Riemannian manifold $\tilde{M}$. It is equipped with the induced Riemannian metric: the inner product of tangent vectors to M equals the inner product of these vectors considered as tangent vectors to $\tilde{M}$. As a result, we obtain two Levi-Cività connections: ∇ on the *Riemannian submanifold M* and $\tilde{\nabla}$ on $\tilde{M}$.

The tangent space $T_p\tilde{M}$ to $\tilde{M}$ at a point $p \in M$ can be represented as the direct sum of the tangent space T_pM and the normal space N_pM. Therefore, for each tangent vector X in the space $T_p\tilde{M}$, we can consider its projections X^T on the tangent space T_pM and $X^\perp$ on the normal space N_pM.

Let X and Y be vector fields on M. We can extend them to vector fields on $\tilde{M}$, take the covariant derivative $\tilde{\nabla}_X Y$, and represent it in the form $(\tilde{\nabla}_X Y)^T + (\tilde{\nabla}_X Y)^\perp$. We will see shortly that the tangential component $(\tilde{\nabla}_X Y)^T$ is $\nabla_X Y$. The normal component $(\tilde{\nabla}_X Y)^\perp$ is, by definition, the *second fundamental form* $\mathbf{II}(X, Y)$. With the second fundamental form we can associate the *second quadratic form* $\mathbf{II}(X, X)$. For hypersurfaces, we identify the normal space with numbers, and therefore the second quadratic form takes numerical values. In the general case, the second quadratic form takes values in a linear space of dimension equal to the difference of the dimensions of the manifolds $\tilde{M}$ and M.

The symmetry of the Levi-Cività connection implies that of the second fundamental form:

$$\mathbf{II}(X, Y) - \mathbf{II}(Y, X) = (\tilde{\nabla}_X Y)^{\perp} - (\tilde{\nabla}_Y X)^{\perp} = [X, Y]^{\perp} = 0,$$

because the commutator of vector fields on a manifold M is a vector field on M.

Theorem 5.14 (Gauss' Formula)   *If vector fields X and Y on M are extended to $\tilde{M}$, then*

$$\tilde{\nabla}_X Y = \nabla_X Y + \mathbf{II}(X, Y).$$

Proof We must prove that the tangential component $(\tilde{\nabla}_X Y)^T$ equals $\nabla_X Y$. Consider the map $\nabla^T : T_p M \times T_p M \to T_p M$ defined by $\nabla^T_X Y = (\tilde{\nabla}_X Y)^T$. Let us show that $\nabla^T = \nabla$. In view of the uniqueness of the Levi-Cività connection, it suffices to check that ∇^T is a symmetric connection compatible with the metric.

First, we check that ∇^T is a connection:

$$(\tilde{\nabla}_{fX} Y)^T = f(\tilde{\nabla}_X Y)^T = f \nabla^T_X Y,$$
$$(\tilde{\nabla}_X fY)^T = (\partial_X f)Y^T + f(\tilde{\nabla}_X Y)^T$$
$$= (\partial_X f)Y^T + f \nabla^T_X Y.$$

Then we check the symmetry of ∇^T:

$$\nabla^T_X Y - \nabla^T_Y X = (\tilde{\nabla}_X Y - \tilde{\nabla}_Y X)^T$$
$$= [X, Y]^T = [X, Y].$$

Finally, we check that the connection ∇^T is compatible with the metric:

$$\partial_X (Y, Z) = (\tilde{\nabla}_X Y, Z) + (Y, \tilde{\nabla}_X Z)$$
$$= ((\tilde{\nabla}_X Y)^T, Z) + (Y, (\tilde{\nabla}_X Z)^T)$$
$$= (\nabla_X Y, Z) + (Y, \nabla_X Z).$$

$\square$

For a surface in $\mathbb{R}^3$, the second quadratic form makes it possible to calculate the covariant derivative of a normal vector field (by the Weingarten formula); a similar formula also exists for hypersurfaces. In the general case, the following theorem holds.

Theorem 5.15 (Weingarten Formula) *Let X and Y be tangent vector fields on a Riemannian submanifold $M \subset \tilde{M}$, and let N be a normal vector field. Suppose that they are extended to vector fields on $\tilde{M}$. Then*

$$(\tilde{\nabla}_X N, Y) = -(N, \mathbf{II}(X, Y)).$$

Proof Clearly, $(N, Y) = 0$. Therefore,

$$0 = \partial_X(N, Y) = (\tilde{\nabla}_X N, Y) + (N, \tilde{\nabla}_X Y)$$

$$= (\tilde{\nabla}_X N, Y) + (N, \nabla_X Y + \mathbf{II}(X, Y)) = (\tilde{\nabla}_X N, Y) + (N, \mathbf{II}(X, Y)),$$

because the vectors N and $\nabla_X Y$ are perpendicular. $\qquad\square$

The second fundamental form also makes it possible to calculate the difference between the curvature tensors on the manifolds M and $\tilde{M}$. We denote these tensors by $R(X, Y, Z, W)$ and $\tilde{R}(X, Y, Z, W)$. Here X, Y, Z, and W are vectors in $T_p M$.

Theorem 5.16 (Gauss Equation) *The curvature tensors on a manifold and a submanifold are related by*

$$\tilde{R}(X, Y, Z, W) = R(X, Y, Z, W) - (\mathbf{II}(X, W), \mathbf{II}(Y, Z)) + (\mathbf{II}(X, Z), \mathbf{II}(Y, W)).$$

Proof Let us extend vectors X, Y, Z, and W first to vector fields in a neighborhood of a point p in the manifold M and then to a neighborhood of this point in the manifold $\tilde{M}$. According to the Gauss formula, we have

$$\tilde{R}(X, Y, Z, W) = (\tilde{\nabla}_X \tilde{\nabla}_Y Z - \tilde{\nabla}_Y \tilde{\nabla}_X Z - \tilde{\nabla}_{[X,Y]} Z, W)$$

$$= (\tilde{\nabla}_X(\nabla_Y Z + \mathbf{II}(Y, Z)) - \tilde{\nabla}_Y(\nabla_X Z + \mathbf{II}(X, Z)), W)$$

$$- (\nabla_{[X,Y]} Z + \mathbf{II}([X, Y], Z), W).$$

The second fundamental form takes values in the normal space, and W is a tangent vector; therefore, $(\mathbf{II}([X, Y], Z), W) = 0$. To the two other terms containing the second fundamental form the Weingarten formula applies:

$$\tilde{R}(X, Y, Z, W) = (\tilde{\nabla}_X \nabla_Y Z, W) - (\mathbf{II}(Y, Z), \mathbf{II}(X, W))$$

$$- (\tilde{\nabla}_Y \nabla_X Z, W) + (\mathbf{II}(X, Z), \mathbf{II}(Y, W))$$

$$- (\nabla_{[X,Y]} Z, W).$$

Now we decompose $\tilde{\nabla}$ into a normal and a tangential component. All inner products of the normal component and tangent vectors vanish, so that only the tangential component remains:

$$\tilde{R}(X, Y, Z, W) = (\nabla_X \nabla_Y Z, W) - (\nabla_Y \nabla_X Z, W) - (\nabla_{[X,Y]} Z, W)$$
$$- (\mathbf{II}(Y, Z), \mathbf{II}(X, W)) + (\mathbf{II}(X, Z), \mathbf{II}(Y, W))$$
$$= R(X, Y, Z, W) - (\mathbf{II}(X, W), \mathbf{II}(Y, Z)) + (\mathbf{II}(X, Z), \mathbf{II}(Y, W)).$$

□

HISTORICAL COMMENT In 1827 Gauss derived an equation expressing the difference between the Riemann curvature tensor of a manifold and that of a submanifold in terms of the second quadratic form in the case of a surface in three-dimensional space. In this case, $\tilde{R}(X, Y, Z, W) = 0$ and the Gauss equation expresses $R(X, Y, Z, W)$ in terms of the second quadratic form.

Theorem 5.17

(a) *Let V be a vector field given along a curve γ on a Riemannian submanifold $M \subset \tilde{M}$. Then $\tilde{\nabla}_{\gamma'} V = \nabla_{\gamma'} V + \mathbf{II}(\gamma', V)$.*

(b) *Let γ be a geodesic on a Riemannian submanifold $M \subset \tilde{M}$. Then $\tilde{\nabla}_{\gamma'} \gamma' = \mathbf{II}(\gamma', \gamma')$.*

Proof

(a) Consider a moving frame e_i ($i = 1, \ldots, \dim M$) along the curve γ on the manifold M. Let us expand the vector $V(t)$ at a point $\gamma(t)$ in this frame: $V(t) = V^i(t) e_i$. We have

$$\tilde{\nabla}_{\gamma'} V = \tilde{\nabla}_{\gamma'} V^i e_i = (V')^i e_i + V^i \tilde{\nabla}_{\gamma'} e_i$$
$$= (V')^i e_i + V^i \nabla_{\gamma'} e_i + V^i \mathbf{II}(\gamma', e_i) = \nabla_{\gamma'} V + \mathbf{II}(\gamma', V).$$

(b) Let us apply the equation in (a) to a vector field V equal to the velocity of the curve γ:

$$\tilde{\nabla}_{\gamma'} \gamma' = \nabla_{\gamma'} \gamma' + \mathbf{II}(\gamma', \gamma').$$

It remains to note that $\nabla_{\gamma'} \gamma' = 0$ for a geodesic.

□

Now we are ready to revisit sectional curvature and prove its property which was stated but not proved. Consider a two-dimensional plane Π in the tangent space at a point $p \in M^n$. Take a neighborhood of zero in this tangent space, small enough for the map $\exp_p$ to be a diffeomorphism in this neighborhood. Let S_Π be the two-dimensional surface being the image under $\exp_p$ of the part of Π contained in the chosen neighborhood.

Theorem 5.18 *The sectional curvature K_Π equals the Gaussian curvature K of the surface S_Π at the point p.*

Proof First, we prove that the second fundamental form of the surface S_Π regarded as a submanifold of M^n vanishes at p. Given any vector $V \in T_p S_\Pi$, consider a geodesic $\gamma = \gamma_V$ on M^n for which $\gamma(0) = p$ and $\gamma'(0) = V$. According to Theorem 5.17, we have

$$0 = \tilde{\nabla}_{\gamma'} \gamma' = \nabla_{\gamma'} \gamma' + \mathbf{II}(\gamma', \gamma').$$

The vector $\nabla_{\gamma'} \gamma'$ lies in the tangent space, and $\mathbf{II}(\gamma', \gamma')$ lies in the normal space; therefore, each of them vanishes. Thus, $\mathbf{II}(V, V) = 0$ for any vector $V \in T_p S_\Pi$, and since the second fundamental form is symmetric, it follows that it vanishes at the point p.

The Gauss equation shows that at the point p, for tangent vectors to the surface S_Π, the Riemann tensor for the surface S_Π coincides with that for the manifold M^n. Therefore, the sectional curvature corresponding to the plane Π equals the sectional curvature of the surface S_Π, i.e., the Gaussian curvature of S_Π at the point p.   □

5.10   Totally Geodesic Submanifolds

A Riemannian submanifold $M \subset \tilde{M}$ is said to be *totally geodesic* at a point $p \in M$ if a sufficiently small part of any geodesic line on M passing through the point p is also geodesic on the manifold $\tilde{M}$. An example of a submanifold totally geodesic at a point p was considered in the preceding section: this is the two-dimensional surface S_Π being the image of a neighborhood of zero in a plane Π under the map $\exp_p$.

A Riemannian submanifold $M \subset \tilde{M}$ is said to be *totally geodesic* if it is totally geodesic at each point. This means that any geodesic line on M is also geodesic on $\tilde{M}$. Examples of geodesic submanifolds are geodesic lines on a Riemannian manifold and the intersection of the sphere with a subspace through its center.

In the proof of Theorem 5.18 we showed that, for the surface S_Π, the second fundamental form at a point p vanishes. This property is characteristic of totally geodesics submanifolds.

Theorem 5.19 *A Riemannian submanifold $M \subset \tilde{M}$ is totally geodesic if and only if its second fundamental form $\mathbf{II}(X, Y)$ is zero.*

Proof Let γ be a geodesic on a Riemannian submanifold $M \subset \tilde{M}$. Then, according to Theorem 5.17, we have $\tilde{\nabla}_{\gamma'} \gamma' = \mathbf{II}(\gamma', \gamma')$. Therefore, any geodesic on the manifold M is also a geodesic on the manifold $\tilde{M}$ if and only if the second quadratic form $\mathbf{II}(X, X)$ is zero. The second fundamental form $\mathbf{II}(X, Y)$ is symmetric, and hence it is zero if and only if so is the second quadratic form.   □

Theorem 5.20 *Let γ be a curve on a totally geodesic submanifold $M \subset \tilde{M}$. Then*

(a) *the parallel transports of a tangent vector to M along γ inside M and inside $\tilde{M}$ coincide;*
(b) *the transport along γ of a vector orthogonal to M is orthogonal to M.*

Proof

(a) Let V be a vector field on the manifold M parallel along the curve γ, i.e., such that $\tilde{\nabla}_{\gamma'} V = 0$. We must prove that V is also parallel on the manifold $\tilde{M}$. According to Theorem 5.17, we have $\tilde{\nabla}_{\gamma'} V = \nabla_{\gamma'} V + \mathbf{II}(\gamma', V) = 0$, because the second fundamental form is zero for a totally geodesic manifold.
(b) The parallel transport along γ of a vector V orthogonal to M and a vector W tangent to M preserves the angle between these vectors, and it takes W to a tangent vector to M. Therefore, the transported vector V is orthogonal to all tangent vectors to M.

$\square$

Problem 5.11 Suppose that the fixed point set of an isometry σ of a Riemannian manifold $\tilde{M}$ is a connected submanifold of M. Prove that the submanifold $M \subset \tilde{M}$ is totally geodesic.

5.11   Jacobi Fields and Conjugate Points

The first and second variation formulas and their proofs in the many-dimensional case are the same as in the two-dimensional case (see Sects. 3.18 and 3.19). The definition of Jacobi vector fields and conjugate points is precisely the same as well (see Sect. 3.20). However, in the many-dimensional case, it makes sense to consider the multiplicity of conjugate points. Recall that we have proved that, in the two-dimensional case, their multiplicity is always 1. The same argument shows that, in the n-dimensional case, the multiplicity of conjugate points is at most $n-1$. Namely, the dimension of the space of Jacobi fields vanishing at $t = 0$ equals n; moreover, there exists a Jacobi field which vanishes at $t = 0$ and does not vanish at $t = 1$ (an example of such a field is $J(t) = t V(t)$, where $V(t) = \gamma'(t)$ is the velocity vector of a geodesic).

Let us show that the multiplicity of a conjugate point on an n-dimensional Riemannian manifold may equal $n-1$. Consider diametrically opposite points p and q on the sphere $\mathbb{S}^n$. Jacobi fields can be obtained by using variations in the class of geodesics with loose endpoints. Moreover, in the case where the endpoints are fixed, Jacobi fields vanishing at these endpoints are obtained. The one-parameter families of motions of the sphere $\mathbb{S}^n$ which leave the points p and q fixed determine Jacobi fields vanishing at p and q. Such motions of the sphere $\mathbb{S}^n$ correspond to motions of the equatorial sphere $\mathbb{S}^{n-1}$. The sphere $\mathbb{S}^{n-1}$ can be rotated in $n - 1$ independent directions; therefore, we obtain $n - 1$ independent Jacobi fields vanishing at the points p and q.

The proof that a geodesic containing two conjugate points is not a shortest curve with given endpoints remains the same.

The proof of the theorem about a relationship between conjugate points and critical points of the map exp needs no substantial changes: it suffices to take a basis consisting of n vectors $X_1, \ldots, X_n$ instead of a basis consisting of two vectors X_1, X_2.

In the study of Jacobi fields, of most interest are those orthogonal to geodesics. The point is that, given any geodesic γ, the tangent vector field $(a + bt)\gamma'(t)$, where a and b are constants, is a Jacobi field. Such Jacobi fields are of little use, and it is better to exclude them from consideration.

Theorem 5.21 *If a Jacobi field J is orthogonal to a geodesic γ at two points, then it is orthogonal to γ at all points.*

Proof Since $\nabla_{\gamma'}\gamma' = 0$, it follows that $\frac{d}{dt}(J, \gamma') = (\nabla_{\gamma'}J, \gamma') + (J, \nabla_{\gamma'}\gamma') = (\nabla_{\gamma'}J, \gamma')$ and $\frac{d^2}{dt^2}(J, \gamma') = (\nabla_{\gamma'}^2 J, \gamma')$. Using the Jacobi equation $\nabla_{\gamma'}^2 J - R(\gamma', J)\gamma' = 0$, we obtain $\frac{d^2}{dt^2}(J, \gamma') = (R(\gamma', J)\gamma', \gamma') = R(\gamma', J, \gamma', \gamma') = 0$ by virtue of the symmetry properties of the curvature tensor. Thus, $(J(t), \gamma'(t))$ is a linear function of t. If it vanishes at two points, then it also vanishes at all other points.   □

For calculating Jacobi fields normal coordinates are convenient. Normal coordinates on a Riemannian manifold M^n are defined in precisely the same way as for a surface. They are introduced in a sufficiently small neighborhood of a point $p \in M^n$. Normal coordinates correspond to the image of a Cartesian coordinate system in the tangent space at p under the map $\exp_p$. Let $e_1, \ldots, e_n$ be an orthonormal basis in $T_p M^n$, and let $q = \exp_p(x^1 e_1 + \cdots + x^n e_n)$. Then $(x^1, \ldots, x^n)$ are the *normal coordinates* of the point q.

The differential of the map $\exp_p$ at the origin is the identity map; therefore, in normal coordinates the first quadratic form at a point p is $(dx^1)^2 + \cdots + (dx^n)^2$, i.e., $g_{ij} = \delta_{ij}$, and the geodesics through p are given by the $x^1 = tc^1, \ldots, x^n = tc^n$.

Exactly the same argument as in the case of a surface proves that all Christoffel symbols at a point p vanish in normal coordinates (see the solution of Problem 3.28).

Theorem 5.22 *Let $x^1, \ldots, x^n$ be normal coordinates in a neighborhood of a point p, and let $\gamma(t)$ be a geodesic through p $(\gamma(0) = p)$. Then, for any vector $W = W^i \partial_i \in T_p M^n$, a Jacobi field J along γ for which $\nabla_{\gamma'}J(0) = W$ is given by $J(t) = t W^i \partial_i$ in normal coordinates.*

Proof Consider the vector field on $\gamma(t)$ defined by $J(t) = t W^i \partial_i$ in normal coordinates. At the origin all Christoffel symbols are zero; therefore, $\nabla_{\gamma'}J(0) = (t W^i)'\partial_i = W^i \partial_i = W$. Thus, the vector field J satisfies the initial condition, and hence it suffices to check that J is a Jacobi field. Let us construct a variation with variation vector field J in the class of geodesics. The geodesic $\gamma(t)$ is given by $\gamma(t) = (t V^1, \ldots, t V^n)$, where $V = \gamma'(0)$. Consider the variation

$$\gamma(s, t) = (t V^1 + st W^1, \ldots, t V^n + st W^n).$$

This is a variation in the class geodesics, so that $\frac{\partial}{\partial s}\gamma(0, t)$ is a Jacobi field. Clearly, $\frac{\partial}{\partial s}\gamma(0, t) = (t W^1, \ldots, t W^n) = J(t)$. $\qquad\square$

Using Theorem 5.22, we can obtain the following expression for the metric in terms of the tensor curvature in normal coordinates.

Theorem 5.23 *Let $x^1, \ldots, x^n$ be the normal coordinates of a point x. Then, in normal coordinates,*

$$g_{ij}(x) = \delta_{ij} + \frac{1}{3} R_{jkli} x^k x^l + o(\|x\|^2).$$

Proof Let p be the origin. To each point x we assign the vector X with the same coordinates and consider the geodesic $\gamma(t) = \exp_p(tX)$. Consider also the Jacobi fields J_i ($i = 1, \ldots, n$) along this geodesic for which $J_i(0) = 0$ and $(\nabla_{\gamma'} J_i)(0) = \partial_i$. These Jacobi fields have the form $J_i(t) = t\partial_i$. We are interested in the inner product (∂_i, ∂_j) at the point $\gamma(1)$. We begin with calculating the derivative of the inner product $(J_i(t), J_j(t))$ with respect to t.

Clearly, $\frac{df(\gamma(t))}{dt} = \partial_{\gamma'} f$, whence

$$\frac{d}{dt}(J_i, J_j) = \partial_{\gamma'}(J_i, J_j) = (\nabla_{\gamma'} J_i, J_j) + (J_i, \nabla_{\gamma'} J_j).$$

For brevity, we will denote the derivative with respect to t and the covariant derivative $\nabla_{\gamma'}$ by a prime. The calculation of derivatives can easily be continued:

$$(J_i, J_j)'' = (J_i'', J_j) + 2(J_i', J_j') + (J_i, J_j'');$$
$$(J_i, J_j)''' = (J_i''', J_j) + 3(J_i'', J_j') + 3(J_i', J_j'') + (J_i, J_j''');$$
$$(J_i, J_j)'''' = (J_i'''', J_j) + 4(J_i''', J_j') + 6(J_i'', J_j'') + 4(J_i', J_j''') + (J_i, J_j'''').$$

We are interested in the values of these derivatives at $t = 0$. For $t = 0$, the Jacobi equation has the form $(\nabla_X)^2 J_i = R(X, J_i)X$. By assumption, $J_i(0) = 0$, whence $J_i''(0) = 0$. Thus, for $t = 0$, all terms containing J or J'' vanish. Therefore, the first and third derivatives are zero:

$$(J_i, J_j)''(0) = 2(J_i', J_j')(0) = 2(\partial_i, \partial_j),$$
$$(J_i, J_j)''''(0) = 4(J_i'''(0), \partial_j) + 4(\partial_i, J_j'''(0)).$$

To calculate $J'''(0)$, we use the Jacobi equation $J'' = R(X, J)X$. As a result, we obtain

$$J'''' = R'(X, J)X + R(X', J)X + R(X, J')X + R(X, J)X'.$$

Taking into account the relations $J(0) = 0$, $X' = 0$, and $J'_i(0) = \partial_i$, we see that $J'''_i(0) = R(X, J'_i(0))X = R(X, \partial_i)X$, and hence

$$(J'''_i(0), \partial_j) = (R(X, \partial_i)X, \partial_j) = (\partial_i, J'''_j(0)) = R_{jkli}x^k x^l.$$

It is required to calculate (∂_i, ∂_j) at the point $\gamma(1)$. At this point $\partial_i = J_i$; thus, we must calculate $(J_i, J_j)(1)$. We have

$$(J_i, J_j)(t) = (\partial_i, \partial_j)(0)t^2 + \frac{1}{3}R_{jkli}x^k x^l t^4 + \cdots = \delta_{ij}t^2 + \frac{1}{3}R_{jkli}x^k x^l t^4 + \dots.$$

For $t = 1$, we obtain the required relation. ☐

Theorem 5.22 gives an expression for a Jacobi field on any Riemannian manifold. In the case of a space of constant curvature, we can obtain another explicit expression for a Jacobi field orthogonal to a geodesic.

Theorem 5.24 *Let M^n be a Riemannian manifold of constant sectional curvature c, and let γ be a geodesic with natural parameterization. Then a Jacobi field along γ orthogonal to γ and vanishing at $t = 0$ has the form $J(t) = u(t)n(t)$, where n is a field of vectors orthogonal to γ which is parallel along γ and the function $u(t)$ has one of the following forms (depending on the sign of c): t (for $c = 0$), $R\sin(t/R)$ (for $c = \frac{1}{R^2} > 0$), and $R\sinh(t/R)$ (for $c = -\frac{1}{R^2} < 0$).*

Proof Recall that the Jacobi equation for a Jacobi field J along a geodesic γ has the form $(\nabla_V)^2 J - R(V, J)V = 0$, where $V = \gamma'$ is the velocity vector of the geodesic. In the proof of Schur's theorem (Theorem 5.13) we showed that, for a space of constant sectional curvature c, the curvature tensor has the form $R(X, Y)Z = c((Y, Z)X - (X, Z)Y)$. Substituting this expression into the Jacobi equation, we obtain

$$0 = (\nabla_V)^2 J - c((J, V)V - (V, V)J)$$

$$= (\nabla_V)^2 J + cJ,$$

because the vectors J and V are orthogonal and $\|V\| = 1$.

Let us find out what fields of the form $J(t) = u(t)n(t)$, where $u(t)$ is a function and $n(t)$ is a field parallel along γ and orthogonal to γ, are solutions of the obtained equation. Clearly, $(\nabla_V)^2(u(t)n(t)) = u''(t)n(t)$, because the vector field $n(t)$ is parallel along γ, i.e., $\nabla_V n = 0$. Thus, we arrive at the equation $u''(t) + cu(t) = 0$ with initial condition $u(0) = 0$. Up to proportionality the solutions of this equation coincide with those in the statement of the theorem.

The dimension of the space of vector fields parallel along γ and orthogonal to γ equals $n - 1$. The dimension of the space of Jacobi fields orthogonal to γ and vanishing at $t = 0$ is at most $n - 1$. Therefore, in fact, the particular solutions found above exhaust all solutions.   □

Employing both expressions for a Jacobi field, we can describe the structure of the metric of a space of constant curvature in a sufficiently small neighborhood and, thereby, prove that a space of constant curvature is locally isomorphic to the sphere, Euclidean space, or hyperbolic space.

Theorem 5.25 *Let M^n be a Riemannian manifold of constant sectional curvature c, and let $x^1, \ldots, x^n$ be normal coordinates in a neighborhood of a point p. Suppose that a point q of M^n lies on a sphere $(x^1)^2 + \cdots + (x^n)^2 = r^2$ and a tangent vector V at q is represented in the form $V = V^\perp + V^T$, where $V^\perp$ is a vector perpendicular to the sphere (with respect to the normal coordinates) and V^T is a vector tangent to it. Then the metric g of the Riemannian manifold is expressed in terms of the Euclidean norm $\| \cdot \|$ in normal coordinates as follows:*

$$g(V, V) = \begin{cases} \|V^\perp\|^2 + \|V^T\|^2 = \|V\|^2 & \text{if } c = 0, \\ \|V^\perp\|^2 + \frac{R^2}{r^2}\left(\sin^2 \frac{r}{R}\right)\|V^T\|^2 & \text{if } c = \frac{1}{R^2} > 0, \\ \|V^\perp\|^2 + \frac{R^2}{r^2}\left(\sinh^2 \frac{r}{R}\right)\|V^T\|^2 & \text{if } c = -\frac{1}{R^2} < 0. \end{cases}$$

Proof Let $\|V\|_g^2 = g(V, V)$. According to Gauss' lemma, the vectors $V^\perp$ and V^T are orthogonal with respect to the metric g, so that $\|V\|_g^2 = \|V^\perp\|_g^2 + \|V^T\|_g^2$. The length of the velocity vector of a geodesic from the point p is the same in both metrics; therefore, $\|V^\perp\|_g^2 = \|V^\perp\|^2$. It remains to calculate $\|V^T\|_g^2$.

Let γ be a radial geodesic with natural parameterization joining the points $p = \gamma(0)$ and $q = \gamma(r)$, and let $W = V^T$. According to Theorem 5.22, a Jacobi field J along γ vanishing at p and equal to W at q is given by $J(t) = \frac{t}{r}W^i \partial_i$. Since the Jacobi field J is orthogonal to the geodesic γ at the points p and q, it follows that it is also orthogonal to it at all other points (see Theorem 5.21).

Now we apply Theorem 5.24 and represent a Jacobi field J on a Riemannian manifold of constant sectional curvature c which is orthogonal to γ and vanishes at $t = 0$ in the form $J(t) = u(t)n(t)$, where n is a field of vectors orthogonal to γ which is parallel along γ and $u(t)$ is a function of one of the three forms specified in the statement of the theorem. Then $\nabla_{\gamma'} J(0) = u'(0)n(0) = n(0)$, because $u'(0) = 1$ in all of the three cases. Since the field $n(t)$ is parallel, it follows that the length of the vector $n(t)$ is constant and equals the length of $n(0)$. Therefore,

$$\|W\|_g^2 = \|J(r)\|_g^2 = \|u(r)n(0)\|_g^2 = |u(r)|^2 \cdot \|\nabla_{\gamma'} J(0)\|_g^2.$$

Next, we have $\nabla_{\gamma'} J(0) = \frac{1}{r} W^i \partial_i \in T_p M^n$. At the point p both metrics coincide, so that

$$\|\nabla_{\gamma'} J(0)\|_g = \frac{1}{r} \|W^i \partial_i\|_g = \frac{1}{r} \|W\|.$$

Thus, $\|V^T\|_g^2 = \left|\frac{u(r)}{r}\right|^2 \|V^T\|^2$, where $\frac{u(r)}{r} = 1$ for $c = 0$, $\frac{u(r)}{r} = \frac{R}{r} \sin \frac{r}{R}$ for $c > 0$, and $\frac{u(r)}{r} = \frac{R}{r} \sinh \frac{r}{R}$ for $c < 0$. $\square$

Corollary *Riemannian manifolds M and M' of dimension n which have the same constant sectional curvature are locally isometric, i.e., any points $p \in M$ and $p' \in M'$ have isometric neighborhoods.*

Proof Given points p and p', consider geodesic balls of sufficiently small radius ε. Using normal coordinates in these balls, we can map one of the balls onto the other, and in normal coordinates the Riemannian metrics on these balls are the same. $\square$

A Riemannian manifold locally isometric to Euclidean space is said to be *flat*. A manifold is flat if and only if one of the following conditions holds: (1) all sectional curvatures are zero; (2) the curvature tensor is zero. Indeed, the curvature tensor of a flat manifold is the same as that of Euclidean space, i.e., zero. If the curvature tensor is zero, then so are all sectional curvatures, and therefore the manifold is locally isometric to Euclidean space.

5.12   Product of Riemannian Manifolds

Consider a manifold $M = M_1 \times M_2$ which is the direct product of manifolds M_1 and M_2. Let X_1 and X_2 be vector fields on M_1 and on M_2. They can be assigned the vector field $X_1 + X_2$ on M defined by

$$(X_1 + X_2)_{(p,q)} = (X_1)_p \oplus (X_2)_q.$$

To each vector $X_{(p,q)} \in T_{(p,q)} M$ we can assign vectors $(X_1)_p \in T_p M_1$ and $(X_2)_q \in T_q M_2$ so that $X_{(p,q)} = (X_1 + X_2)_{(p,q)}$. Therefore, given any vector field X on M, at each point (p, q) we can write the expression $X_{(p,q)} = (X_1 + X_2)_{(p,q)}$, where X_1 and X_2 are some vectors at the points p and q. But a vector field X can be represented in the form $X = X_1 + X_2$ only in the case where $(X_1)_p \in T_p M_1$ in this expression does not depend on q and $(X_2)_q \in T_q M_2$ does not depend on p.

The product of Riemannian manifolds M_1 and M_2 is the manifold $M = M_1 \times M_2$ with Riemannian metric equal to the sum of the Riemannian metrics of the manifolds M_1 and M_2. In other words, for $X_i, Y_i \in T M_i$,

$$g_M(X_1 + X_2, Y_1 + Y_2) = g_{M_1}(X_1, Y_1) + g_{M_2}(X_2, Y_2).$$

It suffices to define inner product in each tangent space. For arbitrary vector fields X and Y on M, $g_M(X, Y)$ is calculated at each point separately.

Clearly, the tangent spaces to the manifolds M_1 and M_2 are orthogonal in the tangent space to the manifold M.

In the rest of this section, we assume that $X_i, Y_i, Z_i, W_i \in TM_i$.

Let us show that, for any connections ∇, ∇^1, and ∇^2 on the Riemannian manifolds M, M_1, and M_2, we have

$$\nabla_{X_1+X_2}(Y_1 + Y_2) = \nabla^1_{X_1}Y_1 + \nabla^2_{X_2}Y_2. \tag{5.7}$$

First, we check that $\nabla_{X_1}Y_2 = 0$ and $\nabla_{X_2}Y_1 = 0$. We prove only the first relation (the proof of the second is similar). We apply Koszul's formula

$$2(\nabla_X Y, Z) = \partial_X(Y, Z) + \partial_Y(X, Z) - \partial_Z(X, Y)$$
$$+ ([X, Y], Z) - ([X, Z], Y) - ([Y, Z], X).$$

Note beforehand that $[X_1, Y_2] = 0$. Here X_1 and Y_2 are regarded as vector fields on M; the coordinates of the vector X_1 corresponding to M_2 are zero, and its coordinates corresponding to M_1 do not depend on those coordinates of the point which correspond to M_2. We check that $[X_1, Y_2] = 0$ coordinatewise. Namely, we assume that $X_1 = fe_i$ and $Y_2 = ge_j$, where f does not depend on the jth coordinate and g does not depend on the ith one. Then

$$[fe_i, ge_j] = fg[e_i, e_j] + f(\partial_i g)e_j - g(\partial_j f)e_i = 0,$$

because each of the three terms is zero.

It suffices to prove that (5.7) holds at each point. Thus, we can assume that a vector field Z has the form $Z = Z_1 + Z_2$ (we can represent any vector Z in this form at a given point and then extend this vector to a vector field constant in each of the two components). Let us write Koszul's formula for $X = X_1$, $Y = Y_2$, and $Z = Z_1 + Z_2$. Clearly, we have $(X_1, Y_2) = 0$ and $[X_1, Y_2] = 0$. Therefore,

$$2(\nabla_{X_1}Y_2, Z) = \partial_{X_1}(Y_2, Z) + \partial_{Y_2}(X_1, Z) - ([X_1, Z], Y_2) - ([Y_2, Z], X_1)$$
$$= \partial_{X_1}(Y_2, Z_2) + \partial_{Y_2}(X_1, Z_1) - ([X_1, Z_1], Y_2) - ([Y_2, Z_2], X_1).$$

All terms in this expression are zero. For example, $\partial_{X_1}(Y_2, Z_2) = 0$ because the coordinates of the vectors Y_2 and Z_2 do not depend on the coordinates of the point in the first component, and $([X_1, Z_1], Y_2) = 0$ because the vectors $[X_1, Z_1]$ and Y_2 are orthogonal.

Let us check that $\nabla_{X_1}Y_1 = \nabla^1_{X_1}Y_1$ (the proof of the equation $\nabla_{X_2}Y_2 = \nabla^2_{X_2}Y_2$ is similar). Again applying Koszul's formula, we see that

$$(\nabla^1_{X_1}Y_1, Z_1 + Z_2) = (\nabla^1_{X_1}Y_1, Z_1) = (\nabla_{X_1}Y_1, Z_1) = (\nabla_{X_1}Y_1, Z_1 + Z_2).$$

Theorem 5.26 $R(X_1 + X_2, Y_1 + Y_2, Z_1 + Z_2, W_1 + W_2) = R(X_1, Y_1, Z_1, W_1) + R(X_2, Y_2, Z_2, W_2)$.

Proof We use formula (5.7), the definition

$$R(X, Y)Z = \nabla_X \nabla_Y Z - \nabla_Y \nabla_X Z - \nabla_{[X,Y]} Z$$

of the curvature tensor, and the fact that $[X_1, Y_2] = 0$. As a result, we obtain

$$R(X_1 + X_2, Y_1 + Y_2)(Z_1 + Z_2) = R(X_1, Y_1)Z_1 + R(X_2, Y_2)Z_2.$$

Therefore,

$$R(X_1 + X_2, Y_1 + Y_2, Z_1 + Z_2, W_1 + W_2)$$
$$= (R(X_1 + X_2, Y_1 + Y_2)(Z_1 + Z_2), W_1 + W_2)$$
$$= R(X_1, Y_1, Z_1, W_1) + R(X_2, Y_2, Z_2, W_2).$$

$\square$

In particular, the sectional curvature of the plane spanned by the vectors X_1 and Y_2 is zero.

Problem 5.12 Prove that M_1 and M_2 are totally geodesic submanifolds in $M_1 \times M_2$.

Problem 5.13 Prove that a curve on a Riemannian product of manifolds is geodesic if and only if its projection on each factor is geodesic.

5.13   Holonomy

The parallel transport of a vector along a piecewise smooth curve $\gamma(t)$, $t \in [a, b]$, is defined in a natural way. Let $\gamma(t)$ be smooth everywhere except at points $t_1 < t_2 < \cdots < t_n$. Then we first transport the vector from the point $\gamma(a)$ to $\gamma(t_1)$ along a smooth curve, then from $\gamma(t_1)$ to $\gamma(t_2)$, and so on.

The parallel transport along a piecewise smooth *loop*, i.e., a curve which begins and ends at the same point p, defines a self-isometry of the space $T_p M^n$. Given two loops, we can compose a single loop which goes first along one of the loops and then along the other. (It is for this purpose that we consider piecewise smooth rather than smooth loops: a loop composed of two smooth loops is only piecewise smooth.) Thus, the isometries under consideration form a group Hol_p, which is a subgroup of $O(n)$; the inverse element corresponds to the loop traversed in the reverse direction. This group is called the *holonomy group*. Considering only contractible loops, we obtain the so-called *restricted holonomy group* Hol_p^0.

The holonomy groups Hol_p and Hol_q of a connected manifold at different points p and q are isomorphic. The isomorphism can be constructed as follows. Choose any smooth path α with initial point p and terminal point q. To a loop γ with initial and terminal point p we assign the loop $\omega = \alpha^{-1}\gamma\alpha$ with initial and terminal point q (first, is goes along α^{-1}, then along γ, and finally along α). This map is a group homomorphism, because the parallel transport along the paths $\alpha^{-1}\gamma_1\gamma_2\alpha$ and $\alpha^{-1}\gamma_1\alpha\alpha^{-1}\gamma_2\alpha$ yields the same result. Indeed, the transport along $\alpha\alpha^{-1}$ is the identity transformation, because as the path α is traversed in the reverse direction, the vector repeats its motion backwards. The homomorphism $\mathrm{Hol}_q \to \mathrm{Hol}_p$ defined by $\omega \mapsto \alpha\omega\alpha^{-1}$ is inverse to the homomorphism $\mathrm{Hol}_p \to \mathrm{Hol}_q$. The restricted holonomy groups Hol_p^0 and Hol_q^0 are isomorphic as well; the proof is exactly the same. In what follows, we usually do not specify the point at which the holonomy group is calculated.

A connected manifold is orientable if and only if its holonomy group is contained in $\mathrm{SO}(n)$. Indeed, given any path, we can define the parallel transport of the tangent space orientation along this path. A manifold is orientable if and only if the orientation is preserved by the parallel transport along any loop. On the other hand, the holonomy group is contained in $\mathrm{SO}(n)$ if and only if the orientation is preserved by the transport along any loop.

Problem 5.14 Prove that $\mathrm{Hol}(\mathbb{S}^2) = \mathrm{SO}(2)$.

Problem 5.15

(a) Prove that $\mathrm{Hol}(\mathbb{S}^3) = \mathrm{SO}(3)$.
(b) Prove that $\mathrm{Hol}(\mathbb{S}^n) = \mathrm{SO}(n)$.

Problem 5.16

(a) Prove that the holonomy group of a cone with a flat metric is the cyclic group whose generator is the rotation throught the angle at the vertex of the cone net. (We consider a cone without vertex in order that the surface be smooth.)
(b) Prove that the holonomy group of the Möbius band with flat metric consists of the identity map and the map $V \mapsto -V$.

Problem 5.17 Prove that the holonomy group of a product of Riemannian manifolds is the product of holonomy groups.

5.14   Commutator and Curvature

Curvature can be interpreted as infinitesimal holonomy. We begin with a similar interpretation in the simpler case of a commutator of vector fields. Our exposition follows [Fa1].

Consider vector fields X and Y. We can locally define an integral curve (trajectory) of the vector field X, that is, a curve γ for which the velocity vector at a point $\gamma(t)$ equals $X(\gamma(t))$. Using trajectories, we can define the map X_t, which equals the displacement along a trajectory during time t (this map is not defined if the trajectory cannot be extended to t).

Fig. 5.2 The map σ

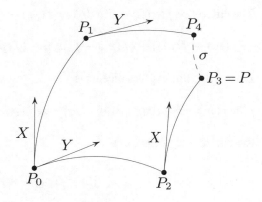

For a function $f(\gamma(t))$ on a trajectory γ, the Taylor series has the form

$$f(X_t p) - f(p) = \sum_{k=1}^{n} \frac{t^k}{k!}(\partial_X)^k f(p) + o(t^n).$$

Indeed, we have $\frac{df}{dt} = X^i \frac{\partial f}{\partial x^i} = \partial_X f$, whence $\frac{d^k f}{dt^k} = (\partial_X)^k f$.

Given a sufficiently small t, we set $\sigma(t) = Y_t X_t Y_{-t} X_{-t} p$. The map σ is schematically shown in Fig. 5.2. As t changes from 0 to some value t_0, the point $\sigma(t)$ traces a curve joining the points $p = p_3$ and p_4.

Theorem 5.27 *For any smooth function* f,

$$f(\sigma(t)) - f(\sigma(0)) = t^2 \partial_{[X,Y]} f(p) + o(t^2),$$

i.e.,

$$\lim_{t \to 0} \frac{f(\sigma(\sqrt{t})) - f(\sigma(0))}{t} = \partial_{[X,Y]} f(p).$$

Proof Let us write the Taylor series for all sides of the parallelogram:

$$f(p_4) - f(p_1) = t\partial_Y f(p_1) + \frac{t^2}{2}(\partial_Y)^2 f(p_1) + o(t^2),$$

$$f(p_1) - f(p_0) = t\partial_X f(p_0) + \frac{t^2}{2}(\partial_X)^2 f(p_0) + o(t^2),$$

$$f(p_3) - f(p_2) = t\partial_X f(p_2) + \frac{t^2}{2}(\partial_X)^2 f(p_2) + o(t^2),$$

$$f(p_2) - f(p_0) = t\partial_Y f(p_0) + \frac{t^2}{2}(\partial_Y)^2 f(p_0) + o(t^2).$$

The difference $f(\sigma(t)) - f(\sigma(0)) = f(p_4) - f(p_3)$ can be represented in the form

$$(f(p_4) - f(p_1)) + (f(p_1) - f(p_0)) - (f(p_3) - f(p_2)) - (f(p_2) - f(p_0)).$$

Up to $o(t^2)$ this expression equals

$$t(\partial_Y f(p_1) - \partial_Y f(p_0)) - t(\partial_X f(p_2) - \partial_X f(p_0)) = t^2(\partial_X \partial_Y - \partial_Y \partial_X)f(p_0) + o(t^2);$$

the differences of the form

$$\frac{t^2}{2}((\partial_Y)^2 f(p_1) - (\partial_Y)^2 f(p_0))$$

are of order $o(t^2)$, and they can be ignored.   □

Now we proceed to the curvature. In calculating the curvature it is convenient to consider commuting vector fields, and therefore we assume that the vector fields X and Y commute. In this case, we have $\sigma(t) = Y_t X_t Y_{-t} X_{-t} p = p$, i.e., the path along the sides of the parallelogram is closed (see Fig. 5.3).

We need the following version of Taylor's formula for the parallel transport of a vector along a curve. Let X be a vector field in a neighborhood of a curve γ. We set $V = \gamma'(0)$ and denote the transport of a vector along the curve γ to a point $\gamma(t)$ by τ_t. Then

$$\tau_0 X(\gamma(t)) - X(\gamma(0)) = \sum_{k=1}^{n} \frac{t^k}{k!} \nabla_V^k X + o(t^n).$$

Fig. 5.3 The closed path

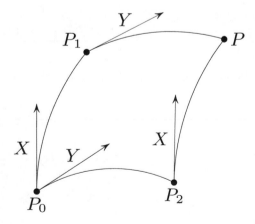

Indeed, let us apply Taylor's usual formula to the function $f(t) = \tau_0 X(\gamma(t))$. To calculate

$$f'(t) = \lim_{h \to 0} \frac{\tau_0 X(\gamma(t+h)) - \tau_0 X(\gamma(t))}{h},$$

we first transport the vector $X(\gamma(t+h))$ to the point $\gamma(t)$ and then transport both vectors $\tau_t X(\gamma(t+h))$ and $X(\gamma(t))$ to the point $\gamma(0)$. As a result, we obtain

$$f'(t) = \tau_0 \lim_{h \to 0} \frac{\tau_t X(\gamma(t+h)) - X(\gamma(t))}{h} = \tau_0 \nabla_{\gamma'(t)} X.$$

Therefore, $f^k(t) = \tau_0(\nabla^k_{\gamma'(t)} X)$ and $f^k(0) = \nabla^k_V X$.

Theorem 5.28 *Let X and Y be commuting vector fields. Then the displacement during time t along trajectories of the fields $-X$, $-Y$, X, and Y (i.e., under the traversal $p \to p_2 \to p_0 \to p_1 \to p$) of a tangent vector Z at the point p is*

$$\Delta Z = t^2 R(Y, X)Z + o(t^2),$$

and therefore

$$\lim_{t \to 0} \frac{\Delta Z}{t^2} = R(Y, X)Z.$$

Proof Let us denote the transport of a vector to the point p_i along the corresponding arc by τ_i and the vector of a vector field Z at the point p_i by Z_i. As in the proof of Theorem 5.27, consider four differences:

$$\tau_1 Z - Z_1 = t \nabla_Y Z_1 + \frac{t^2}{2} \nabla_Y^2 Z_1 + o(t^2), \tag{5.8}$$

$$\tau_0 Z_1 - Z_0 = t \nabla_X Z_0 + \frac{t^2}{2} \nabla_X^2 Z_0 + o(t^2), \tag{5.9}$$

$$\tau_2 Z - Z_2 = t \nabla_X Z_2 + \frac{t^2}{2} \nabla_X^2 Z_2 + o(t^2), \tag{5.10}$$

$$\tau_0 Z_2 - Z_0 = t \nabla_Y Z_0 + \frac{t^2}{2} \nabla_Y^2 Z_0 + o(t^2). \tag{5.11}$$

Let us apply the parallel transport τ_0 to both sides of Eqs. (5.8) and (5.10):

$$\tau_0 \tau_1 Z - \tau_0 Z_1 = t \tau_0(\nabla_Y Z_1) + \frac{t^2}{2} \tau_0(\nabla_Y^2 Z_1) + o(t^2), \tag{5.12}$$

$$\tau_0 \tau_2 Z - \tau_0 Z_2 = t \tau_0(\nabla_X Z_2) + \frac{t^2}{2} \tau_0(\nabla_X^2 Z_2) + o(t^2). \tag{5.13}$$

Summing (5.12) and (5.9) and subtracting the sum of (5.13) and (5.11), we obtain, after cancellations, $\tau_0\tau_1 Z - \tau_0\tau_2 Z$ on the left-hand side. Next,

$$t\tau_0\nabla_Y Z_1 - t\nabla_Y Z_0 = t^2\nabla_X\nabla_Y Z_0 + o(t^2), \tag{5.14}$$

$$t\tau_0\nabla_X Z_2 - t\nabla_X Z_0 = t^2\nabla_Y\nabla_X Z_0 + o(t^2). \tag{5.15}$$

The differences $\frac{t^2}{2}(\tau_0\nabla_Y^2 Z_1 - \nabla_Y^2 Z_0)$ and $\frac{t^2}{2}(\tau_0\nabla_X^2 Z_2 - \nabla_X^2 Z_0)$ are of order $o(t^2)$, and we ignore them. As a result, we arrive at

$$\tau_0\tau_1 Z - \tau_0\tau_2 Z = t^2[\nabla_X, \nabla_Y]Z_0 + o(t^2).$$

The transports $\tau_0\tau_1 Z$ and $\tau_0\tau_2 Z$ correspond to the traversals $p \to p_1 \to p_0$ and $p \to p_2 \to p_0$, and we therefore denote them by $\tau(p \to p_1 \to p_0)Z$ and $\tau(p \to p_2 \to p_0)Z$. After we apply the transport $\tau(p \to p_1 \to p_0)^{-1} = \tau(p_0 \to p_1 \to p)$ to both sides of the obtained equation, the left-hand side takes the form

$$Z - \tau(p \to p_2 \to p_0 \to p_1 \to p)Z = -\Delta Z,$$

and the right-hand side becomes, up to $o(t^2)$,

$$\tau(p_0 \to p_1 \to p)t^2[\nabla_X, \nabla_Y]Z_0 = t^2[\nabla_X, \nabla_Y]Z = t^2 R(X, Y)Z$$
$$= -t^2 R(Y, X)Z.$$

□

Remark Not only do Theorems 5.27 and 5.28 have similar proofs, but also Theorem 5.28 can be derived from Theorem 5.27; see [Sa1].

Theorem 5.29 *A Riemannian manifold is flat if and only if its restricted holonomy group is trivial.*

Proof First, suppose that the restricted holonomy group of a Riemannian manifold is trivial. Then by Theorem 5.28 we have $\Delta Z = 0$ for all transports along parallelograms; therefore, the curvature tensor is zero. Thus, all sectional curvatures vanish, and hence the manifold is locally isometric to Euclidean space.

Now suppose that a manifold is locally isometric to Euclidean space. A contraction of a loop can be represented as a sequence of its transformations in neighborhoods isometric to domains of Euclidean space, and hence parallel transport along a contractible loop is the identity transformation. □

5.15   Solutions of Problems

5.1 Recall (see p. 95) that, for any function f, we have

$$[fW, V] = f[W, V] - (\partial_V f)W, \quad [W, fV] = f[W, V] + (\partial_W f)V.$$

We need these relations to find out the behavior of the right-hand side of (5.1) under the replacement of X, Y, and Z by fX, fY, and fZ (we are not interested in the terms which are obtained by multiplication by f). Namely, to begin with, we must check that under the replacement of Z by fZ there arise no terms containing $\partial_X f$, $\partial_Y f$, or $\partial_Z f$, i.e., the vector $\nabla_X Y$ defined by (5.1) does not depend on Z for fixed X and Y. Then, we must verify that ∇ is a connection, i.e., first, the assignment of the vector field $\nabla_X Y$ to vector fields X and Y is linear in X and Y and, second, $\nabla_{fX} Y = f\nabla_X Y$ and $\nabla_X(fY) = f\nabla_X Y + (\partial_X f)Y$ for any function f.

Under the replacement of Z by fZ the terms containing $\partial_X f$ arise in the first and fifth summands. These summands are $(Y, (\partial_X f)Z)$ and $((-\partial_X f)Z, Y)$. Their sum vanishes. The calculation for $\partial_Y f$ is similar. Terms containing $\partial_Z f$ arise in no summand.

The linearity of the operation ∇ in X and Y is obvious. Under the replacement of X by fX, in addition to terms obtained by multiplication by f, there appear the terms

$$(\partial_Y f)(X, Z) - (\partial_Z f)(X, Y) - (\partial_Y f)(Z, X) + (\partial_Z f)(X, Y) = 0.$$

This agrees with the formula $\nabla_{fX} Y = f\nabla_X Y$.

Under the replacement of Y by fY, there appear the terms

$$(\partial_X f)(Y, Z) - (\partial_Z f)(X, Y) + (\partial_X f)(Y, Z) + (\partial_Z f)(Y, X) = 2(\partial_X f)(Y, Z).$$

This agrees with the formula $\nabla_X(fY) = f\nabla_X Y + (\partial_X f)Y$.

The relations

$$(\nabla_X Y, Z) + (Y, \nabla_X Z) = \partial_X (Y, Z)$$

and

$$(\nabla_X Y, Z) - (\nabla_Y X, Z) = ([X, Y], Z)$$

are easy to verify directly from the definition. Therefore, the connection ∇ is compatible with the metric and has zero torsion.

5.2 Obviously, $d(x, y)$ thus defined is nonnegative and symmetric. The triangle inequality holds because, given a curve joining points p and q and a curve joining points q and r, we can compose a curve joining p and r. It only remains to check that $d(x, y) > 0$ if $x \neq y$. To this end, we choose a chart $U \to \mathbb{R}^n$ at the point x.

Let (V, W) be the inner product of vectors V and W in $\mathbb{R}^n$, and let $g(y)(V, W) = g(y)_{ij} V^i W^j$ be the inner product of these vectors in the tangent space at the point $y \in U$ with respect to the given Riemannian metric. Consider a closed ball D of radius r centered at x which is contained in U. On the set of all pairs (y, V) with $y \in D$ and $\|V\|^2 = (V, V) = 1$ the function $g(y)(V, V)$ attains a positive minimum C^2. For any vector $V \neq 0$ and any number $\lambda \neq 0$, the ratios $g(V, V)/\|V\|^2$ and $g(\lambda V, \lambda V)/\|\lambda V\|^2$ are equal; therefore, given any nonzero vector in the tangent space at any point of the ball D, the ratio of its length in the Riemannian metric to its Euclidean length is at least C. Thus, if the point y lies inside this ball, then $d(x, y) \geq C\|x - y\|$, and if it lies outside, then $d(x, y) \geq Cr$.

5.3 It suffices to prove the statement for $\mathbb{R}^n$. Let d_1 be the Euclidean metric of $\mathbb{R}^n$, and let d_2 be a Riemannian metric on $\mathbb{R}^n$. If we prove that the metrics d_1 and d_2 are related by $c_1 d_1 \leq d_2 \leq c_2 d_1$, where c_1 and c_2 are positive constants, then it will follow that they induce the same topology. In the solution of Problem 5.2 we showed that $d_2 \geq c_1 d_1$ in a closed ball D, using the fact that the function $g(y)(V, V)$ has a positive minimum on the set of all pairs (y, V) with $y \in D$ and $(V, V) = 1$. To prove the inequality $d_2 \leq c_2 d_1$, it suffices to note that this function also has a positive maximum.

5.4 No. Take, for example, points x and y in the plane $\mathbb{R}^2$ which are symmetric about the origin. Then these points cannot be joined by a smooth curve of minimum length on the connected Riemannian manifold $\mathbb{R}^2 \setminus \{0\}$.

5.5 Suppose that a geodesic γ on the sphere $\mathbb{S}^n$ passes through some point z and its velocity vector at this point equals V. Consider the plane which contains the vector V and passes through the center of the sphere and the point z. We can choose a coordinate system so that this plane is the coordinate plane Ox^1x^2, i.e., so that it is determined by the equations $x^3 = \cdots = x^{n+1} = 0$. We must prove that the geodesic γ lies entirely in the plane Ox^1x^2. Suppose that some point w of γ has a nonzero coordinate x^i, $i > 2$. Consider the transformation which leaves all coordinates but x^i intact and changes the sign of x^i. The sphere, the point z, and the velocity vector of the geodesic at z are invariant under this transformation; therefore, so is the geodesic γ. Thus, the transformation leaves the point w fixed. We have arrived at a contradiction.

5.6

(a) In the proof of Theorem 5.5 we have used only the fact that any geodesic through x can be extended without bound.

(b) Let A be a closed bounded subset of M^n. It follows from the boundedness of A that the distance from the point x to any point y in A does not exceed some number d. The points x and y can be joined by a geodesic of length at most d. Therefore, the set A is contained in the image of a closed ball of radius d under the map $\exp_x$, which is compact. The set A is closed, and hence it is compact as well.

5.7 According to Theorem 5.9, there exists a bounded Riemannian metric on M^n. By assumption this metric is geodesically complete; hence the closure of a geodesic ball of any finite radius is compact. Since the metric is bounded, the whole manifold M^n coincides with a geodesic ball of some finite radius. Therefore, M^n is compact.

5.8 Clearly,

$$\frac{\partial}{\partial \alpha} R(X + \alpha Z, Y + \beta W, X + \alpha Z, Y + \beta W)$$

$$= R(Z, Y + \beta W, X + \alpha Z, Y + \beta W) + R(X + \alpha Z, Y + \beta W, Z, Y + \beta W).$$

Therefore, at the point $\alpha = 0$, $\beta = 0$,

$$\frac{\partial^2}{\partial \alpha \partial \beta} \left(R(X + \alpha Z, Y + \beta W, X + \alpha Z, Y + \beta W) \right)$$

$$= R(Z, W, X, Y) + R(Z, Y, X, W) + R(X, W, Z, Y) + R(X, Y, Z, W)$$

$$= 2R(X, Y, Z, W) + 2R(Z, Y, X, W),$$

$$\frac{\partial^2}{\partial \alpha \partial \beta} \left(R(X + \alpha W, Y + \beta Z, X + \alpha W, Y + \beta Z) \right)$$

$$= R(W, Z, X, Y) + R(W, Y, X, Z) + R(X, Z, W, Y) + R(X, Y, W, Z)$$

$$= 2R(X, Y, W, Z) + 2R(X, Z, W, Y).$$

Moreover, $R(Z, Y, X, W) - R(X, Z, W, Y) = R(Z, Y, X, W) + R(Z, X, W, Y) = -R(Z, W, Y, X) = R(X, Y, Z, W)$.

5.9 The coordinates of the vector $X = e_n$ are $X^1 = \cdots = X^{n-1} = 0$ and $X^n = 1$; therefore, $\mathrm{Ric}(X, X) = R^k_{ikj} X^i X^j = R^k_{nkn} = \sum_{k=1}^{n} R_{knkn} = \sum_{k=1}^{n-1} R_{knkn}$, because $R_{nnnn} = 0$. It is also clear that $K(e_j, e_n) = R_{jnjn}$, whence $\sum_{j-1}^{n-1} K(e_j, e_n) = \sum_{j=1}^{n-1} R_{jnjn}$.

5.10 In an orthonormal basis we have $S = \sum_{k=1}^{n} \mathrm{Ric}(e_k, e_k)$. Expressing the terms $\mathrm{Ric}(e_k, e_k)$ by the formula of Problem 5.9, we obtain the required result.

5.11 Consider a geodesic γ on the manifold M. Let V be the velocity vector of this geodesic at some point p. Under the isometry σ all points of the curve γ remain fixed, and hence V remains fixed under the map σ_*.

Now consider a geodesic $\tilde{\gamma}$ on the manifold $\tilde{M}$ passing through the point p and having the velocity vector V at this point. The isometry σ leaves the vector V fixed, and hence it leaves fixed all points of the geodesic $\tilde{\gamma}$. Thus, the curve $\tilde{\gamma}$ lies in M. This curve is locally minimal in M, because it is locally minimal even in $\tilde{M}$. Since the geodesics $\tilde{\gamma}$ and γ in M have the same velocity vector at p, it follows that they coincide.

5.12 Consider, e.g., a geodesic γ on the manifold M_1. Let X_1 be the velocity vector of this geodesic. Then $\nabla^1_{X_1} X_1 = 0$. Therefore, (5.7) implies $\nabla_{X_1} X_1 = 0$. Thus, the curve γ is also geodesic on M.

5.13 Consider a curve γ on the manifold $M = M_1 \times M_2$. Let us decompose its velocity vector into two components: $X = X_1 + X_2$. Clearly, X_1 and X_2 are the velocity vectors of the projections of γ on M_1 and on M_2. By virtue of (5.7), we have

$$\nabla_{X_1+X_2}(X_1 + X_2) = \nabla^1_{X_1} X_1 + \nabla^2_{X_2} X_2.$$

Therefore, a curve γ on the manifold M is geodesic if and only if its projection on each of the factors is geodesic.

5.14 Consider a triangle composed of arcs of great circles on the sphere. The parallel transport along its sides rotates a vector through an angle of $\alpha + \beta + \gamma - \pi$, where α, β, and γ are the angles of the triangle. The sum of angles of a spherical triangle is larger than π but less than 3π; therefore, the angle of rotation can be arbitrary.

5.15

(a) Consider the tangent space to $\mathbb{S}^3$ at a point p. Any direct motion of this vector space is a rotation about some axis. Let α be the plane perpendicular to this axis and passing through p. The intersection of the sphere $\mathbb{S}^3$ with the hyperplane containing α and passing through the center of the sphere is the submanifold $\mathbb{S}^2$ of $\mathbb{S}^3$. According to Theorem 5.20, the parallel transport along a loop lying in $\mathbb{S}^2$ leaves fixed any vector tangent to $\mathbb{S}^3$ at p and perpendicular to the plane α, and it transforms any vector in the plane α as if it were transported inside $\mathbb{S}^2$. According to Problem 5.14, we can realize rotation through any angle by a parallel transport along some curve on the sphere $\mathbb{S}^2$. On the sphere $\mathbb{S}^3$ the same transport realizes the initial motion of the tangent space.

(b) Consider the tangent space to $\mathbb{S}^n$ at a point p. Any direct motion of this space can be represented as the composition of an even number of symmetries about hyperplanes. The composition of two such symmetries is a direct motion which leaves fixed a subspace of dimension $n - 2$. Let us show that the group $\mathrm{Hol}(\mathbb{S}^n)$ contains a subgroup consisting of direct motions leaving fixed this subspace of dimension $n - 2$. To this end, just as in the solution of part (a) of the problem, we take a plane α perpendicular to this subspace and consider the sphere $\mathbb{S}^2$ being the intersection of $\mathbb{S}^n$ with the subspace of dimension 3 containing α and passing through the center of the sphere.

5.16

(a) On a surface with flat metric, running round a sufficiently small loop is the identity map. A contraction of a loop can be represented as a sequence of its transformations in small domains. Therefore, on a surface with flat metric contractible loops correspond to the identity element in the holonomy group.

Fig. 5.4 Parallel transport
along a generator of the
fundamental group

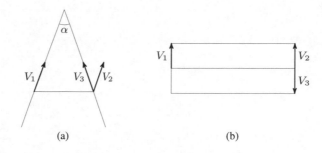

(a)   (b)

On the net of a cone consider parallel transport along a generator of the
fundamental group of the cone (see Fig. 5.4a). Such a transport takes the vector
V_1 to the vector V_2, while on the net the vector V_1 is identified with V_3. The
angle between the vectors V_2 and V_3 equals the angle α at the vertex of the net.
(b) On the net of a Möbius band consider parallel transport along a generator of the
fundamental group (see Fig. 5.4b). Such a transport takes the vector V_1 to the
vector V_2, while on the net V_1 is identified with V_3.

5.17 It is sufficient to consider the product of two Riemannian manifolds. Let γ be
a curve on a manifold $M = M_1 \times M_2$, and let V be a vector field on this curve.
We decompose the velocity vector X of γ and the vector V into two components:
$X = X_1 + X_2$ and $V = V_1 + V_2$. We have

$$\nabla_X V = \nabla^1_{X_1} V_1 + \nabla^2_{X_2} V_2.$$

Clearly, X_1 and X_2 are the velocity vectors of the projections of the curve γ on M_1
and on M_2. Therefore, the vector field V is parallel along the curve γ if and only if
the projections of each vector V on the tangent spaces to M_1 and to M_2 are parallel
along the projections of the curve γ. This immediately implies the statement.

Chapter 6
Lie Groups

In this chapter we discuss the differential geometry of Riemannian manifolds equipped with a group structure, i.e., the differential geometry of Lie groups with a Riemannian metric. On a Lie group of most interest are Riemannian connections consistent with multiplication, and above all, left-invariant metrics.

6.1 Lie Groups and Algebras

A *Lie group* G is a group which is simultaneously a smooth manifold on which the maps $G \times G \to G$ and $G \to G$ given by $(x, y) \mapsto xy$ and $x \mapsto x^{-1}$ are smooth.

Problem 6.1 Prove that if the map $(x, y) \mapsto x^{-1}y$ is smooth, then so are both maps $(x, y) \mapsto xy$ and $x \mapsto x^{-1}$.

The algebraic and geometric structures of a Lie group are involved in an object called a left-invariant vector field, which is important for many purposes. A left-invariant vector field is constructed by using left translations. The *left translation* by an element g in a Lie group is the map $L_g(x) = gx$.

Any smooth map $h \colon M \to N$ of manifolds induces a map of certain objects associated with these manifolds. Moreover, objects can be transferred in two opposite directions. The induced map may transfer them in the same direction, i.e., from M to N. This is the case for vectors: a curve γ on M is mapped to the curve $h(\gamma)$ on N. In such cases, the induced map is usually denoted by h_*. But the induced map may also transfer objects in the opposite direction, from N to M. This is the case for functions: a function $f(n)$ on N is mapped to the function $f(h(m))$ on M. In such cases, the induced map is denoted by h^*. Differential forms are transferred from N to M as well: a form $\omega(X_1, \ldots, X_k)$, where $X_1, \ldots, X_k$ are vectors on N, is mapped to the form $h^*\omega(Y_1, \ldots, Y_k) = \omega(h_*Y_1, \ldots, h_*Y_k)$, where $Y_1, \ldots, Y_k$ are vectors on M.

© The Author(s), under exclusive license to Springer Nature Switzerland AG 2022
V. V. Prasolov, *Differential Geometry*, Moscow Lectures 8,
https://doi.org/10.1007/978-3-030-92249-8_6

A left translation, as well as any other map, induces actions on various objects associated with the Lie group G. For example, the action on the functions on the Lie group is defined by $(L_g^* f)(x) = (f \circ L_g)(x) = f(L_g(x)) = f(gx)$. The action on the tangent vectors is defined as follows. Suppose that a tangent vector V_x at a point $x \in G$ is determined by a curve $\gamma(t)$, i.e., $\gamma(0) = x$ and $\gamma'(0) = V_x$. Then $L_{g*}V_x$ is the tangent vector at the point gx determined by the curve $g\gamma(t)$, i.e., $L_{g*} = dL_g$. These two actions are consistent with the action of vectors on functions, which is defined by $V_x f = \partial_{V_x} f = \frac{d}{dt} f(\gamma(t))|_{t=0}$. Indeed, $(L_{g*}V_x)f = \frac{d}{dt} f(g\gamma(t))|_{t=0} = V_x(L_g^* f)$.

A vector field V is said to be *left-invariant* if any left translation leaves this vector field fixed, i.e., $L_{g*}V_x = V_{gx}$ for all $x \in G$ and all $g \in G$. Each left-invariant vector field is completely determined by the vector V_e at the identity element e. Indeed, the vector of this left-invariant vector field at any point $x \in G$ must equal $L_{x*}V_e$. And conversely, if we take a tangent vector V_e at the identity element e and consider the vector field equal to $L_{x*}V_e$ at the point x, then this vector field will be left-invariant. Indeed, $L_{g*}V_x = L_{g*}L_{x*}V_e = L_{gx*}V_e = V_{gx}$.

Problem 6.2 Prove that on a Lie group of dimension n there exist n vector fields linearly independent at each point.

Consider the tangent space $T_e G$ at the identity element e of a Lie group G. Let X_e be a vector in this space, and let X be the corresponding left-invariant vector field. The *Lie algebra* $\mathfrak{g}$ of the Lie group G is the space $T_e G$ endowed with the operation $[X_e, Y_e] = [X, Y]_e$; thus, to calculate the product of vectors X_e and Y_e in the Lie algebra, we must extend them to left-invariant vector fields X and Y, calculate their commutator $[X, Y]$, and take the value of this commutator at e. The multiplication operation $[\cdot, \cdot]$ in the Lie algebra $\mathfrak{g}$ is called *commutator*, or *Lie bracket*.

Problem 6.3 Prove that the commutator of left-invariant vector fields X and Y is a left-invariant vector field.

In addition to left translations L_g, we can consider right translations $R_h(x) = xh$. The action of a right translation on the tangent vectors is defined as follows. Suppose that a tangent vector W_x at a point $x \in G$ is determined by a curve $\gamma(t)$, i.e., $\gamma(0) = x$ and $\gamma'(0) = W_x$. Then $R_{h*}W_x$ is the tangent vector at xh determined by the curve $\gamma(t)h$, i.e., $R_{h*} = dR_h$. The vector field W is said to be *right-invariant* if any right translation leaves fixed this vector field.

Let X_e be a vector in the space $T_e G$. We denote the corresponding left-invariant vector field by X^L and the corresponding right-invariant vector field by X^R.

It is easy to check that $L_g \circ L_h = L_{gh}$ and $R_g \circ R_h = R_{hg}$. Thus, the map $g \mapsto L_g$ is a homomorphism of the group G to the group $\mathrm{Aut}(G)$, and the map $g \mapsto R_g$ is an antihomomorphism. It is also clear that $L_g \circ R_h = R_h \circ L_g$: each of these operations takes x to gxh.

An operation on the tangent space $T_e G$ at the identity element e of the Lie group G can be defined by using not only left-invariant vector fields but also right-invariant ones. These are different operations: they differ in sign. In the solution of

Problem 6.4 this is proved for matrix Lie groups, i.e., for Lie groups consisting of matrices. For arbitrary Lie groups, this is proved in the solution of Problem 6.8.

Problem 6.4 Given a matrix Lie group G, prove that the operation on the tangent space $T_I G$ at the identity matrix I which is defined by using left-invariant vector fields is taking the commutator of matrices, and the operation defined by using right-invariant vector fields is taking the commutator with the negative sign.

Problem 6.5 Prove that the differential of the multiplication $\mu: G \times G \to G$ is related to the differentials of left and right translations as $\mu_*(V_g, W_h) = R_{h*}V_g + L_{g*}W_h$.

Consider the map $\mathrm{Inv}: G \to G$ given by $\mathrm{Inv}(g) = g^{-1}$.

Problem 6.6

(a) Prove that $\mathrm{Inv}_* V_g = -g^{-1}V_g g^{-1}$ for a matrix Lie group G.
(b) Prove that $\mathrm{Inv}_* V_g = -(L_{g*})^{-1}R_{g^{-1}*}V_g$ for any Lie group.

Problem 6.7 Let X^L and X^R be the left- and right-invariant vector fields corresponding to a vector $X_e \in T_e G$. Prove that $(\mathrm{Inv}_* X^L)_g = -X_g^R$ for any point $g \in G$.

Problem 6.7 establishes a relationship between left- and right-invariant vector fields: the map Inv takes any left-invariant vector field to a right-invariant one.

Problem 6.8 Prove that the right commutator differs from the left commutator in sign, i.e., $[X^L, Y^L]_e = -[X^R, Y^R]_e$.

Let F^t be the flow of a vector field V on a Lie group G, which means that $F^t: G \to G$ is the displacement during time t along a trajectory of V.

Theorem 6.1 *The flow of a left-invariant vector field is the right translation by the image of the identity element e, and the flow of a right-invariant vector field is the left translation; i.e., $F^t = R_{F^t(e)}$ for a left-invariant vector field and $F^t = L_{F^t(e)}$ for a right-invariant vector field.*

Proof Let V be a left-invariant vector field (the case of a right-invariant field is treated in a similar way). It is required to prove that $F^t(x) = R_{F^t(e)}(x)$. Let us fix a point x and consider the curve

$$x(t) = R_{F^t(e)}(x) = x F^t(e) = L_x(F^t(e)).$$

By assumption $\frac{d}{dt}F^t(e) = V(F^t(e))$. The left invariance of V implies $L_{x*}(V(F^t(e)) = V(L_x(F^t(e)))$. Therefore, $x(0) = x$ and

$$\frac{dx}{dt}(t) = L_{x*}(V(F^t(e)) = V(L_x(F^t(e))) = V(x(t)).$$

Thus, $x(t)$ is the trajectory of the vector field V starting at x, i.e., $x(t) = F^t(x)$. $\quad\square$

Corollary *The flow of any left-invariant vector field commutes with the flow of any right-invariant vector field (and, therefore, the commutator of a left-invariant vector field and a right-invariant one vanishes).*

Proof Let F^t be the flow of a left-invariant vector field V, and let H^s be the flow of a right-invariant vector field W. Then

$$(F^t \circ H^s)(x) = (R_{F^t(e)} \circ L_{H^s(e)})(x) = H^s(e) \cdot x \cdot F^t(e)$$
$$= (L_{H^s(e)} \circ R_{F^t(e)})(x) = (H^s \circ F^t)(x).$$

$\square$

A tangent vector X_e at the identity element $e \in G$ determines a left-invariant vector field X^L and a right-invariant vector field X^R. Through each point of the Lie group G a trajectory of the left-invariant field X^L and a trajectory of the right-invariant field X^R pass. These trajectories are generally different. But the trajectories passing through the identity element e coincide and determine (for sufficiently small t) a one-parameter subgroup, namely, the displacement of the identity element e during time t along the trajectory.

To prove the coincidence of the two trajectories, we use Theorem 6.1. First, we show that the trajectories of both vector fields are defined not only for sufficiently small t but also for any t. For the flow of a left-invariant vector field, we have $F^{t+s}(e) = F^s(F^t(e)) = R_{F^s(e)}(F^t(e)) = F^t(e) \cdot F^s(e)$. A similar equation holds for the flow of a right-invariant vector field. This equation shows, in particular, that the flow $F^t(e)$ is defined not only for sufficiently small t, but also for any t. Indeed, given any t, the flow $F^t(e)$ can be specified by $F^t(e) = (F^{t/n}(e))^n$, where n is large enough. The formula $F^t(x) = R_{F^t(e)}(x)$ shows that, given any t, trajectories of a left-invariant vector field passing not only through the identity element e but also through any point x are defined.

Let F^t be the flow constructed from a left-invariant vector field X^L or from a right-invariant vector field X^R. We have proved that the map $h: \mathbb{R} \to G$ given by $h(t) = F^t(e)$ is a homomorphism. In either case, $\frac{dh}{dt}(0) = X_e$. Differentiating the equations $h(t + s) = h(t)h(s)$ and $h(t + s) = h(s)h(t)$ with respect to s at $s = 0$, we obtain, respectively,

$$\frac{dh}{dt}(t) = L_{h(t)*}(X_e) = X^L(h(t))$$

and

$$\frac{dh}{dt}(t) = R_{h(t)*}(X_e) = X^R(h(t)).$$

Therefore, $h(t)$ is the trajectory of both X^L and X^R through $h(0) = e$.

Using the one-parameter group $F^t(e)$, we can define the exponential map $\exp: \mathfrak{g} \to G$ by setting $\exp(X_e) = F^1(e)$, where $F^1(e)$ is the displacement of the identity element e during time $t = 1$ along the flow F^t of the left-invariant vector field X^L (or of the right-invariant vector field X^R) generated by the vector X_e.

The differential of the map $\exp: \mathfrak{g} \to G$ at the point 0 is the identity map $\mathfrak{g} \to \mathfrak{g}$. Therefore, the map $\exp$ is a diffeomorphism between some neighborhood of the point $0 \in \mathfrak{g}$ and some neighborhood of the point $e \in G$.

Example For the Lie group GL_n of nonsingular matrices of order n, the Lie algebra is the algebra $\mathfrak{gl}_n$ of all matrices of order n, and

$$\exp(A) = I + A + \frac{A^2}{2!} + \cdots + \frac{A^k}{k!} + \dots .$$

Proof It is easy to check that the function

$$\theta_A(t) = I + At + \frac{A^2 t^2}{2!} + \cdots + \frac{A^k t^k}{k!} + \dots$$

is the one-parameter subgroup corresponding to the tangent vector A.   □

By using the exponential map, it is possible to define multiplication in a Lie algebra in a different way. Let us identify a neighborhood of zero in the given Lie algebra and a neighborhood of the identity element in the Lie group by means of the map $\exp$. Consider coordinates with origin at zero in the chosen neighborhood of zero; they induce coordinates in the chosen neighborhood of the identity element. In these coordinates multiplication in the Lie group can be decomposed as

$$xy = x + y + c(x, y) + \dots ,$$

where $c(x, y)$ is some bilinear function and the dots denote higher-order terms. Indeed, we have $xy = A + Bx + Cy + \dots$. Setting $x = e$, we obtain $y = A + Cy + \dots$, whence $A = 0$ and $C = 1$. A similar argument proves that $B = 1$.

The bilinear function c is also defined on the Lie algebra. The commutation operation in the Lie algebra can be specified by $[X, Y]_1 = 2c(X, Y)$. A complete definition of this operation, with all identifications expressed explicitly, is as follows:

$$\exp X \cdot \exp Y = \exp(X + Y + \frac{1}{2}[X, Y]_1 + \dots).$$

It is easy the check that $[X, Y]_1 = -[Y, X]_1$. Indeed,

$$\exp(0) = \exp X \cdot \exp(-X) = \exp(X - X + \frac{1}{2}[X, -X]_1 + \dots),$$

whence $[X, X]_1 = 0$. Therefore, $[X + Y, X + Y]_1 = 0$, which means that $[X, Y]_1 + [Y, X]_1 = 0$.

Theorem 6.2 *For sufficiently small* $X, Y \in \mathfrak{g}$,

$$\exp(-X)\exp(-Y)\exp X \exp Y = \exp([X, Y]_1 + \ldots), \tag{6.1}$$

$$\exp X \exp Y \exp(-X) = \exp(Y + [X, Y]_1 + \ldots). \tag{6.2}$$

Proof Clearly,

$$\exp(-X)\exp(-Y) = \exp(-X - Y + \frac{1}{2}[X, Y]_1 + \ldots) = \exp A,$$

$$\exp X \exp Y = \exp(X + Y + \frac{1}{2}[X, Y]_1 + \ldots) = \exp B,$$

$$\exp A \exp B = \exp(A + B + \frac{1}{2}[A, B]_1 + \ldots).$$

Moreover,

$$A + B = -X - Y + \frac{1}{2}[X, Y]_1 + \cdots + X + Y + \frac{1}{2}[X, Y]_1 + \cdots = [X, Y]_1 + \ldots$$

and $[A, B]_1 = [-X - Y, X + Y]_1 + \cdots = 0 + \ldots$. Therefore,

$$\exp(-X)\exp(-Y)\exp X \exp Y = \exp A \exp B$$

$$= \exp(A + B + \frac{1}{2}[A, B]_1 + \ldots)$$

$$= \exp([X, Y]_1 + \ldots).$$

Next,

$$\exp X \exp Y \exp(-X) = \exp B \exp(-X) = \exp(B - X - \frac{1}{2}[B, X]_1 + \ldots).$$

Moreover, $B - X = Y + \frac{1}{2}[X, Y]_1 + \ldots$ and $[B, X]_1 = [X + Y, X]_1 + \cdots = [Y, X]_1 + \ldots$. Therefore,

$$\exp X \exp Y \exp(-X) = \exp(Y + \frac{1}{2}[X, Y]_1 - \frac{1}{2}[Y, X]_1 + \ldots)$$

$$= \exp(Y + [X, Y]_1 + \ldots).$$

$\square$

Now the equivalence of the two definitions of commutator in a Lie algebra, i.e., the identity $[X, Y]_1 = [X, Y]$, follows directly from Theorem 6.3.

Theorem 6.3 *For left-invariant vector fields X and Y,*

$$\exp(-tX)\exp(-tY)\exp(tX)\exp(tY) = \exp(t^2[X, Y] + o(t^2)).$$

Proof We use Theorem 5.27 (see p. 201), taking the identity element e for p. In this case, we have $X_t p = p\exp(tX)$ for $p \in G$, and therefore

$$\sigma(t) = \exp(-tX)\exp(-tY)\exp(tX)\exp(tY).$$

For a function f on the Lie group with $f(e) = 0$, Theorem 5.27 gives

$$f(\exp(-tX)\exp(-tY)\exp(tX)\exp(tY)) = t^2\partial_{[X,Y]}f(e) + o(t^2)$$

$$= f(\exp(t^2[X, Y] + o(t^2))),$$

which implies the required equation. □

Corollary *For left-invariant vector fields X and Y, the velocity vector of the curve*

$$\exp(-\sqrt{t}X)\exp(-\sqrt{t}Y)\exp(\sqrt{t}X)\exp(\sqrt{t}Y)$$

at t = 0 equals [X, Y].

For a right-invariant vector field, we have $X_t p = \exp(tX)p$, so that

$$\exp(tY)\exp(tX)\exp(-tY)\exp(-tX)) = \exp(t^2[X, Y] + o(t^2)).$$

For right-invariant vector fields X and Y, the velocity vector of the curve

$$\exp(\sqrt{t}Y)\exp(\sqrt{t}X)\exp(-\sqrt{t}Y)\exp(-\sqrt{t}X)$$

at $t = 0$ equals $[X, Y]$.

Example The Lie algebra of the group of nonsingular matrices is the algebra of matrices with commutator $[X, Y] = XY - YX$.

Proof We must show that the decompositions of $\exp X \exp Y$ and $\exp(X + Y + \frac{1}{2}(XY - YX))$ coincide up to second-order terms. It is easy to check that, up to second-order terms, both of them equal

$$I + X + Y + \frac{X^2}{2} + XY + \frac{Y^2}{2}.$$

□

HISTORICAL COMMENT The theory of Lie groups and algebras was developed by Sophus Lie (1842–1899). He came to this theory through the study of partial differential equations. He hoped to find a theory of such equations similar to Galois' theory of algebraic equations. In 1873–1874 Lie began to develop the theory of continuous transformations of groups, already independently of his initial goal.

6.2 Adjoint Representation and the Killing Form

The map $I_g = R_{g^{-1}} \circ L_g = L_g \circ R_{g^{-1}}$, which takes x to gxg^{-1}, is an automorphism of a Lie group G. Such an automorphism is said to be *inner*.

An *adjoint representation* of a Lie group G is the map $\mathrm{Ad}: G \to \mathrm{Aut}(\mathfrak{g})$ defined by $\mathrm{Ad}_g = (dI_g)_e$, that is, the differential of an inner automorphism at the identity element.

For a matrix Lie group, $\mathrm{Ad}_A(X_e) = AX_eA^{-1}$. Indeed, the map I_A takes each curve $X(t)$ representing a tangent vector X_e to the curve $AX(t)A^{-1}$. The tangent vector to this curve at $t = 0$ is AX_eA^{-1}.

It is easy to check that

$$g \exp(t X_e)g^{-1} = \exp(t\, \mathrm{Ad}_g(X_e)). \tag{6.3}$$

Indeed, both maps $t \mapsto g\exp(tX_e)g^{-1}$ and $t \mapsto \exp(t\,\mathrm{Ad}_g(X_e))$ are homomorphisms, and their derivatives at $t = 0$ equal $\mathrm{Ad}_g(X_e)$. One-parameter subgroups of a Lie group having the same tangent vectors at $t = 0$ coincide.

Problem 6.9 Prove that, for matrices, $Ae^{tX}A^{-1} = e^{tAXA^{-1}}$.

An adjoint representation of a Lie group G is a group homomorphism (a homomorphism from the Lie group G to the automorphism group of the Lie algebra $\mathfrak{g}$). Indeed, differentiating the equation $I_{gh} = I_g \circ I_h$ by the chain rule, we obtain $\mathrm{Ad}_{gh} = \mathrm{Ad}_g \circ \mathrm{Ad}_h$.

An adjoint representation of a Lie group relates the left-invariant vector fields to the right-invariant ones as follows: the left translation of a vector $X_e \in T_eG$ to a point g equals the right translation to this point of the vector $\mathrm{Ad}_g(X_e)$. Indeed, we have $L_{g*}X_e = R_{g*}R_{g^{-1}*}L_{g*}X_e = R_{g*}(R_{g^{-1}*}L_{g*}X_e) = R_{g*}(\mathrm{Ad}_g(X_e))$.

The *adjoint representation* of a Lie algebra $\mathfrak{g}$ is the homomorphism ad of the Lie algebra $\mathfrak{g}$ to the Lie algebra of endomorphisms of the space $\mathfrak{g}$ defined by $\mathrm{ad}_X = (d\,\mathrm{Ad})_e(X)$. For example, in the case of a matrix Lie group, as a curve representing a tangent vector $X \in T_eG$ we can take e^{tX}. Then

$$\mathrm{ad}_X(Y) = \frac{d}{dt}(\mathrm{Ad}_{e^{tX}}(Y))\Big|_{t=0} = \frac{d}{dt}(e^{tX}Ye^{-tX})\Big|_{t=0} = XY - YX = [X, Y].$$

Let us prove this formula in the general case.

Theorem 6.4 *The identity* $\mathrm{ad}_X(Y) = [X, Y]$ *holds for all* $X, Y \in \mathfrak{g}$.

Proof Using the decomposition of Theorem 6.2, we obtain

$$\mathrm{Ad}_{\exp(tX)}(\exp(sY)) = \exp(tX)\exp(sY)\exp(-tX) = \exp(sY + st[X, Y] + \dots).$$

Therefore,

$$
\begin{aligned}
\mathrm{ad}_X(Y) &= \frac{d}{dt}\frac{d}{ds}\mathrm{Ad}_{\exp(tX)}(\exp(sY))\Big|_{s=0}\Big|_{t=0}\\
&= \frac{d}{dt}\frac{d}{ds}\exp(sY + st[X, Y] + \dots)\Big|_{s=0}\Big|_{t=0}\\
&= \frac{d}{dt}(Y + t[X, Y] + \dots)\Big|_{t=0} = [X, Y].
\end{aligned}
$$

$\square$

Problem 6.10 Given an $X \in \mathfrak{g}$, prove that $\mathrm{Ad}_{\exp X} = \exp(\mathrm{ad}_X)$, where $\exp X$ is the exponential map from the Lie algebra to the Lie group and $\exp(\mathrm{ad}_X)$ is the exponential of a linear operator.

The *Killing form* is the symmetric bilinear form $B(X, Y)$ on a Lie algebra $\mathfrak{g}$ defined by $B(X, Y) = \mathrm{tr}(\mathrm{ad}_X \circ \mathrm{ad}_Y)$. Bilinearity follows from the linearity of the map $X \mapsto \mathrm{ad}_X$, and symmetry follows from $\mathrm{tr}(AB) = \mathrm{tr}(BA)$.

Theorem 6.5 *The Killing form has the following properties:*

(a) *Each operator Ad_g is orthogonal with respect to B, i.e., $B(X, Y) = B(\mathrm{Ad}_g(X), \mathrm{Ad}_g(Y))$ for all $g \in G$ and $X, Y \in \mathfrak{g}$.*

(b) *Each operator ad_Z is skew-symmetric with respect to B, i.e., $B(\mathrm{ad}_Z(X), Y) = -B(X, \mathrm{ad}_Z(Y))$ (this identity can also be written as $B([X, Z], Y) = B(X, [Z, Y])$).*

Proof

(a) If $\sigma : \mathfrak{g} \to \mathfrak{g}$ is an automorphism of a Lie algebra $\mathfrak{g}$, i.e., σ is a linear isomorphism and $\sigma[X, Y] = [\sigma X, \sigma Y]$, then $\mathrm{ad}_{\sigma X} \circ \sigma = \sigma \circ \mathrm{ad}_X$, i.e., $\mathrm{ad}_{\sigma X} = \sigma \circ \mathrm{ad}_X \circ \sigma^{-1}$. Let $\sigma = \mathrm{Ad}_g$. Then

$$
\begin{aligned}
B(\mathrm{Ad}_g(X), \mathrm{Ad}_g(Y)) &= \mathrm{tr}(\mathrm{ad}_{\mathrm{Ad}_g}(X) \circ \mathrm{ad}_{\mathrm{Ad}_g}(Y))\\
&= \mathrm{tr}(\mathrm{Ad}_g \circ \mathrm{ad}_X \circ \mathrm{Ad}_g^{-1} \circ \mathrm{Ad}_g \circ \mathrm{ad}_Y \circ \mathrm{Ad}_g^{-1})\\
&= \mathrm{tr}(\mathrm{ad}_X \circ \mathrm{ad}_Y) = B(X, Y).
\end{aligned}
$$

(b) Twice applying the Jacobi identity, we obtain

$$
\begin{aligned}
[Z, [X, [Y, W]]] &= [[Z, X], [Y, W]] + [X, [Z, [Y, W]]]\\
&= [[Z, X], [Y, W]] + [X, [[Z, Y], W]] + [X, [Y, [Z, W]]],
\end{aligned}
$$

whence

$$\text{ad}_Z \circ \text{ad}_X \circ \text{ad}_Y = \text{ad}_{\text{ad}_Z(X)} \circ \text{ad}_Y + \text{ad}_X \circ \text{ad}_{\text{ad}_Z(Y)} + \text{ad}_X \circ \text{ad}_Y \circ \text{ad}_Z,$$

i.e.,

$$[\text{ad}_Z, \text{ad}_X \circ \text{ad}_Y] = \text{ad}_{\text{ad}_Z(X)} \circ \text{ad}_Y + \text{ad}_X \circ \text{ad}_{\text{ad}_Z(Y)}.$$

Since $\text{tr}([A, B]) = 0$, it follows that $B(\text{ad}_Z(X), Y) + B(X, \text{ad}_Z(Y)) = 0$. $\square$

HISTORICAL COMMENT Wilhelm Killing (1847–1923) classified semisimple Lie algebras between 1888 and 1890. For studying Lie algebras, he introduced a symmetric bilinear form on a Lie algebra; this form has subsequently become known as the Killing form.

6.3 Connections and Metrics on Lie Groups

In the context of Lie groups of special interest are connections and metrics consistent with the group structure in one sense or another, e.g., left-invariant connections and metrics. First, we discuss connections which are constructed without employing metric and may have nonzero torsion, and then we proceed to the Levi-Cività connections constructed for various metrics.

On a Lie group G a canonical left-invariant covariant differentiation $_L\nabla$ is defined. Its curvature tensor is zero, while torsion is not always zero. The one-parameter subgroups of the Lie group correspond to geodesics through the identity element. We can also define a symmetric covariant differentiation which leads to the same geodesics.

The canonical left-invariant covariant differentiation $_L\nabla$ is defined by setting $_L\nabla_X Y = 0$ for any left-invariant vector field Y (and any, not necessarily left-invariant, vector field X). This completely determines covariant differentiation. Indeed, any vector field Z on G can be represented as a linear combination of the basis left-invariant vector fields e_i: $Z = f^i e_i$. Covariant differentiation obeys Leibniz' rule; therefore,

$$_L\nabla_X Z = {}_L\nabla_X f^i e_i = f^i {}_L\nabla_X e_i + (\partial_X f^i) e_i = (\partial_X f^i) e_i.$$

For the canonical left-invariant covariant differentiation, the curvature tensor $_L R$ is zero, and the torsion tensor $_L T(X, Y)$ equals $-[X, Y]$, provided that the vector fields X and Y are left-invariant. To prove this, note that any vectors X_p, Y_p, and Z_p given at a point p can be extended to left-invariant vector fields X, Y, and Z.

Therefore, in calculating the curvature and torsion tensors at a given point, we can take left-invariant vector fields, and then all covariant derivatives vanish:

$$_LT(X, Y) = {_L}\nabla_X Y - {_L}\nabla_Y X - [X, Y] = -[X, Y],$$

$$_LR(X, Y)Z = ({_L}\nabla_X)({_L}\nabla_Y)Z - ({_L}\nabla_Y)({_L}\nabla_X)Z - {_L}\nabla_{[X,Y]}Z = 0.$$

For the canonical left-invariant covariant differentiation $_L\nabla$ on a Lie group G, an integral curve $\gamma(t)$ of a left-invariant vector field X is geodesic. Indeed, we have $_L\nabla_{\gamma'}\gamma' = {_L}\nabla_{\gamma'}X = 0$. The curve $g\gamma(t)$ is an integral curve of X as well:

$$(g\gamma(t))' = (L_g)_*\gamma'(t) = (L_g)_*X(\gamma(t)) = X(g\gamma(t)).$$

At $t = 0$ the curve $\gamma_e(t) = \gamma(0)^{-1}\gamma(t)$ passes through the identity element e. The geodesics $\gamma_e(t)$ through the identity element e are one-parameter subgroups. Indeed, for a fixed t_0, we have $\gamma_e(t_0)\gamma_e(t) = \gamma(t_0 + t)$, because both curves are integral curves of the vector field X, so that they coincide at $t = 0$. It follows, in particular, that the curve $\gamma_e(t)$ can be extended without bound, i.e., it is defined for all t. Any geodesic has the form $g\gamma_e(t)$; therefore, any geodesic can be extended without bound.

A symmetric covariant differentiation ∇ on a Lie group G can be defined as

$$\nabla_X Y = {_L}\nabla_X Y - \frac{1}{2}{_L}T(X, Y).$$

Theorem 6.6

(a) *The geodesics with respect to the covariant differentiations ∇ and $_L\nabla$ coincide.*
(b) *The torsion tensor T of the connection ∇ is zero.*
(c) *The curvature tensor R for left-invariant vector fields X, Y, and Z is calculated by*

$$R(X, Y)Z = \frac{1}{4}[Z, [X, Y]].$$

Proof (a) Clearly, $\nabla_{\gamma'}\gamma' = {_L}\nabla_{\gamma'}\gamma'$, because $T(\gamma', \gamma') = 0$.

(b) and (c) The curvature and torsion tensors can be calculated by using left-invariant vector fields. For left-invariant vector fields X and Y, we have, by definition,

$$\nabla_X Y = {_L}\nabla_X Y - \frac{1}{2}{_L}T(X, Y) = -\frac{1}{2}{_L}T(X, Y) = \frac{1}{2}[X, Y].$$

Therefore,

$$T(X, Y) = \nabla_X Y - \nabla_Y X - [X, Y]$$

$$= \frac{1}{2}[X, Y] - \frac{1}{2}[Y, X] - [X, Y] = 0.$$

Next,

$$R(X, Y)Z = \nabla_X \nabla_Y Z - \nabla_Y \nabla_X Z - \nabla_{[X,Y]} Z$$

$$= \frac{1}{2} \nabla_X [Y, Z] - \frac{1}{2} \nabla_Y [X, Z] - \frac{1}{2} [[X, Y], Z]$$

$$= \frac{1}{4} [X, [Y, Z]] - \frac{1}{4} [Y, [X, Z]] - \frac{1}{2} [[X, Y], Z] = \frac{1}{4} [Z, [X, Y]].$$

To derive the last equality, we have used the Jacobi identity.   □

A left-invariant Riemannian metric on a Lie group can be constructed as follows. We choose a Euclidean metric (positive definite inner product) on the tangent space at the identity element (i.e., on the Lie algebra) and spread it over the entire group by using the left translations $L_g \colon (L_{g*}X, L_{g*}Y)_g = (X, Y)_e$. For such a metric, any left translation is an isometry. Indeed, if follows directly from the definition that L_{g*} is an isometry from $T_e G$ to $T_g G$ and $L_{g^{-1}*}$ is an isometry from $T_g G$ to $T_e G$. The map $L_{g*}|_h \colon T_h G \to T_{gh} G$ can be represented as the composition of the isometries $T_h G \to T_e G$ and $T_e G \to T_{gh} G$; therefore, it is an isometry as well.

A Riemannian metric on a Lie group G is said to be *bi-invariant* if it is both left-invariant and right-invariant.

Theorem 6.7 *A left-invariant metric on a Lie group G is bi-invariant if and only if, for each $g \in G$, the map Ad_g preserves the restriction of this metric to the Lie algebra* $\mathfrak{g}$.

Proof First, suppose that a metric on a Lie group is both left-invariant and right-invariant. Then both maps L_g and $R_{g^{-1}}$ are isometries of the Lie group. Their composition $I_g = L_g \circ R_{g^{-1}}$ is an isometry of the Lie group as well. The map Ad_g is the differential of I_g at the identity element, and hence the adjoint action on the Lie algebra is isometric.

Now suppose that a metric on a Lie group is left-invariant and the adjoint action on the Lie algebra preserves this metric. Let us prove that the right translations preserve it as well. The map of the tangent space at a point h induced by the right translation R_g can be represented as the composition of the map of the tangent space at the identity element e induced by this right translation and the inverse of a map of the same form:

$$(R_g)_*|_h = (R_{hg})_*|_e \circ ((R_h)_*|_e)^{-1}.$$

Indeed, we can map the curve $\gamma(t)$ for which $\gamma(0) = h$ first to the curve $\gamma(t)h^{-1}$ and then to the curve $(\gamma(t)h^{-1})hg = \gamma(t)g$. Therefore, it suffices to prove that the right translation induces an isometry of the tangent space at the identity element. This is evident from the representation

$$(R_g)_* = (L_g)_* \circ \mathrm{Ad}_g^{-1}.$$

□

Thus, the left-invariant metrics on a Lie group G are in one-to-one correspondence with the inner products in its Lie algebra $\mathfrak{g}$, and the bi-invariant metrics on G are in one-to-one correspondence with the Ad-invariant inner products in $\mathfrak{g}$.

Theorem 6.8 *If an inner product in a Lie algebra is* Ad-*invariant, then* $([X, Y], Z) = -(Y, [X, Z])$ *for all* $X, Y, Z \in \mathfrak{g}$.

Proof The Ad-invariance of the given inner product implies

$$(\mathrm{Ad}_g(Y), Z) = (\mathrm{Ad}_g(Y), \mathrm{Ad}_g \, \mathrm{Ad}_{g^{-1}}(Z)) = (Y, \mathrm{Ad}_{g^{-1}}(Z)).$$

Therefore,

$$([X, Y], Z) = (\mathrm{ad}_X(Y), Z) = \left(\frac{d}{dt} \mathrm{Ad}_{\exp t X}(Y) \Big|_{t=0}, Z \right)$$

$$= \frac{d}{dt}(\mathrm{Ad}_{\exp t X}(Y), Z) \Big|_{t=0} = \frac{d}{dt}(Y, \mathrm{Ad}_{\exp(-t X)}(Z)) \Big|_{t=0}$$

$$= (Y, -\mathrm{ad}_X(Z)) = -(Y, [X, Z]).$$

$\square$

Theorem 6.8 can also be formulated as follows: If all maps Ad_g are orthogonal with respect to some inner product in a Lie algebra, then all maps ad_X are skew-symmetric with respect to this inner product. For a connected Lie group, the converse is also true; we leave its proof to the reader as a problem.

Problem 6.11 Prove that if a Lie group is connected and all maps ad_X are skew-symmetric, then all maps Ad_g are orthogonal.

Problem 6.12 Prove that, for a Lie group G with a bi-invariant metric, the map $\mathrm{Inv} \colon G \to G$ defined by $\mathrm{Inv}(g) = g^{-1}$ is an isometry.

6.4 Maurer–Cartan Equations

Consider a basis in a Lie algebra $\mathfrak{g}$. It consists of tangent vectors at the identity element $e \in G$. Extending these vectors to left-invariant vector fields, we obtain a basis X_i of left-invariant vector fields. It is associated with a basis ω^i in the dual space: $\omega^i(X_j) = \delta^i_j$.

The elements of the space dual to the space of left-invariant vector fields are differential 1-forms constant on each left-invariant vector field. Indeed, if a 1-form ω lies in the dual space, then it can be written as $\omega = a_i \omega^i$, and a left-invariant vector field can be written as $X = b^j X_j$ (a_i and b^j are some constants). Therefore, $\omega(X) = a_i b^j \delta^i_j = a_i b^i$. Conversely, if a 1-form ω is constant on each left-invariant vector field, then $\omega(X_i) = a_i$, whence $\omega = a_i \omega^i$.

In a different way, the elements of the space dual to the space of left-invariant vector fields can also be described as left-invariant differential 1-forms on a Lie group. Before defining a left-invariant 1-form, we recall how maps act on 1-forms. Let $f: M \to N$ be a smooth map. If X is a vector field on M and ω is a 1-form on N, then f_*X is a vector field on N, and $f^*\omega$ is the 1-form on M defined by $(f^*\omega)(X) = \omega(f_*X)$. Thus, the value of the form $f^*\omega$ at a vector $X \in T_x M$ equals the value of the form ω at the vector $f_*X \in T_{f(x)}N$.

A differential 1-form ω is *left-invariant* if the left translation L_g takes it to itself: $L_g^*\omega = \omega$, i.e., the value of the form ω at a vector $X \in T_x G$ equals its value at the vector $L_{g*}X \in T_{gx}G$.

Given a left-invariant vector field X, its value at a point gx equals the value of the vector field $L_{g*}X$ at gx. Therefore, for a left-invariant 1-form ω and a left-invariant vector field X, $\omega(X)$ takes the same value at all points. Now suppose given a 1-form ω such that $\omega(X)$ is constant for any left-invariant vector field X. Then the value of ω at the vector $L_{g*}X \in T_{gx}G$ equals the value of ω at $X \in T_x G$, and therefore the form ω left-invariant.

The *Maurer–Cartan form* on a Lie group G with Lie algebra $\mathfrak{g}$ is the left-invariant $\mathfrak{g}$-valued differential 1-form ω on G which takes the constant value X_e at each left-invariant vector field X (X_e is the value of X at the point e). At an arbitrary vector $X_g \in T_g G$ the form ω takes the value $(L_{g^{-1}})_*X_g \in T_e G$.

In addition to the left-invariant Maurer–Cartan form ω^L, for which $\omega^L(X_g) = (L_{g^{-1}})_*X_g$, we can consider the right-invariant Maurer–Cartan form ω^R, for which $\omega^R(X_g) = (R_{g^{-1}})_*X_g$.

Using the Maurer–Cartan form ω (left or right), we can explicitly specify a trivialization of the tangent bundle T_*G (i.e., a mapping $T_*G \to G \times \mathfrak{g}$). Namely, to each tangent vector $V_g \in T^*G$ we can assign the pair $(g, \omega(V_g))$.

Problem 6.13 Consider the composition of the map $G \times \mathfrak{g} \to T_*G$ inverse to the left trivialization and the right trivialization $T_*G \to G \times \mathfrak{g}$. Prove that this composition is the map $(g, V) \mapsto (g, \mathrm{Ad}_g(V))$.

Problem 6.14 Suppose given a basis X_i of the space of left-invariant vector fields and a basis ω^i of the dual space. Let X_{ei} be the vector of the field X_i at the point e. Prove that $X_{ei} \otimes \omega^i = \omega$ (the Maurer–Cartan form).

For a matrix Lie group G, the Maurer–Cartan form ω_G at a point $g = (x_j^i)$ can be written as $g^{-1}dg = (x_j^i)^{-1}d(x_j^i)$, where dg is the identity map of the tangent bundle. Indeed, we have $\omega_G(V_g) = L_{g^{-1}*}V_g = g^{-1}V_g$.

If both vector fields X and Y are left-invariant, then $\omega([X, Y]) = [X, Y]_e = [X_e, Y_e] = [\omega(X), \omega(Y)]$; in the last two expressions, the brackets denote commutator in the Lie algebra.

The differential of the 1-form ω can be expressed in terms of its derivatives in directions of vector fields and the commutator as follows (see Appendix, p. 259):

$$d\omega(X, Y) = \partial_X(\omega(Y)) - \partial_Y(\omega(X)) - \omega([X, Y]).$$

For left-invariant vector fields, $\partial_X(\omega(Y)) = \partial_Y(\omega(X)) = 0$ (the derivative of a constant quantity vanishes); therefore,

$$d\omega(X, Y) = -\omega([X, Y]) = -[\omega(X), \omega(Y)].$$

To derive the equation

$$d\omega(X, Y) + [\omega(X), \omega(Y)] = 0, \tag{6.4}$$

we used the left-invariance of the vector fields X and Y. To calculate the commutator of vector fields, we need the whole fields, rather that their values at a point. But the left-hand side of the resulting Eq. (6.4) depends only on the values of vector fields at a point. Therefore, this equation holds for any vector fields X and Y, rather than only for left-invariant ones. Equation (6.4) is called the *Maurer–Cartan equation*.

For a matrix Lie group, we have $(d\omega)^i_j = d(\omega^i_j)$ and

$$[\omega(X), \omega(Y)]^i_j = (\omega(X)\omega(Y) - \omega(Y)\omega(X))^i_j$$

$$= \omega(X)^i_k \omega(Y)^k_j - \omega(Y)^i_k \omega(X)^k_j$$

$$= (\omega^i_k \wedge \omega^k_j)(X, Y).$$

Therefore, for a matrix Lie group, the Maurer–Cartan equation can be written in the form $d\omega = -\omega \wedge \omega$.

HISTORICAL COMMENT Ludwig Maurer (1859–1927) obtained the Maurer–Cartan equation in 1899. Élie Joseph Cartan (1869–1951) obtained this equation in 1904 and applied it in the method of moving frame, which he began to develop in 1910.

6.5 Invariant Integration on a Compact Lie Group

In this section we use some basic facts concerning the integration of n-forms over n-dimensional manifolds.

The volume form on a manifold M^n is a nowhere vanishing n-form ω. This form determines the orientation of M^n. The volume form can be integrated over the manifold; i.e., for the volume form ω, the integral $\int_{M^n} \omega$ can be defined. This integral obeys the change-of-variable rule: if $\varphi: M^n \to M^n$ is a diffeomorphism of the manifold M^n, then

$$\int_{M^n} \omega = \pm \int_{M^n} \varphi^*\omega,$$

where the plus sign corresponds to an orientation-preserving diffeomorphism and the minus sign, to an orientation-reversing one.

Using the volume form, we can integrate functions on a manifold: $\int_{M^n} f = \int_{M^n} f\omega$. When the volume form ω is replaced by $c\omega$, the integral of the function is multiplied by $|c|$ (for negative c, orientation switches together with the sign of the form against which f is integrated).

We will be interested in the case where the manifold M^n is a compact Lie group G. All n-forms on an n-dimensional space are proportional, and hence there exists a unique, up to proportionality, left-invariant n-form on an n-dimensional Lie group G. On a compact Lie group G a left-invariant volume form ω for which $\int_G \omega = 1$ is determined up to multiplication by ± 1. We choose one of the two such forms and set

$$\int_G f(g)\, dg = \int_G f = \int_G f\omega$$

for every smooth function f.

Theorem 6.9 *The measure dg on a compact Lie group is left- and right-invariant, i.e.,*

$$\int_G f(hg)\, dg = \int_G f(gh)\, dg = \int_G f(g)\, dg$$

for any $h \in G$.

Proof The left invariance of the measure dg follows from that of the form ω. To prove it, we need the equation $(f \circ L_h)(L_h^*\omega) = L_h^*(f\omega)$, which holds because both these forms take the value $f(hg)\omega(L_{h*}X_g)$ at the vector X_g. Thus,

$$\int_G f(hg)\, dg = \int_G (f \circ L_h)\omega = \int_G (f \circ L_h)(L_h^*\omega)$$

$$= \int_G L_h^*(f\omega) = \int_G f\omega = \int_G f(g)\, dg.$$

In addition to the equation given above, we have also used the change-of-variable formula (clearly, L_h is an orientation-preserving diffeomorphism, because it preserves the volume form).

Now let us prove the right invariance of the measure dg. First, we note that the transformations L_g and R_g commute and the left-invariance of ω implies that of $R_g^*\omega$. The left-invariant form is unique up to proportionality; therefore, $R_g^*\omega = c(g)\omega$, where $c(g)$ is a nonzero real number. Since $R_g \circ R_h = R_{hg}$, it follows that the map $g \mapsto c(g)$ is a homomorphism from the compact group G to the multiplicative group of real numbers. A compact subgroup of the multiplicative group of real numbers consists of either the unique element 1 or the two elements ± 1; hence $|c(g)| = 1$. The map R_h preserves orientation if and only if $c(h) > 0$.

Thus,

$$\int_G f(gh)\, dg = \int_G (f \circ R_h)\omega = c(h) \int_G (f \circ R_h)(R_h^*\omega)$$

$$= c(h) \int_G R_h^*(f\omega) = c(h)\, \mathrm{sgn}(c(h)) \int_G f\omega = \int_G f\omega.$$

□

Example A left-invariant n-form on a noncompact Lie group of dimension n may not be right-invariant.

Proof Consider the group of affine transformations of the straight line $T_{(a,\lambda)}(t) = a + \lambda t$, where $\lambda \neq 0$. We have

$$T_{(a,\lambda)}T_{(b,\mu)}(t) = T_{(a,\lambda)}(b + \mu t) = a + \lambda b + \lambda \mu t,$$

whence $(a, \lambda)(b, \mu) = (a + \lambda b, \lambda \mu)$.

Let us introduce coordinates (x, y) on this group and consider the 2-form $\omega = \frac{dx \wedge dy}{y^2}$. Clearly,

$$L_{(a,\lambda)}(x, y) = (a, \lambda)(x, y) = (a + \lambda x, \lambda y),$$

$$R_{(a,\lambda)}(x, y) = (x, y)(a, \lambda) = (x + ya, y\lambda).$$

Therefore, $L_{(a,\lambda)}^*\omega = \frac{(\lambda dx) \wedge (\lambda dy)}{(\lambda y)^2} = \omega$ and $R_{(a,\lambda)}^*\omega = \frac{(dx + a\,dy) \wedge (\lambda dy)}{(\lambda y)^2} = \frac{1}{\lambda}\omega$, i.e., the 2-form ω is left-invariant but not right-invariant.   □

In the proof of Theorem 6.9 we defined a homomorphism $c\colon G \to \mathbb{R}$ for a Lie group G with a left-invariant form ω by $R_g^*\omega = c(g)\omega$. Given a compact connected Lie group G, we have $c(g) = 1$ for all $g \in G$ (for a disconnected compact Lie group, $c(g) = \pm 1$). For the group of affine transformations of the straight line, $c((a, \lambda)) = \frac{1}{\lambda}$.

In the proof of Theorem 6.10 we use the fact that a closed subgroup of a Lie group is itself a Lie group. A proof of this fact can be found in the book [Ad, pp. 17–19].

Theorem 6.10 *On a Lie group G a bi-invariant Riemannian metric can be defined if and only if the set $\mathrm{Ad}(G)$ (the image of G under the map Ad) is relatively compact in $\mathrm{GL}(\mathfrak{g})$ (i.e., the closure of this set is compact).*

Proof First, suppose that there exists a bi-invariant Riemannian metric on the Lie group G. Then, according to Theorem 6.7, the map Ad_g is orthogonal with respect to this metric for each $g \in G$. Therefore, the set $\mathrm{Ad}(G)$ lies in the group $\mathrm{O}(\mathfrak{g})$ of orthogonal transformations of the space $\mathfrak{g}$. This group is compact, and therefore so is the closure of $\mathrm{Ad}(G)$.

Now suppose that the closure of the set $\mathrm{Ad}(G)$ is compact. By virtue of Theorem 6.7, it suffices to construct an Ad-invariant inner product in the space $\mathfrak{g}$.

Indeed, the left-invariant Riemannian metric determined by this inner product is also right-invariant.

The closure of the set $\mathrm{Ad}(G)$ in $\mathrm{GL}(\mathfrak{g})$ is a compact Lie group H. The group H acts on $\mathfrak{g}$, i.e., for any $h \in H$ and $X \in \mathfrak{g}$, an element $hX \in \mathfrak{g}$ is defined. Choose an arbitrary inner product $g(X, Y)$ in $\mathfrak{g}$ and its average over H, i.e., consider the following inner product in $\mathfrak{g}$:

$$(X, Y) = \int_H (hX, hY) dh.$$

The invariance of integral follows from that of this inner product with respect to the action of any element $t \in H$:

$$(tX, tY) = \int_H (thX, thY) dh = \int_H (hX, hY) dh = (X, Y).$$

In particular, this inner product is invariant with respect to the action of any element $\mathrm{Ad}_g \in H$. $\qquad\square$

Corollary *On any compact Lie group G a bi-invariant Riemannian metric can be defined.*

Proof The image of a compact group G under the map Ad is compact. $\qquad\square$

Problem 6.15 Prove that, for a compact Lie group G, the Killing form is negative definite.

6.6   Lie Derivative

The Lie derivative is the operator of differentiation of a tensor field along a vector field. The Lie derivative, as opposed to covariant derivative, does not depend on a Riemannian metric. In the next section, we will need the Lie derivative of a Riemannian metric.

A vector field X on a manifold M^n determines a transformation X_t of M^n for sufficiently small t, namely, the displacement during time t along an integral curve of the vector field X. To calculate the derivative of a tensor field S in the direction of a vector field X, we need some method for transporting the value of S in the space of tensors at a point $X_t(x)$ to the space of tensors at the point $x = X_0(x)$. Vectors are transported in same direction in which the map acts, and differential forms are transported in the reverse direction. Therefore, in the case of a vector field, we can use the transformation $(X_{-t})_*$ for transporting spaces of tensors, and in the case of a differential form, we can use the transformation $(X_t)^*$.

In the general case, to a tensor

$$\frac{\partial}{\partial x^{i_1}} \otimes \cdots \otimes \frac{\partial}{\partial x^{i_p}} \otimes dx^{j_1} \otimes \cdots \otimes dx^{j_q}$$

we apply the transformation

$$(X_{-t})_* \frac{\partial}{\partial x^{i_1}} \otimes \cdots \otimes (X_{-t})_* \frac{\partial}{\partial x^{i_p}} \otimes (X_t)^* dx^{j_1} \otimes \cdots \otimes (X_t)^* dx^{j_q}.$$

We denote this transformation of a tensor S by $X_t^* S$. For functions, we set $(X_t^* f)(x) = f(X_t(x))$.

At a point x the *Lie derivative* $(L_X S)|_x$ of a tensor field S in the direction of a vector field X is defined by

$$(L_X S)|_x = \lim_{t \to 0} \frac{X_t^* S(X_t(x)) - S(x)}{t} = \frac{d}{dt} X_t^* S(X_t(x)) \Big|_{t=0}.$$

Using this definition, we calculate the Lie derivative first for a function and then for a vector field and a 1-form. We assume that local coordinates x^i are chosen and $X = X^i \frac{\partial}{\partial x^i}$ in these coordinates.

The Lie derivative of a function is the directional derivative: $L_X f = \partial_X f$. Indeed,

$$L_X f = \frac{d}{dt} f(X_t(x)) \Big|_{t=0} = \frac{\partial f}{\partial x^i} \cdot \frac{d(X_t(x))^i}{dt} \Big|_{t=0} = \frac{\partial f}{\partial x^i} X^i.$$

The Lie derivative of a vector field is the commutator of vector fields: $L_X Y = [X, Y]$. Indeed, if $Y = Y^i \frac{\partial}{\partial x^i}$, then

$$L_X Y = \frac{d}{dt} (X_{-t})_* \left(Y^i \frac{\partial}{\partial x^i} \right) \Big|_{t=0} = \frac{d}{dt} \left(Y^i(X_t(x)) \frac{\partial X_{-t}^j}{\partial x^i} \frac{\partial}{\partial x^j} \right) \Big|_{t=0}.$$

Since X_0 is the identity transformation, it follows that $\frac{\partial X_{-t}^j}{\partial x^i} \Big|_{t=0} = \delta_i^j$. Moreover, $\frac{d}{dt} X_{-t} \big|_{t=0} = -X$. Therefore,

$$L_X Y = \frac{\partial Y^i}{\partial x^k} X^k \delta_i^j \frac{\partial}{\partial x^j} + Y^i \left(-\frac{\partial X^j}{\partial x^i} \right) \frac{\partial}{\partial x^j}$$

$$= \left(X^k \frac{\partial Y^j}{\partial x^k} - Y^k \frac{\partial X^j}{\partial x^k} \right) \frac{\partial}{\partial x^j} = [X, Y].$$

Problem 6.16 Let X be a left-invariant vector field on a Lie group, and let Y be any vector field. Prove that

$$[X, Y]_e = \lim_{t \to 0} \frac{(R_{\exp(-tX_e)})_* Y_{\exp(tX_e)} - Y_e}{t}.$$

The Lie derivative of a differential form $\omega = \omega_j dx^j$ is

$$L_X \omega = \left(\frac{\partial \omega_j}{\partial x^i} X^i + \frac{\partial X^i}{\partial x^j} \omega_i \right) dx^j.$$

This is proved in approximately the same way as in the case of vector fields; the only substantial difference is that the proof uses the relation $\frac{d}{dt} X_t \big|_{t=0} = X$. By definition,

$$L_X \omega = \frac{d}{dt} (X_t)^* \left(\omega_j dx^j \right) \Big|_{t=0} = \frac{d}{dt} \left(\omega_j (X_t(x)) \frac{\partial X_t^j}{\partial x^k} dx^k \right) \Big|_{t=0}.$$

We also have $\frac{\partial X_t^j}{\partial x^k} \big|_{t=0} = \delta_k^j$ and $\frac{d}{dt} X_t \big|_{t=0} = X$. Therefore,

$$L_X \omega = \frac{\partial \omega_j}{\partial x^i} X^i \delta_k^j dx^k + \omega_j \frac{\partial X^j}{\partial x^k} dx^k = \left(\frac{\partial \omega_j}{\partial x^i} X^i + \frac{\partial X^i}{\partial x^j} \omega_i \right) dx^j.$$

In calculating $L_X S$, is it sometimes more convenient to deal with series expansions rather than with derivatives, i.e., take the following expression for the definition of the Lie derivative:

$$X_t^* S(X_t(x)) = S(x) + t(L_X S)|_x + o(t).$$

Let us calculate, e.g., the Lie derivative of a vector field Y in the direction of a vector field X by using the series expansion. For this purpose, consider the expression

$$X_t^* Y(X_t(x)) = Y(x) + t(L_X Y)|_x + o(t),$$

where $X_t^* = (X_{-t})_*$. We apply the operation $(X_t)_*$ to both sides of this expression and take the first summand on the right-hand side to the left-hand side:

$$Y(X_t) - (X_t)_* Y = t(X_t)_* (L_X Y) + o(t).$$

Consider the derivative of a function f in the direction of the vector field $Y(X_t) - (X_t)_* Y$:

$$
\begin{aligned}
\partial_{Y(X_t)-(X_t)_* Y} f &= \partial_{Y(X_t)} f - \partial_{(X_t)_* Y} f \\
&= (\partial_Y f) \circ X_t - \partial_Y (f \circ X_t) \\
&= \partial_Y f + t \partial_X \partial_Y f + o(t) - \partial_Y (f + t \partial_X f + o(t)) \\
&= t (\partial_X \partial_Y f - \partial_Y \partial_X f) + o(t).
\end{aligned}
$$

Thus, we again obtain the formula $L_X Y = [X, Y]$.

If T is a tensor of type $(0, k)$, then, given any vector fields $X, Y_1, \ldots, Y_k$, we have

$$
(L_X T)(Y_1, \ldots, Y_k) = \partial_X (T(Y_1, \ldots, Y_k)) - \sum_{i=1}^{k} T(Y_1, \ldots, L_X Y_i, \ldots, Y_k).
$$

In particular, the Lie derivative of a metric tensor has the form

$$
L_X g(Y, Z) = \partial_X g(Y, Z) - g([X, Y], Z) - g(Y, [X, Z]).
$$

We have already proved this relation for a tensor of type $(0, 1)$, i.e., for a 1-form. Now we give a proof using the series expansion. In calculating the Lie derivative of a vector field by using the series expansion we have obtained the equation

$$
Y(X_t) = (X_t)_* Y + t(X_t)_* (L_X Y) + o(t).
$$

This equation implies

$$
\begin{aligned}
((X_t)^* T)(Y) = T((X_t)_* Y) &= T(Y(X_t) - t(X_t)_* (L_X Y)) + o(t) \\
&= T(Y) \circ X_t - tT((X_t)_* (L_X Y)) + o(t) \\
&= T(Y) + t \partial_X (T(Y)) - tT((X_t)_* (L_X Y)) + o(t).
\end{aligned}
$$

Therefore,

$$
\begin{aligned}
(L_X T)(Y) &= \lim_{t \to 0} \frac{((X_t)^* T)(Y) - T(Y)}{t} \\
&= \lim_{t \to 0} (\partial_X (T(Y)) - T((X_t)_* (L_X Y))) \\
&= \partial_X (T(Y)) - T(L_X Y).
\end{aligned}
$$

The general case differs only in that the notation is more complex.

HISTORICAL COMMENT Sophus Lie introduced only a special case of the Lie derivative, the commutator of vector fields. The Lie derivative in the general case was introduced by Władysław Ślebodziński (1884–1972) in 1931 .

6.7 Infinitesimal Isometries

A vector field X on a Riemannian manifold M is called a *Killing field* if $L_X g(Y, Z) = 0$ for any vector fields Y and Z, i.e., the Lie derivative of the metric tensor in the direction of X vanishes. Recall that the Lie derivative of the metric tensor has the form

$$L_X g(Y, Z) = \partial_X g(Y, Z) - g([X, Y], Z) - g(Y, [X, Z])$$

(see p. 231). In terms of the Levi-Città connection ∇, the equation $L_X g(Y, Z) = 0$ can be written as

$$g(\nabla_Y X, Z) + g(Y, \nabla_Z X) = 0$$

for any vector fields Y and Z. Indeed, the Levi-Città connection is torsion-free and compatible with the metric, so that

$$\partial_X g(Y, Z) = g(\nabla_X Y, Z) + g(Y, \nabla_X Z)$$
$$= g(\nabla_Y X + [X, Y], Z) + g(Y, \nabla_Z X + [X, Z]).$$

Therefore,

$$\partial_X g(Y, Z) - g([X, Y], Z) - g(Y, [X, Z]) = g(\nabla_Y X, Z) + g(Y, \nabla_Z X).$$

Thus, if X is a Killing field, then $\nabla X(Y) = \nabla_Y X$ is a linear operator of Y skew-symmetric with respect to the metric g.

A vector field X on a manifold M^n is a Killing field if and only if the local one-parameter group of transformations X_t (displacements by t along integral trajectories of X) consists of isometries. Indeed, if all transformations X_t are isometries, then $g(X_t^* Y, X_t^* Z) = g(Y, Z)$, and therefore $L_X g = 0$. Now suppose that $L_X g = 0$. Then

$$\lim_{t \to 0} \frac{1}{t}(g(X_t^* Y, X_t^* Z) - g(Y, Z)) = 0.$$

Since $X_t X_s = X_{t+s}$, we have

$$\lim_{t \to 0} \frac{1}{t}(g(X_{s+t}^* Y, X_{s+t}^* Z) - g(X_s^* Y, X_s^* Z)) = 0.$$

Therefore, the map $s \mapsto g(X_s^* Y, X_s^* Z)$ is constant, which means that X_s is an isometry.

In view of this, Killing vector fields are also called *infinitesimal isometries*.

Theorem 6.11 *The commutator* $[X, Y]$ *of Killing fields* X *and* Y *is a Killing field.*

Proof Recall that $[X, Y] = L_X Y = \frac{d}{dt}((X_{-t})_* Y(X_t(x)))\big|_{t=0}$. Therefore, for a fixed t, the diffeomorphisms $X_{-t} Y_s X_t$ generate the vector field $[X, Y]$. The diffeomorphism $X_{-t} Y_s X_t$ is a composition of isometries, and hence it is an isometry as well. It follows that

$$
L_{[X,Y]} g = \frac{\partial^2}{\partial s \partial t}(X_{-t} Y_s X_t)^* g \bigg|_{s=t=0} = 0.
$$

$\square$

Thus, the Killing fields form a Lie algebra.

Problem 6.17 Prove that the right-invariant vector fields on a Lie group with a left-invariant metric are Killing fields.

A Killing field X on a manifold M^n is a Jacobi field along any geodesic on this manifold. Indeed, a Killing field X generates a local one-parameter group of isometries. Isometries take geodesics to geodesics. Therefore, any Killing field X generates a variation (with loose endpoints) of a geodesic in the class of geodesics.

Recall that a Jacobi vector field J on a geodesic γ is completely determined by the initial conditions $J(0)$ and $\nabla_V J(0)$, where $V = \gamma'$ is the velocity vector of the geodesic. Therefore, to define a Killing field, it is sufficient to specify its value at one point and the values of its covariant derivatives in all directions at this point. In particular, if a Killing field vanishes together with its covariant derivatives in all directions at some point, then it vanishes everywhere.

The operator $\nabla X(Y) = \nabla_Y X$ is a linear skew-symmetric operator of Y. Therefore, to define a Killing field, it suffices to specify a vector and a skew-symmetric operator. It follows that the dimension of the Lie algebra of Killing fields on an n-dimensional Riemannian manifold is at most $n + \frac{n(n-1)}{2} = \frac{n(n+1)}{2}$.

The isometry group of a Riemannian manifold has many nice properties. This group is always a Lie group (by the Myers–Steenrod theorem, 1939). A generalization of this theorem was proved in the book [Pa]. In the same book it was proved that the natural action of the isometry group on the manifold is smooth and, moreover, a homomorphism β from the group of real numbers to a Lie group of isometries is smooth, provided that so is the map $(t, p) \mapsto \beta(t)p$. We will not prove these theorems; instead, we will assume that all Riemannian manifolds under consideration have all these properties.

Under this assumption, we establish a relationship between the Lie algebra of the isometry group of a Riemannian manifold M^n and the Lie algebra of Killing fields on this manifold.

We use the following notation: G is the Lie group of isometries of a Riemannian manifold M^n and $\mathfrak{g}$ is its Lie algebra; X and Y are left-invariant vector fields on the Lie group G; and X_e and Y_e are the vectors of these vector fields at the identity element, in particular, X_e and Y_e are elements of the Lie algebra $\mathfrak{g}$.

To an element $X_e \in \mathfrak{g}$ we can assign the one-parameter subgroup $\exp(tX_e)$ in G. The isometry $\exp(tX_e)$ of the manifold M^n takes each point $p \in M^n$ to the point $\exp(tX_e)p \in M^n$. We associate the element X_e of the Lie algebra $\mathfrak{g}$ with the vector field $X_p^+ = \frac{d}{dt}(\exp(tX_e)p)\big|_{t=0}$ on the manifold M^n. The one-parameter group X_t^+ of displacements during time t along integral trajectories of the vector field X^+ consists of the isometries $\exp(tX_e)$, so that X^+ is a Killing field.

For each point $p \in M^n$, we consider the map $\pi_p \colon G \to M^n$ that takes every element $g \in G$ to the point $g(p) \in M^n$. It is easy to check that $(\pi_p)_* X_e = X_p^+$. Indeed, the vector $X_e \in T_e G$ can be represented by a curve $\gamma(t) = \exp(tX_e)$ for which $\gamma'(0) = X_e$. Therefore, the vector $(\pi_p)_* X_e \in T_p M^n$ can be represented by the curve $(\pi_p \circ \gamma)(t) = \gamma(t)p = \exp(tX_e)p$, i.e., it equals $\frac{d}{dt}(\exp(tX_e)p)\big|_{t=0} = X_p^+$.

Let us show that the map $X_e \mapsto X^+$ is a one-to-one linear map of the Lie algebra of the Lie group of isometries of the manifold M^n to the Lie algebra of Killing fields on M^n. This map is an anti-isomorphism with respect to the structure of the Lie algebra, i.e., $[X^+, Y^+] = -[X, Y]^+$.

The linearity of the map $X_e \mapsto X^+$ follows from $(\pi_p)_* X_e = X_p^+$. Let us check that it has trivial kernel. Indeed, suppose that a vector field X^+ is zero. Then its integral curves $\exp(tX_e)p$ are constant for all $p \in M^n$. Therefore, $\exp(tX_e)$ is the identity element of the Lie group for all t, which means that $X_e = 0$. Now let us check that any Killing field Z can be represented in the form $Z = X^+$ for some $X_e \in \mathfrak{g}$. The Killing vector field Z determines the one-parameter isometry group Z_t. The one-parameter subgroup $t \mapsto Z_t$ of the Lie group G corresponds to some element X_e of the Lie algebra $\mathfrak{g}$; clearly, $X^+ = Z$.

Now let us prove that $[X^+, Y^+] = -[X, Y]^+$. Each vector $X_e \in \mathfrak{g}$ determines the one-parameter group $\exp(tX_e)$. This group coincides with the one-parameter group X_t^+ of displacements during time t along integral trajectories of the vector field X^+. This one-parameter group can be used to calculate the commutator vector fields as the Lie derivative:

$$[X^+, Y^+]_p = \lim_{t \to 0} \frac{(X_{-t}^+)_* Y_q^+ - Y_p^+}{t} = \lim_{t \to 0} \frac{\exp(-tX_e)_* Y_q^+ - Y_p^+}{t},$$

where $q = \exp(tX_e)p$.

Recall that $Y_q^+ = (\pi_q)_*(Y_e)$. Therefore,

$$\exp(-tX_e)_* Y_q^+ = \exp(-tX_e)_*(\pi_q)_* Y_e = (\exp(-tX_e)\pi_q)_* Y_e.$$

Let us find out the structure of the map $\exp(-tX_e)\pi_q$:

$$(\exp(-tX_e)\pi_q)(g) = (\exp(-tX_e)\pi_{\exp(tX_e)p})(g)$$
$$= \exp(-tX_e)g\exp(tX_e)p = (\pi_p R_{\exp(tX_e)} L_{\exp(-tX_e)})(g);$$

therefore, $\exp(-tX_e)\pi_q = \pi_p R_{\exp(tX_e)} L_{\exp(-tX_e)}$.

The left invariance of the vector field Y implies $(L_{\exp(-tX_e)})_* Y_e = Y_{\exp(-tX_e)}$, so that

$$\exp(-tX_e)_* Y_q^+ = (\exp(-tX_e)\pi_q)_* Y_e = (\pi_p R_{\exp(tX_e)} L_{\exp(-tX_e)})_* Y_e$$
$$= (\pi_p)_* (R_{\exp(tX_e)})_* Y_{\exp(-tX_e)}.$$

Thus,

$$\exp(-tX_e)_* Y_q^+ - Y_p^+ = (\pi_p)_* (R_{\exp(tX_e)})_* Y_{\exp(-tX_e)} - (\pi_p)_* Y_e$$
$$= (\pi_p)_* [(R_{\exp(tX_e)})_* Y_{\exp(-tX_e)} - Y_e].$$

According to Problem 6.16,

$$\lim_{t \to 0} \frac{(R_{\exp(tX_e)})_* Y_{\exp(-tX_e)} - Y_e}{t} = [-X, Y]_e.$$

Therefore, the equation $[X^+, Y^+]_p = -[X, Y]_p^+$ follows from $(\pi_p)_*[-X, Y]_e = -[X, Y]_p^+$.

HISTORICAL COMMENT In 1884 Killing studied geometries with certain special properties related to infinitesimal transformations and introduced Killing fields. He proved that Killing fields form a Lie algebra, which had led him to the notion of a Lie algebra independently of Sophus Lie but 15 years later.

6.8 Homogeneous Spaces

Let G be a Lie group, and let K be its closed Lie subgroup. The set

$$G/K = \{gK | g \in G\}$$

of left cosets can be endowed with the structure of a manifold. Such a manifold is called a *homogeneous space*. We denote it by M and the point of M corresponding to the class $eK = K$ by o.

On the manifold M the Lie group G acts by the translations τ_a, where $a \in G$. A translation τ_a takes a point gK to the point agK. Let us check that the point agK

depends only on gK and the element a. If $g_1 K = g_2 K$, then $g_1 = g_2 k$ for some $k \in K$; therefore, $ag_1 K = ag_2 kK = ag_2 K$. This action is transitive, i.e., given any two points of the manifold M, one of them is mapped to the other by some element of the group G.

The *isotropy group* of a point $m \in M$ with respect to the action of the group G is defined as the set of those $a \in G$ for which $am = m$. For the point o, the isotropy group is the group K, and for a point gK, the isotropy group is the group gKg^{-1}.

A *homogeneous Riemannian space* is a Riemannian manifold M on which its Lie group G of isometries acts transitively. A homogeneous Riemannian space has the form G/K, where K is the isotropy group of some point. Indeed, choose a point $m \in M$ and let K be the isotropy group of this point. Then the map $G/K \to M$ which assigns the element $gm \in M$ to every class $gK \in G/K$ is bijective.

Example 6.1 The sphere $\mathbb{S}^n \subset \mathbb{R}^{n+1}$ is a homogeneous Riemannian space of the form $O(n + 1)/O(n) \cong SO(n + 1)/SO(n)$.

Proof The isometry group of the sphere $\mathbb{S}^n$ is the group $O(n + 1)$ of orthogonal transformations. It acts transitively. The isotropy group of the point $(1, 0, \dots, 0) \in \mathbb{S}^n$ is the subgroup $O(n) \subset O(n + 1)$ consisting of all matrices of the form $\begin{pmatrix} 1 & 0 \\ 0 & A \end{pmatrix}$.
□

In Euclidean and hyperbolic spaces and the sphere the isometry group acts transitively not only on points but also on each tangent space. Homogeneous spaces are not so symmetric: for these spaces, only the action of the isometry group on points is required to be transitive. In a homogeneous space all points have identical geometric properties, but the geometries in different directions at a point may be different.

Consider the natural projection $p \colon G \to G/K$ mapping each element $g \in G$ to the point $gK \in G/K$. Let us calculate the differential $p_*|_e \colon \mathfrak{g} \to T_o(G/K)$ of this map at the point e. Let $X \in \mathfrak{g}$. The one-parameter group $\exp(tX)$ represents this vector; therefore,

$$p_* X = \frac{d}{dt}(p \circ \exp(tX))\Big|_{t=0} = \frac{d}{dt}(\exp(tX)K)\Big|_{t=0}.$$

Problem 6.18 Prove that $\operatorname{Ker} p_*|_e = \mathfrak{k}$ is the Lie algebra of the Lie group K.

The map $p_*|_e \colon \mathfrak{g} \to T_o(G/K)$ is surjective; therefore, the relation $\operatorname{Ker} p_*|_e = \mathfrak{k}$ implies the isomorphism $T_o(G/K) \cong \mathfrak{g}/\mathfrak{k}$.

A homogeneous space G/K is a generalization of a Lie group G: a homogeneous space G/K is a Lie group in the special case where K is the trivial subgroup. The geometry of homogeneous spaces is most similar to that of Lie groups in the case of reductive homogeneous spaces. A homogeneous space G/K is said to be *reductive* if there exists a subspace $\mathfrak{m} \subset \mathfrak{g}$ such that $\mathfrak{g} = \mathfrak{k} \oplus \mathfrak{m}$ and $\operatorname{Ad}_k \mathfrak{m} \subset \mathfrak{m}$ for all $k \in K$. If a homogeneous space G/K is reductive, then the map $p_*|_e \colon \mathfrak{g} \to T_o(G/K)$ induces the canonical isomorphism $\mathfrak{m} \cong T_o(G/K)$, which is a generalization of the canonical isomorphism $\mathfrak{g} \cong T_e(G)$.

Problem 6.19

(a) Prove that if a homogeneous space G/K is reductive and $\mathfrak{g} = \mathfrak{k} \oplus \mathfrak{m}$ is the corresponding $\mathrm{Ad}(K)$-invariant decomposition, then $[\mathfrak{k}, \mathfrak{m}] \subset \mathfrak{m}$.
(b) Prove that if $K \subset G$ is a connected closed subgroup and $\mathfrak{m} \subset \mathfrak{g}$ is a subspace for which $\mathfrak{g} = \mathfrak{k} \oplus \mathfrak{m}$ and $[\mathfrak{k}, \mathfrak{m}] \subset \mathfrak{m}$, then the homogeneous space G/K is reductive.

For example, a homogeneous space G/K is reductive if the subgroup K is compact. For $\mathfrak{m}$ we can be take the orthogonal complement to $\mathfrak{k}$ with respect to an $\mathrm{Ad}(K)$-invariant inner product on $\mathfrak{g}$. An $\mathrm{Ad}(K)$-invariant inner product on $\mathfrak{g}$ can be constructed by using invariant integration over the compact group K. Namely, we can take any inner product $g(X, Y)$ on $\mathfrak{g}$ and average it over the group K, i.e., consider the integral

$$(X, Y) = \int_K g(\mathrm{Ad}_k(X), \mathrm{Ad}_k(Y))dk.$$

The inner product thus obtained is invariant with respect to Ad_k for all $k \in \mathfrak{k}$.

Yet another example of a reductive homogeneous space arises in the following situation.

Example 6.2 Let K be a closed subgroup of a connected Lie group G, and let σ be an involutive automorphism of G leaving all elements of K fixed and such that the connected component of the fixed point set containing the identity element lies in K. Then G/K is a reductive homogeneous space.

Proof Let us show that (1) $\mathfrak{k} = \{X \in \mathfrak{g} | d\sigma(X) = X\}$, (2) $\mathfrak{g} = \mathfrak{k} \oplus \mathfrak{m}$, where $\mathfrak{m} = \{X \in \mathfrak{g} | d\sigma(X) = -X\}$, and (3) $\mathrm{Ad}_k(\mathfrak{m}) \subset \mathfrak{m}$ for all $k \in K$.

(1) Since the map σ leaves all elements of K fixed, it follows that $d\sigma(X) = X$ for $X \in \mathfrak{k}$. Suppose that $d\sigma(X) = X$ and consider the one-parameter subgroup $\gamma(t)$ determined by the vector X. The curves γ and $\sigma \circ \gamma$ have the same initial velocity vector; therefore, $\sigma \circ \gamma = \gamma$, i.e., all points of the curve γ are fixed. Since the curve γ passes through the identity element of G, it follows that it lies in the connected component of the fixed point set containing the identity element, and hence it lies in K. Thus, $X \in \mathfrak{k}$.
(2) Any element $X \in \mathfrak{g}$ can be represented in the form $X = X_{\mathfrak{k}} + X_{\mathfrak{m}}$, where $X_{\mathfrak{k}} = \frac{1}{2}(X + d\sigma(X))$ and $X_{\mathfrak{m}} = \frac{1}{2}(X - d\sigma(X))$. Since the map σ is involutive, it follows that so is $d\sigma$, which implies $d\sigma(X_{\mathfrak{k}}) = X_{\mathfrak{k}}$ and $d\sigma(X_{\mathfrak{m}}) = -X_{\mathfrak{m}}$.
(3) We must show that, given $X \in \mathfrak{m}$ and $k \in K$, we have $d\sigma(\mathrm{Ad}_k(X)) = -\mathrm{Ad}_k(X)$. The equation $\sigma(k) = k$ implies $\sigma(kxk^{-1}) = k\sigma(x)k^{-1}$; therefore, $d\sigma$ and Ad_k commute. Thus,

$$d\sigma(\mathrm{Ad}_k(X)) = \mathrm{Ad}_k(d\sigma(X)) = \mathrm{Ad}_k(-X) = -\mathrm{Ad}_k(X).$$

$\square$

The reductive homogeneous spaces in Example 6.2 are symmetric (see Sect. 6.9).

Let $M = G/K$ be a reductive homogeneous space, and let $\mathfrak{g} = \mathfrak{k} \oplus \mathfrak{m}$ be the corresponding $\mathrm{Ad}(K)$-invariant decomposition. The translations $\tau_a(gK) = agK$ act on the manifold M. Given $k \in K$, the point $o \in M$ is fixed under the action τ_k; therefore, the map $d\tau_k$ acts on $T_o(M)$. It follows from the $\mathrm{Ad}(K)$-invariance of the decomposition that the group $\mathrm{Ad}(K)$ acts on $\mathfrak{m}$. Let us show that the projection $p \colon G \to G/K$ takes the group $\mathrm{Ad}(K)$ acting on $\mathfrak{m}$ to the group of transformations of the form $d\tau_k$, $k \in K$. It is easy to check that $\tau_k \circ p = p \circ I_k$, where $I_k(g) = kgk^{-1}$. Indeed, $\tau_k(p(g)) = \tau_k(gK) = kgK$ and $p(I_k(g)) = p(kgk^{-1}) = kgK$. Differentiating the equation $\tau_k \circ p = p \circ I_k$ at the point $e \in G$, we obtain $d\tau_k \circ dp = dp \circ \mathrm{Ad}_k$.

Theorem 6.12 *Let $M = G/K$ be a reductive homogeneous space, and let $\mathfrak{g} = \mathfrak{k} \oplus \mathfrak{m}$ be the corresponding $\mathrm{Ad}(K)$-invariant decomposition. Then the requirement that the map $dp \colon \mathfrak{m} \to T_o(M)$ is a linear isometry establishes a one-to-one correspondence between the $\mathrm{Ad}(K)$-invariant inner products on $\mathfrak{m}$ and the G-invariant metrics on M.*

Proof Consider an $\mathrm{Ad}(K)$-invariant inner product $(\,,\,)$ on $\mathfrak{m}$. The condition that the map $dp \colon \mathfrak{m} \to T_o(M)$ is a linear isometry defines an inner product $(\,,\,)_o$ on $T_o(M)$. The projection $p \colon G \to G/K$ takes the group $\mathrm{Ad}(K)$ acting on $\mathfrak{m}$ by linear isometries to the group of transformations of the form $d\tau_k$, $k \in K$. Therefore, for each $k \in K$, the transformation $\tau_k \colon T_o(M) \to T_o(M)$ is a linear isometry.

We use this fact to extend the inner product $(\,,\,)_o$ to the whole manifold M in a G-invariant way. Let us check that if $\tau_a(o) = \tau_b(o) = x$ for some $a, b \in G$, then the linear isomorphisms $\tau_{a^{-1}}, \tau_{b^{-1}} \colon T_x(M) \to T_o(M)$ transform the inner product $(\,,\,)_o$ into the same inner product $(\,,\,)_x$ on $T_x(M)$. Indeed, it follows from $\tau_a(o) = \tau_b(o)$ that $aK = bK$, i.e., $b^{-1}a = k \in K$. Therefore, $\tau_{b^{-1}} \circ \tau_a = \tau_k$ is a linear isometry of the space $T_o(M)$, and

$$(\tau_{a^{-1}}(U), \tau_{a^{-1}}(V))_o = (d\tau_k \circ \tau_{a^{-1}}(U), d\tau_k \circ \tau_{a^{-1}}(V))_o$$

$$= (\tau_{b^{-1}}(U), \tau_{b^{-1}}(V))_o.$$

The Riemannian metric on M obtained by such a spread of inner product is G-invariant.

Conversely, if on M a G-invariant Riemannian metric $(\,,\,)$ is given, then the transformations $d\tau_k|_o$, where $k \in K$, are linear isometries. The map $dp_\mathfrak{m}$ must be a linear isometry. Therefore, according to the remark before the theorem, the map dp transforms $(\,,\,)_o$ into an $\mathrm{Ad}(K)$-invariant inner product on $\mathfrak{m}$.   □

Theorem 6.12 shows that G-invariant metrics on a reductive homogeneous space G/K are a generalization of left-invariant metrics on the Lie group G. An analogue of Lie groups with a bi-invariant metric is *naturally reductive homogeneous spaces*.

These are reductive homogeneous spaces $M = G/K$ with a G-invariant metric whose restriction to the subspace $\mathfrak{m}$ has the property

$$([X, Y]_{\mathfrak{m}}, Z) = (X, [Y, Z]_{\mathfrak{m}})$$

for all $X, Y, Z \in \mathfrak{m}$.

For a naturally reductive homogeneous space, geodesic lines can be described. We give this description only briefly; detailed proofs can be found in [ON2].

The subspace $\mathfrak{m} \subset \mathfrak{g}$ is equipped with an inner product. We extend it to the whole space $\mathfrak{g} = \mathfrak{m} \oplus \mathfrak{k}$ as follows. Choose an arbitrary inner product in space $\mathfrak{k}$ and assume that $\mathfrak{k} \perp \mathfrak{m}$. This defines a left-invariant metric on G with respect to which all elements of $\mathfrak{k}$ are vertical (tangent to the fibers of the bundle $p: G \to G/K$) and all elements of $\mathfrak{m}$ are horizontal (perpendicular to the fibers).

To describe geodesics, we need the notion of a Riemannian submersion. A submersion is a smooth map of smooth manifolds whose differential is surjective at each point. A submersion $p: M \to B$ of Riemannian manifolds is called a Riemannian submersion if the map $(\operatorname{Ker} dp)^{\perp} \to TB$ induced on each tangent space is an isometry, i.e., the horizontal tangent subspace at each point $x \in M$ is mapped isometrically to the tangent space at the point $p(x) \in B$. Under a Riemannian submersion the horizontal geodesics on the manifold M are mapped to geodesics on B. For a naturally reductive space, the map $G/K \to M$ is a Riemannian submersion. The one-parameter subgroup $\gamma(t)$ of G determined by a tangent vector $X \in \mathfrak{m}$ is horizontal. Therefore, the projection $p: G \to G/K = M$ takes this geodesic to a geodesic on the homogeneous space M.

6.9   Symmetric Spaces

Consider a complete connected Riemannian manifold M^n. Given a point $p \in M^n$, an isometry $s_p: M^n \to M^n$ is called a *geodesic symmetry* centered at p if $s_p(\exp_V(t)) = \exp_V(-t)$ for all $t \in \mathbb{R}$ and $V \in T_p M^n$. The point p is an isolated fixed point of the geodesic symmetry s_p, and s_p^2 is the identity map. The differential of the isometry s_p at the point p is the identity map with the minus sign.

A Riemannian manifold M^n is called a *symmetric space* if, for any point $p \in M^n$, there exists a geodesic symmetry centered at p.

Remark   The equation $s_p(\exp_V(t)) = \exp_V(-t)$ completely determines a self-map s_p of a complete connected Riemannian manifold, provided that there exists a map satisfying this equation. However, such a map does not exist if $\exp_V(t_1) = \exp_V(t_2)$ and $\exp_V(-t_1) \neq \exp_V(-t_2)$. Moreover, even if such a map exits, it is not necessarily an isometry.

Theorem 6.13   *A symmetric space is a Riemannian homogeneous space.*

Proof We will restrict ourselves to the proof that the isometry group of a symmetric space acts on it transitively; we will not prove the Myers–Steenrod theorem that the isometry group of a Riemannian space is a Lie group.

Take any geodesic of finite length with endpoints p and q and consider the symmetry about the point dividing this geodesic into two parts of equal length. This isometry takes the point p to q. By definition any symmetric space is connected, that is, any two of its points can be joined by a curve consisting of several geodesic links. One endpoint of each link can be isometrically mapped to the other one; thus, any point of a symmetric space can be transferred to any other point by an isometry.

□

Theorem 6.14 *The reductive homogeneous space* $M = G/K$ *in Example 6.2 equipped with any* G*-invariant Riemannian metric is a Riemannian symmetric space, and the geodesic symmetry* ζ *centered at o satisfies the relation* $\zeta \circ p = p \circ \sigma$.

Proof A map $\zeta : M \to M$ satisfying the relation $\zeta \circ p = p \circ \sigma$ exists and is unique. Indeed, let us set $\zeta(pg) = p(\sigma g)$ for any $g \in G$ and check that this map is well-defined. If $pg_1 = pg_2$, then $g_1 K = g_2 K$. Since the map σ leaves fixed all elements of the group K, it follows that $\sigma(g_1)K = \sigma(g_2)K$, i.e., $p\sigma(g_1) = p\sigma(g_2)$.

The map ζ is smooth (this follows from the existence of a local section of the submersion p). The map σ is an involution, and hence so is ζ.

Let us show that $d\zeta = -\mathrm{id}$ at the point o. Clearly, $\zeta(o) = o$. If $Y \in T_o(M)$, then, as we showed in considering Example 6.2, there exists a $Y \in \mathfrak{g}$ for which $d\sigma(Y) = -Y$ and $dp(Y) = y$. Therefore,

$$d\zeta(y) = d\zeta(dp(Y)) = dp(d\sigma(Y)) = dp(-Y) = -y.$$

We have $\tau_{\sigma g} = \zeta \tau_g \zeta$ for all $g \in G$. Indeed, for any $x \in G$,

$$\zeta \tau_g p(x) = \zeta p(gx) = p\sigma(gx) = p(\sigma g \cdot \sigma x) = \tau_{\sigma g} p(x) = \tau_{\sigma g} \zeta p(x).$$

The map ζ is an isometry for any G-invariant metric on M. Indeed, setting $V_o = d\tau_{g^{-1}} V \in T_o(M)$ for each $V \in T_g(M)$, we see that

$$(d\zeta(V), d\zeta(V)) = (d\zeta \, d\tau_g(V_o), d\zeta \, d\tau_g(V_o))$$

$$= (d\tau_{\sigma g} d\zeta(V_o), d\tau_{\sigma g} d\zeta(V_o)) = (d\zeta(V_o), d\zeta(V_o))$$

$$= (-V_o, -V_o) = (V, V).$$

It is also clear that if a Riemannian homogeneous space has a geodesic symmetry ζ with respect to o, then it has a geodesic symmetry with respect to any point $p = \tau(o)$, namely, $\tau \zeta \tau^{-1}$.

□

HISTORICAL COMMENT In 1926 Cartan began to study Riemannian symmetric spaces; he classified them by 1932.

6.10   Solutions of Problems

6.1 For the constant $y = e$ (the identity element of the group), the map $(x, e) \mapsto x^{-1}e = x^{-1}$ is smooth; therefore, so is the map $x \mapsto x^{-1}$. The composition $(x, y) \mapsto (x^{-1}, y) \mapsto (x^{-1})^{-1}y = xy$ is smooth as well.

6.2 The required vector field can be obtained by taking n linearly independent vectors in the tangent space at e and extending them to left-invariant vector fields on the Lie group.

6.3 It is required to prove that if vector fields X and Y are left-invariant, then $L_{g*}[X, Y]_p = [X, Y]_{gp}$ for all $p \in G$ and all $g \in G$. To this end, it suffices to check that the actions of both vectors on a function f yield the same result. Clearly,

$$
\begin{aligned}
(L_{g*}[X, Y]_p)f &= [X, Y]_p(f \circ L_g) = X_p(Y(f \circ L_g)) - Y_p(X(f \circ L_g)) \\
&= X_p(Y(f) \circ L_g) - Y_p(X(f) \circ L_g) \\
&= L_{g*}(X_p)(Y(f)) - L_{g*}(Y_p)(X(f)) \\
&= X_{gp}(Y(f)) - Y_{gp}(X(f)) = ([X, Y]_{gp})f.
\end{aligned}
$$

6.4 Suppose that a tangent vector V at the identity matrix I is determined by a curve $V(t)$, i.e., $V(0) = I$ and $V'(0) = V$. For example, we can take the curve $V(t) = I + tV$. At a point $X \in G$ the left and right translations of this vector are determined by the curves $XV(t) = X + tXV$ and $V(t)X = X + tVX$. The left-invariant commutator of vectors V and W is the commutator at the point I of the vector fields given at X by the curves $X + tXV$ and $X + tXW$, and the right-invariant one is the commutator of vector fields determined by the curves $X + tVX$ and $X + tWX$.

Let us calculate these commutators. Note that the action of a vector field on the linear functions completely determines this field, and therefore it suffices to consider linear functions. Take the linear function $f_A(X) = AX$. On the curve $X + tXW$ this function equals $AX + tAXW$; the derivative of this expression with respect to t is AXW. Thus, under the action of the left-invariant field W^L on f_A we obtain the function $\varphi(X) = AXW$. The action on this function of the left-invariant vector field V^L, which is determined by the curves $X + tXV$, is as follows. First, the expression $A(X + tXV)W = AXW + tAXVW$ is obtained; differentiating this expression with respect to t and setting $X = I$, we obtain AVW. It follows that the left-invariant commutator acts on f_A as $f_A \mapsto A(VW - WV)$, which corresponds to the commutator of the matrices V and W.

Now let us calculate the commutator of right-invariant vector fields. The vector field W^R acts of the linear function $f_A(X) = AX$ as follows: first, the expression $f_A(X + tWX) = AX + tAWX$ is obtained, and then its derivative with respect to t is taken, which equals AWX. Thus, the action yields the function $\varphi(X) = AWX$. To find the result of the action of the vector field V^R on this function, we differentiate the expression $AW(X + tVX) = AWX + tAWVX$ with respect to t and set $X = I$.

As a result, we obtain AWV. It follows that the right-invariant commutator acts on A as $f_A \mapsto A(WV - VW)$, which corresponds to the commutator of the matrices V and W with the minus sign.

6.5 Consider the curve $\gamma(t) = (\gamma_1(t), \gamma_2(t))$, where $\gamma_1(0) = g$, $\gamma_1'(0) = V_g$, $\gamma_2(0) = h$, and $\gamma_2'(0) = W_h$. The map μ takes this curve to the curve $\gamma_1(t)\gamma_2(t)$. The tangent vector to this curve at $t = 0$ equals

$$\frac{d}{dt}(\gamma_1(t)\gamma_2(t))\Big|_{t=0} = \frac{d}{dt}(\gamma_1(t)h)\Big|_{t=0} + \frac{d}{dt}(g\gamma_2(t))\Big|_{t=0} = R_{h*}V_g + L_{g*}W_h.$$

6.6

(a) Suppose that a vector V_g is represented by a curve $X(t)$, i.e., $X(0) = g$ and $X'(0) = V_g$. Then the vector $\mathrm{Inv}_* V_g$ is represented by the curve $X^{-1}(t)$. Let us differentiate the equation $X(t)X^{-1}(t) = I$, where I is the identity matrix. For $t = 0$, we obtain $V_g g^{-1} + g\,\mathrm{Inv}_* V_g = 0$, i.e., $\mathrm{Inv}_* V_g = -g^{-1}V_g g^{-1}$.

(b) Let us apply the formula of Problem 6.5 to the case where $h = g^{-1} = \mathrm{Inv}(g)$ and $W_h = \mathrm{Inv}_* V_g$. We have $\mu(g, \mathrm{Inv}(g)) = e$, which is a constant quantity, whence $\mu_*(V_g, \mathrm{Inv}_* V_g) = 0$, i.e., $R_{g^{-1}*}V_g + L_{g*}\,\mathrm{Inv}_* V_g = 0$. Therefore,

$$\mathrm{Inv}_* V_g = -(L_{g*})^{-1}R_{g^{-1}*}V_g.$$

6.7 Using the formula of Problem 6.6, we obtain $(\mathrm{Inv}_* X^L)_g = \mathrm{Inv}_*(X^L_{g^{-1}}) = -(L_{g^{-1}*})^{-1}R_{g*}(X^L_{g^{-1}}) = -(L_{g^{-1}*})^{-1}R_{g*}L_{g^{-1}*}X_e$. Since the operations L_{g*} and R_{h*} commute, it follows that $(L_{g^{-1}*})^{-1}R_{g*}L_{g^{-1}*}X_e = R_{g*}(L_{g^{-1}*})^{-1}L_{g^{-1}*}X_e = R_{g*}X_e = X^R_g$.

6.8 Consider the vector $W_e = [X^L, Y^L]_e$. It generates the left-invariant vector field W^L and the right-invariant vector field W^R. According to Problem 6.7, we have $(\mathrm{Inv}_*[X^L, Y^L])_e = (\mathrm{Inv}_* W^L)_e = -(W^R)_e = -[X^L, Y^L]_e$ and $(\mathrm{Inv}_*[X^L, Y^L])_e = [\mathrm{Inv}_* X^L, \mathrm{Inv}_* Y^L]_e = [-X^R, -Y^R]_e = [X^R, Y^R]_e$. The left-hand sides of these equations are equal, and hence so are the right-hand sides.

6.9 The required identity follows from (6.3), because $\mathrm{Ad}_A(X_e) = AX_e A^{-1}$ for a matrix Lie group. It can also be easily derived directly from the series expansion of the matrix exponential by using the identity $(AXA^{-1})^n = AX^n A^{-1}$.

6.10 We set $A(t) = \mathrm{Ad}_{\exp(tX)}$ and calculate the derivative:

$$A'(t)Y = \frac{d}{dt}(\exp(tX)Y\exp(-tX))$$

$$= X\exp(tX)Y\exp(-tX) + \exp(tX)Y\exp(-tX)(-X)$$

$$= (\mathrm{ad}_X)\,\mathrm{Ad}_{\exp(tX)}\,Y = (\mathrm{ad}_X)A(t).$$

Thus, $A'(t) = UA(t)$, where $U = \text{ad}_X$ and $A(0) = e$. The unique solution of this equation has the form $A(t) = \exp(tU)$. For $t = 1$, we obtain the required identity.

6.11 In a sufficiently small neighborhood U of the identity element of a Lie group G, for each $g \in U$, we can choose a unique $X \in \mathfrak{g}$ satisfying the condition $g = \exp(X)$. Recall that $\text{Ad}_{\exp(X)} = e^{\text{ad}_X}$ according to Problem 6.10. Therefore, $\text{Ad}_g = \text{Ad}_{\exp(X)} = e^{\text{ad}_X}$. The skew symmetry of the matrix ad_X implies the orthogonality of the matrix $\text{Ad}_g = e^{\text{ad}_X}$.

In a connected Lie group any element can be represented as a product of elements in a neighborhood of the identity element. Clearly, the product of orthogonal matrices is an orthogonal matrix.

6.12 It was shown in the solution of Problem 6.6 that $\text{Inv}_* V_g = -(L_{g*})^{-1} R_{g^{-1}*} V_g$. By assumption both maps $R_{g^{-1}*}$ and L_{g*} are isometries.

6.13 The composition under consideration is given by $(g, V) \mapsto (g, (R_{g^{-1}})_*(L_g)_* V) = (g, (R_{g^{-1}} L_g)_* V) = (g, \text{Ad}_g(V))$.

6.14 Consider a left-invariant vector field $a^j X_j$. The values of both the left-invariant form ω and the left-invariant form $X_{ei} \otimes \omega^i$ at this vector field are equal to $a^j X_{ej}$.

6.15 Since the Lie group G is compact, we can choose an Ad-invariant inner product in the space $\mathfrak{g}$. With respect to this inner product each operator ad_X is skew-symmetric. The matrix (a_{ij}) of this operator in an orthonormal basis of the space $\mathfrak{g}$ is skew-symmetric, whence

$$B(X, X) = \text{tr}(\text{ad}_X \circ \text{ad}_X) = \sum_i \sum_j a_{ij} a_{ji} = -\sum_{i,j} a_{ij}^2 < 0.$$

6.16 According to Theorem 6.1, the displacement during time $-t$ along a trajectory of a left-invariant vector field X starting at e is the transformation $R_{\exp(-tX_e)}$. Therefore, the required relation follows directly from the fact that the commutator of vector fields is the Lie derivative.

6.17 According to Theorem 6.1, the flow X_t of a right-invariant vector field X on a Lie group is the left translation by the image of the identity element, i.e., $X_t(g) = \exp(tX)g$. On the Lie group with a left-invariant metric the transformation X_t is an isometry for each t. Therefore, X is a Killing field.

6.18 If $X \in \mathfrak{k}$, then $\exp(tX)K = K$ for all t. Therefore, $p_* X = 0$.

Now suppose that $p_* X = 0$. This means that, for any function f on G/K, we have

$$\left. \frac{d}{dt} f(\exp(tX)K) \right|_{t=0} = 0.$$

In particular, this equation holds for the function $f_s(gK) = f(\exp(sX)gK)$, whence

$$0 = \frac{d}{dt} f_s(\exp(tX)K)\Big|_{t=0}$$

$$= \frac{d}{dt} f(\exp(sX)\exp(tX)K)\Big|_{t=0}$$

$$= \frac{d}{dt} f(\exp((s+t)X)K)\Big|_{t=0}.$$

Let $r = s + t$. Then

$$\frac{d}{dr} f(\exp(rX)K)\Big|_{r=s} = \frac{d}{dt} f(\exp((s+t)X)K)\Big|_{t=0} = 0.$$

Thus, $f(\exp(rX)K)$ does not depend of r, so that $\exp(rX)K = K$ for any r, i.e., the curve $\exp(rX)$ lies in K. This means that $X \in \mathfrak{k}$.

6.19

(a) For $X \in \mathfrak{k}$ and $Y \in \mathfrak{m}$, we have $\exp(tX) \in K$, whence it follows that $\mathrm{Ad}_{\exp(tX)}(Y) \in \mathfrak{m}$ and the curve

$$\gamma(t) = \mathrm{Ad}_{\exp(tX)}(Y) = e^{t\,\mathrm{ad}_X}(Y)$$

$$= (I + t\,\mathrm{ad}_X + \frac{t^2}{2}(\mathrm{ad}_X)^2 + \ldots)Y$$

$$= Y + t[X, Y] + \frac{t^2}{2}[X, [X, Y]] + \ldots$$

lies in the subspace $\mathfrak{m}$. Therefore, $[X, Y] = \gamma'(0) \in \mathfrak{m}$.

(b) If $X \in \mathfrak{k}$ and $Y \in \mathfrak{m}$, then $Y_1 = [X, Y] \in \mathfrak{m}$, whence $(\mathrm{ad}_X)^2(Y) = [X, [X, Y]] = [X, Y_1] \in \mathfrak{m}$. Similarly, $(\mathrm{ad}_X)^n(Y) \in \mathfrak{m}$ for all positive integers n. Therefore, $\mathrm{Ad}_{\exp(tX)}(Y) = e^{t\,\mathrm{ad}_X}(Y) \in \mathfrak{m}$ for all t, and hence $\mathrm{Ad}_k(Y) \in \mathfrak{m}$ for any element k in the subgroup generated by $\exp(\mathfrak{k})$. If the group K is connected, then this subgroup coincides with K.

Chapter 7
Comparison Theorems, Curvature and Topology, and Laplacian

7.1 The Simplest Comparison Theorems

We have already proved some comparison theorems for surfaces by using Sturm's theorem (see Theorem 3.26 on p. 119). Using the same theorem, we can also prove similar comparison theorems in the case of Riemannian manifolds, where the role of curvature is usually played by sectional curvatures. We will need Jacobi fields J on a geodesic γ and the Jacobi equation

$$\nabla^2_{\gamma'} J = R(\gamma', J)\gamma'.$$

We will consider normal Jacobi fields, i.e., those orthogonal to the geodesic.

Theorem 7.1 (comparison Theorem for Jacobi Fields) *Suppose that all sectional curvatures on a Riemannian manifold M are bounded above by a constant c. Let $\gamma(t)$ be a geodesic with natural parameter on M, and let $J(t)$ be a normal Jacobi field on γ for which $J(0) = 0$. Then the following assertions hold:*

- *if $c = 0$, then $\|J(t)\| \geqslant t\|\nabla_{\gamma'} J(0)\|$ for $t \geqslant 0$;*
- *if $c = \frac{1}{R^2} > 0$, then $\|J(t)\| \geqslant R \sin \frac{t}{R}\|\nabla_{\gamma'} J(0)\|$ for $0 \leqslant t \leqslant \pi R$;*
- *if $c = -\frac{1}{R^2} < 0$, then $\|J(t)\| \geqslant R \sinh \frac{t}{R}\|\nabla_{\gamma'} J(0)\|$ for $t \geqslant 0$.*

Proof The function $\|J(t)\|$ is smooth at $t \neq 0$. Let us calculate its second derivative by using the Jacobi equation. First, note that $\frac{d}{dt}(J, J) = 2(\nabla_{\gamma'} J, J)$, and hence

$$\frac{d}{dt}\|J\| = \frac{(\nabla_{\gamma'} J, J)}{(J, J)^{1/2}}.$$

© The Author(s), under exclusive license to Springer Nature Switzerland AG 2022
V. V. Prasolov, *Differential Geometry*, Moscow Lectures 8,
https://doi.org/10.1007/978-3-030-92249-8_7

Therefore,

$$\frac{d^2}{dt^2}\|J\| = \frac{d}{dt}\frac{(\nabla_{\gamma'}J, J)}{(J, J)^{1/2}}$$

$$= \frac{(\nabla_{\gamma'}^2 J, J)}{(J, J)^{1/2}} + \frac{(\nabla_{\gamma'}J, \nabla_{\gamma'}J)}{(J, J)^{1/2}} - \frac{(\nabla_{\gamma'}J, J)^2}{(J, J)^{3/2}}$$

$$= \frac{(R(\gamma', J)\gamma', J)}{\|J\|} + \frac{\|\nabla_{\gamma'}J\|^2}{\|J\|} - \frac{(\nabla_{\gamma'}J, J)^2}{\|J\|^3}.$$

The contribution of the last two terms is nonnegative, because $(\nabla_{\gamma'}J, J)^2 \leqslant \|\nabla_{\gamma'}J\|^2\|J\|^2$ by virtue of Cauchy's inequality.

The vectors J and γ' are orthogonal, and $\|\gamma'\| = 1$. Therefore, the sectional curvature of the plane spanned by J and γ' equals

$$\frac{(R(J, \gamma')\gamma', J)}{\|J\|^2} = -\frac{(R(\gamma', J)\gamma', J)}{\|J\|} \cdot \frac{1}{\|J\|}.$$

Thus, it follows from the assumptions of the theorem that

$$\frac{d^2}{dt^2}\|J\| \geqslant -c\|J\|$$

for $\|J\| > 0$.

Now we can use Sturm's theorem. We set $y(t) = k\|J(t)\|$ and choose a constant k so that $y'(0) = 1$. The function $y(t)$ satisfies the differential equation $y'' + a(t)y = 0$, where $a(t) \leqslant c$. Consider the solution of the differential equation $z'' + cz = 0$ satisfying the conditions $z(0) = 0$ and $z'(0) = 1$. (For $c = 0$, this solution has the form $z(t) = t$, for $c > 0$, it has the form $z(t) = R \sin \frac{t}{R}$, and for $c < 0$, it has the form $z(t) = R \sinh \frac{t}{R}$.) It can be derived from Sturm's theorem that $y(t) \geqslant z(t)$ if $t \geqslant 0$ and $z(t) \neq 0$. The solution $z(t)$ can vanish at a positive t only if $c > 0$; in this case, the minimum t at which $z(t)$ vanishes equals πR. Let us explain how Sturm's theorem can be applied in this situation. To apply it, we must make sure that $y(t)$ cannot vanish on the half-open interval on which $z(t) > 0$. For small t, the positivity of $y(t)$ follows from $y'(0) = 1$, and for larger t, the inequality $y(t) \geqslant z(t)$ shows that $y(t)$ cannot vanish until so does $z(t)$.

To obtain the required result, we must check that $k = \frac{1}{\|\nabla_{\gamma'}J(0)\|}$, i.e., $\frac{d}{dt}\|J(0)\| = \|\nabla_{\gamma'}J(0)\|$. For this purpose, we can use Theorem 5.22: the Jacobi vector field J for which $J(0) = 0$ and $\nabla_{\gamma'}J(0) = W^i\partial_i$ is given in normal coordinates by $J(t) = tW^i\partial_i$. Indeed, it follows from this relation that $\frac{d}{dt}\|J(0)\| = \|W^i\partial_i\|$. □

Theorem 7.2 (comparison Theorem for Conjugate Points) *Suppose that all sectional curvatures on a Riemannian manifold M are bounded above by a constant c. If $c \leqslant 0$, then M has no points conjugate along a geodesic, and if $c = \frac{1}{R^2} > 0$,* *then the distance from any point to a nearest point conjugate to it along some geodesic is at least πR.*

Proof Let $c \leqslant 0$. Then, according to Theorem 7.1, any nontrivial normal Jacobi field $J(t)$ for which $J(0) = 0$ does not vanish at $t > 0$. Moreover, if $c = \frac{1}{R^2} > 0$, then $J(t)$ does not vanish at $0 < t < \pi R$. $\square$

Theorem 5.25 (see p. 196) describes the metric of a space of constant curvature c in normal coordinates. Let $g_c(V, V)$ denote the metric specified in the statement of this theorem for a space of constant curvature c.

Theorem 7.3 (Comparison Theorem for a Metric) *Suppose that all sectional curvatures on a Riemannian manifold M with Riemannian metric g are bounded above by a constant c. Then $g(V, V) \geqslant g_c(V, V)$ in any normal coordinate system.*

Proof As in the proof of Theorem 5.25, we decompose the vector V in two components, a component V^T tangent to the geodesic sphere and a component $V^\perp$ directed along a radial geodesic, i.e., perpendicular to the geodesic sphere. From the orthogonality of these two components it follows that

$$g(V, V) = g(V^T, V^T) + g(V^\perp, V^\perp).$$

The velocity vector of a radial geodesic has the same length in the metrics g and g_c: it equals the length of this vector in Euclidean norm associated with the normal coordinates. Therefore, $g(V^\perp, V^\perp) = g_c(V^\perp, V^\perp)$. The vector V^T coincides with the vector of some normal Jacobi field vanishing at $t = 0$. Applying the comparison theorem for Jacobi fields to this vector, we obtain $g(V^T, V^T) \geqslant g_c(V^T, V^T)$. $\square$

7.2   The Cartan–Hadamard Theorem

Recall that an isometry is a diffeomorphism of Riemannian manifolds transforming the Riemannian metric of one of the manifold into the Riemannian metric of the other. A *local isometry* is a map of Riemannian manifolds which transforms the Riemannian metric of one of the manifolds into the Riemannian metric of the other. Any local isometry is a local diffeomorphism.

Problem 7.1 Suppose that local isometries φ_1 and φ_2 from a connected Riemannian manifold M onto a Riemannian manifold N coincide at a point $x \in M$ and the differentials of φ_1 and φ_2 at this point coincide as well. Prove that then the maps φ_1 and φ_2 coincide.

To prove the Cartan–Hadamard theorem, we need the following lemma about local isometries of geodesically complete manifolds. Recall that, according to the Hopf–Rinow theorem, the geodesic completeness of a Riemannian manifold is equivalent to its topological completeness.

A map $p: M \to N$ of connected manifolds is called a *covering* if each point $x \in N$ has a neighborhood U such that its preimage is homeomorphic to the product of U and a discrete set F and the restriction of p to $U \times F$ is the natural projection $U \times F \to U$. It is easy to show that, in this case, the sets F at different points $x \in N$ have the same cardinality. In particular, the image of the map p coincides with N.

Lemma *If manifolds M and N are connected and M is geodesically complete, then any local isometry $p: M \to N$ is a covering.*

Proof Consider any point $x \in N$ and choose a positive number r so that the map $\exp_x$ is a diffeomorphism from an open ball of radius r in $T_x N$ onto a geodesic ball U of radius r in the manifold N.

The preimage of a point x under p is a set of points y_i in the manifold M. For each point y_i, consider the image U_i of an open ball of radius r in $T_{y_i} M$ under the map $\exp_{y_i}$. The local isometry p takes each geodesic in M to a geodesic in N. Therefore, the radial geodesics in the geodesic ball U_i are mapped to radial geodesics in the geodesic ball U. Taking into account the geodesic completeness of the manifold M, we see that the map $p: U_i \to U$ is a diffeomorphism.

Now let us check that $p^{-1}(U)$ is contained in the union of the sets U_i. Take any point $q \in U$; it can be joined with the center x of the geodesic ball by a unique (up to parameterization) minimal geodesic γ. We assume that $\gamma(0) = q$ and $\gamma(s) = x$ and set $V = \gamma'(0)$. Let $\tilde{q}$ be a point in the preimage of q, i.e., such that $p(\tilde{q}) = q$. The space $T_{\tilde{q}} M$ contains a unique vector $\tilde{V}$ for which $p_* \tilde{V} = V$. Let $\tilde{\gamma}$ be a geodesic from $\tilde{q}$ with initial velocity $\tilde{V}$. Since the manifold M is geodesically complete, it follows that $y = \tilde{\gamma}(s)$ is defined. Clearly, $p(y) = x$, i.e., y is one of the points y_i. Thus, $q \in U_i$. □

The following example shows that the geodesic completeness assumption on the manifold M is essential.

Example Consider a covering $\mathbb{S}^2 \to \mathbb{R}P^2$. Let us introduce a metric on the projective plane $\mathbb{R}P^2$ so that the given covering is a local isometry. The given map restricted to the sphere $\mathbb{S}^2$ minus a point is a local isometry but not a covering.

Theorem 7.4 (Cartan–Hadamard) *Let M be a complete connected Riemannian manifold all of whose sectional curvatures are nonpositive. Then, for any point $y \in M$, the map $\exp_y: T_y M \to M$ is a covering. In particular, the universal covering of the manifold M is diffeomorphic to $\mathbb{R}^n$. Moreover, if M is simply connected, then M is itself diffeomorphic to $\mathbb{R}^n$.*

Proof Since all sectional curvatures are nonpositive, it follows that the manifold has no points conjugate along a geodesic. In turn, this implies that the map $\exp_y: T_y M \to M$ has no critical points. Therefore, the map $\exp_y: T_y M \to M$ is a local diffeomorphism. This makes it possible to construct a Riemannian metric on $T_y M$ as follows. We take a point $a \in T_y M$, consider the point $\exp_y(a) \in M$, and lift the Riemannian metric g at this point by means of the map $\exp_y$. As a result, we obtain a Riemannian metric $\tilde{g}$ on $T_y M$ for which $\exp_y$ is a local isometry.

To apply the lemma, we must verify that the Riemannian manifold $T_y M$ with metric $\tilde{g}$ is geodesically complete. First, note that the geodesics through the origin are straight lines, and the velocity of motion along them is constant. Indeed, these lines are mapped to parameterized geodesics on the manifold M. Now we can apply the fact that a connected Riemannian manifold is geodesically complete if it has a point such that the map exp is defined on the entire tangent space to this point (see the remark to the Hopf–Rinow theorem on p. 178). □

HISTORICAL COMMENT For surfaces in Euclidean space, Theorem 7.4 was first proved by Hans Carl Friedrich von Mangoldt (1854–1925) in 1881. In 1898 it was proved independently by Hadamard. The general form of this theorem was proved by Cartan in 1928.

7.3 Manifolds of Positive Curvature

The simplest comparison theorems of Sect. 7.1 deal with the situation where the sectional curvature is bounded above. In the case where it is bounded below by a positive constant, the following statement can be proved by a similar method.

Theorem 7.5 *Suppose that all sectional curvatures on a Riemannian manifold M are bounded below by a positive constant $c = \frac{1}{R^2}$. Then any part of a geodesic on M which is longer than πR contains conjugate points.*

Proof In the proof of Theorem 7.1 we showed that the function $y(t) = k\|J(t)\|$ dominates the function $z(t) = R \sin t R$. Now we have the reverse inequality, so that the functions y and z should be interchanged: the function $y(t) = R \sin t R$ dominates the function $z(t) = k\|J(t)\|$ (in the region where both these functions are positive). Since the function $y(t)$ vanishes at $t = \pi R$, it follows that $z(t)$ must vanish at $t \leqslant \pi R$. □

We define the *diameter* of a Riemannian manifold to be the least upper bound for the distances between its points. For a sphere of radius R, the diameter thus defined equals πR (the longest distance is between diametrically opposite points).

Theorem 7.6 (Bonnet) *Let M be a complete connected Riemannian manifold all of whose sectional curvatures are bounded below by a positive constant $\frac{1}{R^2}$. Then M is compact and its diameter is at most πR.*

Proof First, we prove that the diameter of M is at most πR. Suppose that the distance between points x and y in M is greater that πR. According to the Hopf–Rinow theorem, they can be joined by a minimal geodesic. Since the length of this geodesic is greater than πR, it follows from Theorem 7.5 that it contains conjugate points. But such a geodesic cannot be minimal. This contradiction shows that the diameter of M is at most πR.

Now let us prove that the manifold M is compact. Take any point $x \in M$. It can be joined with any point $y \in M$ by a geodesic of length at most πR. Therefore, the image of the ball of radius R centered at the origin under the map $\exp_x$ coincides with M. Thus, M is the image of a compact set.   □

Using Bonnet's theorem, we can prove the following statement.

Theorem 7.7 *Let M be a complete connected Riemannian manifold all of whose sectional curvatures are bounded below by a positive constant. Then the fundamental group of M is finite.*

Proof Consider the universal covering $p \colon \tilde{M} \to M$. The preimage of any point $x \in M$ under p is a discrete set of cardinality equal to that of the fundamental group. Therefore, it suffices to prove that the manifold $\tilde{M}$ is compact.

The map p can be used to lift the metric g of M to the manifold $\tilde{M}$: the lifted metric is defined by $\tilde{g} = p^*g$. For this metric, all sectional curvatures are bounded below by the same positive constant. Moreover, it is complete. Indeed, the map p is a local isometry, so that we can take a geodesic on $\tilde{M}$, map it to a geodesic on M, extend the latter (using the completeness of M), and then lift the extended geodesic to $\tilde{M}$.   □

The statement concerning the boundedness of the diameter of a Riemannian manifold remains valid under weaker assumptions than those of Bonnet's theorem. It is sufficient to require the boundedness of the Ricci tensor rather than of all sectional curvatures. The assertions about the compactness of a complete Riemannian manifold and the finiteness of its fundamental group remain valid under these assumptions, and their proofs do not change.

Theorem 7.8 (Myers) *Let M be a complete connected n-dimensional Riemannian manifold for which*

$$\mathrm{Ric}(V, V) \geqslant \frac{n-1}{R^2}\|V\|^2.$$

Then the diameter of M is at most πR.

Proof Consider two points x and y in the manifold M. According to the Hopf–Rinow theorem, they can be joined by a minimal geodesic γ. Suppose that the length of this geodesic equals l. We assume that the parameterization of this geodesic is natural and $\gamma(0)$ is its initial point.

Let us apply the second variation formula. For a geodesic γ of length l with velocity vector γ', the quadratic form E_{**} is calculated by

$$E_{**}(W, W) = -\int_0^l (W, (\nabla_{\gamma'})^2 W - R(\gamma', W)\gamma')dt,$$

where W is the variation of the geodesic. For a minimal geodesic, we have $E_{**}(W, W) \geqslant 0$.

Choose a basis $e_1, \ldots, e_{n-1}$ in the orthogonal complement to the vector $\gamma'(0)$ and consider its parallel transport along the geodesic. We obtain vector fields $X_1(t)$, $\ldots, X_{n-1}(t)$. For $i = 1, \ldots, n-1$, consider the vector fields

$$Y_i(t) = \sin\left(\frac{\pi t}{l}\right) X_i(t).$$

Each vector field X_i is parallel along the geodesic, and hence $\nabla_{\gamma'} X_i = 0$. Therefore, $\nabla_{\gamma'} Y_i = \frac{\pi}{l} \cos\left(\frac{\pi t}{l}\right) X_i$ and $(\nabla_{\gamma'})^2 Y_i = -\frac{\pi^2}{l^2} \sin\left(\frac{\pi t}{l}\right) X_i$. Thus,

$$E_{**}(Y_i, Y_i) = -\int_0^l (Y_i, (\nabla_{\gamma'})^2 Y_i - R(\gamma', Y_i)\gamma')dt$$

$$= \int_0^l \sin^2\left(\frac{\pi t}{l}\right)\left(\frac{\pi^2}{l^2} + R(\gamma', X_i, \gamma', X_i)\right) dt.$$

According to Problem 5.9,

$$\mathrm{Ric}(\gamma', \gamma') = \sum_{i=1}^{n-1} K(X_i, \gamma') = \sum_{i=1}^{n-1} R(X_i, \gamma', \gamma', X_i) = -\sum_{i=1}^{n-1} R(\gamma', X_i, \gamma', X_i).$$

Therefore,

$$\int_0^l \sin^2\left(\frac{\pi t}{l}\right)\left((n-1)\frac{\pi^2}{l^2} - \mathrm{Ric}(\gamma', \gamma')\right) dt = \sum_{i=1}^{n-1} E_{**}(Y_i, Y_i) \geqslant 0.$$

By assumption $\mathrm{Ric}(\gamma', \gamma') \geqslant \frac{n-1}{R^2}$. Therefore, $\frac{\pi^2}{l^2} \geqslant \frac{1}{R^2}$, which means that $l \leqslant \pi R$.

$\square$

HISTORICAL COMMENT Sumner Byron Myers (1910–1955) proved Theorem 7.8 in 1941.

7.4 Manifolds of Constant Curvature

Theorem 5.25 on p. 196 describes the local structure of a space of constant curvature. Now, applying the Cartan–Hadamard theorem and the comparison theorem for conjugate points, we can also describe the global structure of a complete simply connected space of constant curvature (i.e., the universal covering of any complete space of constant curvature).

Theorem 7.9 *A complete simply connected space of constant curvature is isometric to either hyperbolic space, Euclidean space, or the sphere.*

Proof To a space of constant negative or zero curvature we apply the Cartan–Hadamard theorem, and to a space of constant positive curvature, the comparison theorem for conjugate points. Consider these two cases separately.

Let M be a complete simply connected space of constant negative or zero curvature. Then, according to the Cartan–Hadamard theorem, the map $\exp_x : T_x M \to M$ is a covering for each point $x \in M$. Since M is simply connected, it follows that $\exp_x : T_x M \to M$ is a diffeomorphism. Lifting the metric g of M to $T_x M$, we obtain a metric $\tilde{g}$ on $T_x M$, with respect to which the map $\exp_x : T_x M \to M$ is an isometry. The Euclidean coordinates on $T_x M$ are normal coordinates for the metric $\tilde{g}$; therefore, the metric $\tilde{g}$ is defined by one of the formulas in Theorem 5.25. Thus, M is isometric to a hyperbolic or Euclidean space.

Now suppose that M^n is a complete simply connected space of constant positive curvature $c = \frac{1}{R^2}$. Let us show that it is isometric to the sphere $\mathbb{S}^n_R$ of radius R. Take diametrically opposite points a and $-a$ on the sphere $\mathbb{S}^n_R$. The map $\exp_a$ is a diffeomorphism from an open ball of radius πR onto the sphere $\mathbb{S}^n_R$ minus the point $-a$. According to the comparison theorem for conjugate points, the distance between any conjugate points on the manifold M^n is at least πR; hence, for any point $x \in M^n$, the restriction of the map $\exp_x$ to an open ball of radius πR in $T_x M^n$ is a local diffeomorphism.

Take any isometry φ_a from an open ball of radius πR in $T_a \mathbb{S}^n_R$ onto an open ball of radius πR in $T_x M^n$. Consider the lifts to an open ball of radius πR in $T_a \mathbb{S}^n_R$ of the metric on the sphere (by means of the map $\exp_a$) and of the metric on M^n (by means of the map $\exp_x \circ \varphi$). These are two metrics of constant curvature $\frac{1}{R^2}$ on this ball. The Euclidean coordinates on the ball are normal coordinates for both metrics, because the radii of this ball are geodesics. Therefore, these metrics coincide. Thus, the map

$$\Phi_a = \exp_x \circ \varphi_a \circ \exp_a^{-1} : \mathbb{S}^n_R \setminus \{-a\} \to M^n$$

is a local isometry onto its image.

On the sphere $\mathbb{S}^n_R$ choose any point b different from a and $-a$ and consider the map Φ_b defined as follows. Let $\Phi_a(b) = y$. Consider the isometry φ_b from an open ball radius πR in $T_b \mathbb{S}^n_R$ onto an open ball of radius πR in $T_y M^n$ which is the restriction of the isometry $(\Phi_a)_* : T_b \mathbb{S}^n_R \to T_y M^n$. We set

$$\Phi_b = \exp_y \circ \varphi_b \circ \exp_b^{-1} : \mathbb{S}^n_R \setminus \{-b\} \to M^n.$$

This map is a local isometry onto its image.

The maps Φ_a and Φ_b take the point b to the point y. The differentials of these maps at the point b coincide as well: $(\Phi_a)_*(b) = (\varphi_b)_* = (\Phi_b)_*(b)$. Therefore, according to Problem 7.1, the maps Φ_a and Φ_b coincide wherever both maps are defined. Gluing together these maps, we obtain a globally defined local isometry from $\mathbb{S}^n_R$ onto M^n. Therefore, M^n is compact, and we can use the fact that a local diffeomorphism from a connected compact manifold onto a simply connected manifold is a (global) diffeomorphism (see, e.g., [Pr2]). $\square$

7.5 Laplace Operator

Let V be a vector space of dimension n equipped with an inner product and an orientation. Then we can define a linear operator, called the star operator, which maps the pth exterior power of the space V to its $(n-p)$th exterior power. Suppose that vectors $e_1, \ldots, e_p$ have unit length and are pairwise orthogonal. We complete them to a positively oriented orthonormal basis $e_1, \ldots, e_n$ and set

$$*(e_1 \wedge \cdots \wedge e_p) = e_{p+1} \wedge \cdots \wedge e_n.$$

To the whole space $\Lambda^p V$ the star operator is extended by linearity.

Problem 7.2 Prove that the map $**: \Lambda^p V \rightarrow \Lambda^p V$ is multiplication by $(-1)^{p(n-p)}$.

For an orientable Riemannian manifold M^n, the star operator is locally defined by

$$*(\omega_1 \wedge \cdots \wedge \omega_p) = \omega_{p+1} \wedge \cdots \wedge \omega_n,$$

where the basis $\omega_1, \ldots, \omega_n$ is dual to a positively oriented orthonormal basis.

According to Problem 7.2, given a p-form ω, we have $**\omega = (-1)^{p(n-p)}\omega$. Consider the operator δ defined by

$$\delta\omega = (-1)^{np+n+1} * d * \omega$$

(the choice of sign will become clear from Theorem 7.10 and its proof).

The operator δ is defined not only on an orientable, but also on a nonorientable manifold, because the star occurs twice in its definition. Indeed, we can take local coordinates. When the orientation of these coordinates changes, the star operator changes sign and the operator δ remains intact.

Now we can define the Laplace operator, or Laplacian:

$$\Delta = d \circ \delta + \delta \circ d.$$

Suppose that M^n is a closed manifold. Then in the space of p-forms on this manifold we can define (inner) product by

$$(\omega, \eta) = \int_{M^n} \omega \wedge *\eta.$$

This product is linear, and $(\omega, \eta) = (\eta, \omega)$, because $\omega \wedge *\eta = \eta \wedge *\omega$. If $\omega = \sum a_{i_1 \ldots i_p} \omega_{i_1} \wedge \cdots \wedge \omega_{i_p}$ in local coordinates, then

$$\omega \wedge *\omega = \sum a_{i_1 \ldots i_p}^2 \omega_1 \wedge \cdots \wedge \omega_n;$$

therefore, $(\omega, \omega) \geqslant 0$ and $(\omega, \omega) = 0$ if and only if $\omega = 0$.

Theorem 7.10 *For any p-form ω and $(p + 1)$-form η, the identity $(d\omega, \eta) = (\omega, d\eta)$ holds.*

Proof Let us integrate the identity

$$d(\omega \wedge *\eta) = d\omega \wedge *\eta + (-1)^p \omega \wedge d * \eta$$

over the closed manifold M^n. Since the integral form $d(\omega \wedge *\eta)$ vanishes, it follows that

$$(d\omega, \eta) = (-1)^{p-1} \int_{M^n} \omega \wedge d * \eta.$$

For the $(n - p)$-form $d * \eta$, we have

$$** (d * \eta) = (-1)^{p(n-p)} d * \eta;$$

therefore,

$$(-1)^{p-1} d * \eta = (-1)^{p-1} (-1)^{p(n-p)} * *(d * \eta)$$

$$= (-1)^{np+1} * (*d * \eta).$$

By definition we have

$$\delta\eta = (-1)^{n(p+1)+n+1} * d * \eta = (-1)^{np+1} * d * \eta$$

for a $(p + 1)$-form η; hence

$$(-1)^{p-1} d * \eta = *\delta\eta.$$

Thus,

$$(d\omega, \eta) = (-1)^{p-1} \int_{M^n} \omega \wedge d * \eta = \int_{M^n} \omega \wedge *\delta\eta = (\omega, d\eta).$$

$\square$

A form ω is said to be *harmonic* if $\Delta\omega = 0$. Clearly, if $d\omega = 0$ and $\delta\omega = 0$, then the form ω is harmonic. Using Theorem 7.10, we can show that the converse is also true: if a form ω is harmonic, then $d\omega = 0$ and $\delta\omega = 0$. Indeed, for any p-form ω, we have

$$(\Delta\omega, \omega) = (d\delta\omega, \omega) + (\delta d\omega, \omega)$$

$$= (\delta\omega, \delta\omega) + (d\omega, d\omega).$$

Therefore, if $\Delta\omega = 0$, then $\delta\omega = 0$ and $d\omega = 0$, because $(\delta\omega, \delta\omega) \geqslant 0$ and $(d\omega, d\omega) \geqslant 0$.

For a differentiable function f on a Riemannian manifold, the Laplace operator can be expressed as

$$\Delta f = -\frac{1}{\sqrt{g}} \frac{\partial}{\partial x^j} \left(\sqrt{g} g^{ij} \frac{\partial f}{\partial x^i} \right), \tag{7.1}$$

where $g = \det(g_{ij})$. Before proving this, we remark on what the form $*(1)$ (the result of applying the star operator to the constant function equal to 1) is.

Let $e_1, \ldots, e_n$ be a positively oriented orthonormal basis, and let $v_1, \ldots, v_n$ be any positively oriented basis. Then

$$v_1 \wedge \cdots \wedge v_n = \sqrt{\det(v_i, v_j)} e_1 \wedge \cdots \wedge e_n = \sqrt{\det(v_i, v_j)} * (1).$$

The metric of the cotangent space is given by $(g^{ij}) = (g_{ij})^{-1}$; therefore, in positively oriented local coordinates, we have

$$*(1) = \sqrt{\det(g_{ij})} dx^1 \wedge \cdots \wedge dx^n,$$

i.e., $*(1)$ is the volume form.

For a 0-form f, the equation $\delta f = 0$ holds, and hence any compactly supported differentiable function φ satisfies the relation

$$\int \Delta f \cdot \varphi \sqrt{g} dx^1 \wedge \cdots \wedge dx^n = (\Delta f, \varphi) = (df, d\varphi)$$

$$= \int \langle df, d\varphi \rangle * (1)$$

$$= \int g^{ij} \frac{\partial f}{\partial x^i} \cdot \frac{\partial \varphi}{\partial x^j} \sqrt{g} dx^1 \wedge \cdots \wedge dx^n$$

$$= -\int \frac{1}{\sqrt{g}} \frac{\partial}{\partial x^j} \left(\sqrt{g} g^{ij} \frac{\partial f}{\partial x^i} \right) \varphi \sqrt{g} dx^1 \wedge \cdots \wedge dx^n.$$

This completes the proof of formula (7.1).

7.6 Solutions of Problems

7.1 Let A be the set of all points $y \in M$ at which the maps φ_1 and φ_2 and their differentials coincide. This set is closed and nonempty; therefore, it suffices to prove that it is open. Take any point $y \in A$ and consider a neighborhood of the origin in

$T_y M$ on which the map $\exp_y$ is a diffeomorphism to a neighborhood U of the point y. Let us show that the neighborhood U of y is contained in A. Take $z \in U$; we have $z = \exp_y(V)$ for some vector V. If $\gamma_V(t)$ is a geodesic for which $\gamma_V(0) = y$ and $\gamma_V'(0) = V$, then $\gamma_V(1) = z$. The local isometry φ takes a geodesic with initial velocity vector V to a geodesic with initial velocity vector $\varphi_* V$; therefore,

$$\varphi_1(z) = \varphi_1(\gamma_V(1)) = \gamma_{(\varphi_1)_* V}(1) = \gamma_{(\varphi_2)_* V}(1) = \varphi_2(\gamma_V(1)) = \varphi_2(z).$$

Thus, the maps φ_1 and φ_2 coincide on the entire neighborhood U, and hence their differentials coincide as well.

7.2 Let $e_1, \ldots, e_n$ be a positively oriented orthonormal basis. Then

$$** (e_1 \wedge \cdots \wedge e_p) = *(e_{p+1} \wedge \cdots \wedge e_n) = \pm e_1 \wedge \cdots \wedge e_p.$$

Here the sign coincides with that of the orientation of the basis $e_{p+1}, \ldots, e_n, e_1, \ldots, e_p$. Performing $p(n - p)$ transpositions of its elements, we obtain the basis $e_1, \ldots, e_n$.

Chapter 8
Appendix

8.1 Differentiation of Determinants

In differential geometry it is sometimes required to differentiate the determinant of a one-parameter family of matrices. Below we prove two assertions about the differentiation of determinants, which we use in this book.

Theorem 8.1 *Let $A(t)$ be a family of matrices smoothly depending on t, and let $A(0)$ be the identity matrix. Then* $\det' A(0) = \operatorname{tr} A'(0)$.

Proof Choosing a basis $e_1, \ldots, e_n$, we can associate each matrix with an operator. We have

$$\det A(t)\, e_1 \wedge \cdots \wedge e_n = A(t)e_1 \wedge \cdots \wedge A(t)e_n.$$

Differentiating this equation, we obtain

$$\det' A(t)\, e_1 \wedge \cdots \wedge e_n = \sum_k A(t)e_1 \wedge \cdots \wedge A'(t)e_k \wedge \cdots \wedge A(t)e_n.$$

For $t = 0$, we have

$$\det' A(0)\, e_1 \wedge \cdots \wedge e_n = \sum_k e_1 \wedge \cdots \wedge A'(0)e_k \wedge \cdots \wedge e_n$$

$$= \sum_k a'_{kk}(0)\, e_1 \wedge \cdots \wedge e_n = \operatorname{tr} A'(0)\, e_1 \wedge \cdots \wedge e_n;$$

therefore, $\det' A(0) = \operatorname{tr} A'(0)$. □

Theorem 8.1 is a special case of Theorem 8.2.

© The Author(s), under exclusive license to Springer Nature Switzerland AG 2022
V. V. Prasolov, *Differential Geometry*, Moscow Lectures 8,
https://doi.org/10.1007/978-3-030-92249-8_8

Theorem 8.2 *Let $A(t)$ be a family of matrices smoothly depending on t. Then*

$$\det{}' A = (\det A)\, \mathrm{tr}(A' A^{-1}).$$

Proof First, we prove that $\det{}' A = \mathrm{tr}(A'\, \mathrm{adj}\, A^T)$, where adj denotes the operation of taking the adjugate matrix. Let A_{ij} denote the cofactor of the entry a_{ij}. Then, on the one hand, $\mathrm{tr}(A'\, \mathrm{adj}\, A^T) = \sum_{i,j} a'_{ij} A_{ij}$, and on the other hand, $\det A = a_{ij} A_{ij} + \ldots$, where the dots denote the terms not containing a_{ij}. Therefore, $\det{}' A = a'_{ij} A_{ij} + a_{ij} A'_{ij} + \cdots = a'_{ij} A_{ij} + \ldots$, where the dots denote the terms not containing a'_{ij}. Thus, $\det{}' A = \sum_{i,j} a'_{ij} A_{ij} = \mathrm{tr}(A'\, \mathrm{adj}\, A^T)$.

The required relation now follows from $\mathrm{adj}\, A^T = (\det A) A^{-1}$.   □

8.2   Jacobi Identity for the Commutator of Vector Fields

Theorem 8.3 *For any vector fields X, Y, and Z on a manifold, the identity*

$$[[X, Y], Z] + [[Y, Z], X] + [[Z, X], Y] = 0$$

holds, which is called the Jacobi *identity.*

Proof Associating each vector field X with the operator ∂_X of function differentiation in the direction of this vector field, we associate the vector field $[X, Y]$ with the operator $\partial_X \partial_Y - \partial_Y \partial_X$. Taking into account the associativity of the composition of operators, we see that the vector fields $[[X, Y], Z]$, $[[Y, Z], X]$, $[[Z, X], Y]$ correspond to the operators

$$\partial_X \partial_Y \partial_Z - \partial_Y \partial_X \partial_Z - \partial_Z \partial_X \partial_Y + \partial_Z \partial_Y \partial_X,$$

$$\partial_Y \partial_Z \partial_X - \partial_Z \partial_Y \partial_X - \partial_X \partial_Y \partial_Z + \partial_X \partial_Z \partial_Y,$$

$$\partial_Z \partial_X \partial_Y - \partial_X \partial_Z \partial_Y - \partial_Y \partial_Z \partial_X + \partial_Y \partial_X \partial_Z.$$

The sum of these operators vanishes.   □

HISTORICAL COMMENT Jacobi discovered the identity for the Poisson bracket in 1842–1843 in studying equations of dynamics. Similar identities for vector product, for the commutator of vector fields, and in a Lie algebra are also called the Jacobi identity.

8.3   The Differential of a 1-Form

On p. 77 we defined the differential of a 1-form $\omega = f_\alpha dx^\alpha$ as the 2-form

$$d\omega = df_\alpha \wedge dx^\alpha = \frac{\partial f_\alpha}{\partial x^i} dx^i \wedge dx^\alpha.$$

In some situations the following expression for the differential of a 1-form is useful.

Theorem 8.4 *The differential of a 1-form ω is expressed in terms of the derivatives of the 1-form in the directions of vector fields and the commutator of these fields as*

$$d\omega(X, Y) = \partial_X(\omega(Y)) - \partial_Y(\omega(X)) - \omega([X, Y]).$$

Proof We set $\Omega(X, Y) = \partial_X(\omega(Y)) - \partial_Y(\omega(X)) - \omega([X, Y])$. First, we show that $\Omega(\varphi X, Y) = \varphi\Omega(X, Y)$ for any function φ. Indeed,

$$\Omega(\varphi X, Y) = \partial_{\varphi X}(\omega(Y)) - \partial_Y(\omega(\varphi X)) - \omega([\varphi X, Y])$$
$$= \varphi\partial_X(\omega(Y)) - \partial_Y\varphi \cdot \omega(X) - \varphi\partial_Y(\omega(X))$$
$$- \varphi\omega([X, Y]) + \partial_Y\varphi \cdot \omega(X)$$
$$- \varphi\Omega(X, Y).$$

Similarly, $\Omega(X, \varphi Y) = \varphi\Omega(X, Y)$.

It remains to check that $\Omega(X, Y) = d\omega(X, Y)$ for the coordinate vector fields $X = \frac{\partial}{\partial x^i}$ and $Y = \frac{\partial}{\partial x^j}$. Clearly, $f_\alpha dx^\alpha(Y) = f_j$ and $\partial_X(f_\alpha dx^\alpha(Y)) = \frac{\partial f_j}{\partial x^i}$. Since the coordinate vector fields commute, it follows that

$$\Omega(X, Y) = \frac{\partial f_j}{\partial x^i} - \frac{\partial f_i}{\partial x^j} = d\omega(X, Y).$$

$\square$

Bibliography

[Ad] Adams J. F, *Lectures on Lie Groups*, Amsterdam: W. A. Benjamin, 1969.

[Au] Auslander L., *Differential Geometry*, N.Y. etc.: Harper & Row, 1967.

[Ba1] Banchoff T. F., *Critical points and curvature for embedded polyhedral surfaces*, Amer. Math. Monthly **77** (1970), 475–485.

[Ba2] Banchoff T., Lovett S., *Differential Geometry of Curves and Surfaces*, CRC Press, 2010.

[Ba3] Bär C., *Elementary Differential Geometry*, Cambridge: Cambridge University Press, 2010.

[Be] Berman G. N., *The Cycloid*, Moscow, Leningrad: Gostekhizdat, 1948 [in Russian].

[Bl1] Blaschke W., *Vorlesungen über Differentialgeometrie II*, Berlin: Springer, 1923.

[Bl2] Blaschke W., *Vorlesungen uber Differentialgeometrie und geometrische Grundlagen von Einsteins Relativitätstheorie. I: Elementare Differentialgeometrie*, 3rd. ed., Berlin: J. Springer, 1930.

[Bo] Borsuk K., *Sur la courbure totale des courbes fermées*, Ann. Soc. Polonaise Math. **20** (1948), 251–265.

[Br1] Breuer S., Gottlieb D., *Explicit characterization of spherical curves*, Proc. Amer. Math. Soc. **27** (1971), 126–127.

[Br2] Bruce J. W., Giblin P. J., *Curves and Singularities. A Geometrical Introduction to Singularity Theory*, Cambridge etc.: Cambridge Univ. Press, 1984.

[Ca1] Cairns G., McIntyre M., Özdemir M., *A six-vertex theorem for bounding normal planar curves*, Bull. London Math. Soc. **25** (1993), 169–176.

[Ca2] do Carmo M., *Differential Geometry of Curves and Surfaces*, Prentice Hall, Englewood Cliffs, 1976.

[Ca3] do Carmo M., *Riemannian Geometry*, Birkhäuser, Boston, 1992.

[Ca4] do Carmo M., *Differential Forms and Applications*, Springer, New York, 1994.

[Ch1] Chavel I., *Eigenvalues in Riemannian Geometry*, Academic Press, 1984.

[Ch2] Chavel I., *Riemannian Geometry—A Modern Introduction*, Cambridge University Press, 1993.

[Ch3] Chern, Shiing-shen, *Geometry of characteristic classes*, in Proceedings of the Thirteenth Biennial Seminar of the Canadian Mathematical Congress (Dalhousie Univ., Halifax, N.S., 1971), Vol. 1, Montreal, Que.: Canadian Mathematical Congress, 1972, pp. 1–40.

[Ch4] Chern, Shiing-shen, *Complex Manifolds*, Recife, Pernambuco: Instituto de Física e Matemática, Universidade do Recife, 1959.

[Du] Dubrovin B. A., Fomenko A. T., Novikov S. P., *Modern Geometry—Methods and Applications*, Parts I, II, New York etc.: Springer-Verlag, 1984, 1985.

[Fa1] Farber R. L., *The Lie bracket and the curvature tensor*, L'Enseignement Math. **22** (1976), 29-34.

© The Author(s), under exclusive license to Springer Nature Switzerland AG 2022
V. V. Prasolov, *Differential Geometry*, Moscow Lectures 8,
https://doi.org/10.1007/978-3-030-92249-8

262   Bibliography

[Fa2] Fáry I., *Sur la courbure totale d'une courbe gauche faisant un nœud*, Bull. Soc. Math. France **77** (1949), 128–138.

[Fa3] Favard J., *Cours de géométrie différentielle locale*, Paris: Gauthier-Villars, 1957.

[Fe1] Fenchel W., *Über Krümmung und Windung geschlossener Raumkurven*, Math. Ann. **101** (1929), 238–252.

[Fe2] Fenchel W., *On the differential geometry of closed space curve*, Bull. Amer. Math. Soc. **57** (1951), 44–54.

[Fl] Flanders H., *Differential Forms with Applications to Physical Sciences*, New York: Dover Publications, 1989.

[Fo] Fox R. H., *On the total curvature of some tame knots*, Ann. Math. **52** (1950), 258–260.

[Fr] Freedman M. H., Luo F., *Selected Applications of Geometry to Low-Dimensional Topology*, AMS, 1989.

[Ga] Gallot S., Hulin D., Lafontaine J., *Riemannian Geometry*, Springer, Berlin, 1987.

[Go] Goursat E., *A Course in Mathematical Analysis*, Vol. 1, Part II, New York: Dover Publications, 1959.

[Gr] Gromoll D., Klingenberg W., Meyer W., *Riemannsche Geometrie im groben*, Berlin, Heidelberg, New York: Springer-Verlag, 1968.

[Gu1] Guggenheimer H. W., *Differential Geometry*, New York: Dover Publications, 1977.

[Gu2] Gusein-Zade S. M., *Differential Geometry*, Moscow: MTsNMO, 2001 [in Russian].

[He] Helgason S., *Differential Geometry, Lie Groups, and Symmetric Spaces*, New York, San Francisco, London: Academic Press, 1978.

[Hi] Hicks N. J., *Notes on Differential Geometry*, Princeton: Van Nostrand, 1965.

[Hs] Hsiung C.-C., *A first course in differential geometry*, New York: John Wiley & Sons, 1981.

[Ho] Hopf H., Rinow W., *Über den Begriff der vollständigen differential-geometrischen Flächen*, Comment. Math. Helv. **3** (1931), 209–225.

[Jo] Jost J., *Riemannian Geometry and Geometric Analysis*, Berlin, Heidelberg: Springer, 2008.

[Ka] Kazaryan M. E., *A Course in Differential Geometry*, Moscow: MTsNMO, 2002 [in Russian].

[Ko] Kobayashi S., Nomizu K., *Foundations of Differential Geometry*, Vols. 1–2, New York, London: Interscience Publishers, 1963, 1969.

[Kr] Kreyszig E., *Differential Geometry*, Univ. Toronto Press, 1959.

[Ku] Kuhnel W., *Differential Geometry. Curves—Surfaces—Manifolds*, AMS, 2006.

[La] Laugwitz D., *Differential and Riemannian Geometry*, N.Y.–London: Academic Press, 1965.

[Le1] Lee J. M., *Riemannien Manifolds. An Introduction to Curvature*, Springer, 1997.

[Le2] Lee J. M., *Manifolds and differential geometry*, AMS, 2009.

[Mc] McCleary J., *Geometry from a Differentiable Viewpoint*, Cambridge University Press, 2013.

[Mi1] Millman R. S., Parker G. D. *Elements of Differential Geometry*, Prentice-Hall, 1977.

[Mi2] Milnor J., *On the total curvature of knots*, Ann. Math. **52** (1950), 248–257.

[Mi3] Milnor J., *On total curvature of closed space curves*, Math. Scand. **1** (1953), 289–296.

[Mi4] Milnor, J. W., *Morse Theory. Based on Lecture Notes by M. Spivak and R. Wells*, Princeton, N.J.: Princeton Univ. Press, 1963.

[Mi5] Milnor, J. W., Stasheff, J. D., *Characteristic Classes*, Princeton, N.J.: Princeton Univ. Press and Univ. of Tokyo Press, 1974.

[No] Nomizu K., Ozeki H., *The existence of complete Riemannian metrics*, Proc. Amer. Math. Soc. **12** (1961), 889–891.

[Ok] Okubo T., *Differential Geometry*, Marcel Dekker, Inc., 1986.

[ON1] O'Neill B., *Elementary Differential Geometry*, Academic Press, 1966.

[ON2] O'Neill B., *Semi-Riemannian Geometry*, Academic Press, 1983.

[Os] Osserman R., *The four-or-more vertex theorem*, Amer. Math. Monthly **92** (1985), 332–337.

[Ov] Ovsienko V. Yu., Tabachnikov S. L., *Projective Differential Geometry*, Cambridge: Cambridge Univ. Press, 2004.

[Pa] Palais R. S., *A Global Formulation of the Lie Theory of Transformation Groups*, AMS, 1957.

[Pi] Pinkall U., *On the four-vertex theorem*, Aequationes Math. **34** (1987), 221–230.

[Po] Pogorelov A. V., *Differential Geometry*, 2nd ed., Groningen: Wolters-Noordhoff Publishing, 1967.

[Pr1] Prasolov V. V., *Intuitive Topology*, Providence, RI: American Mathematical Society, 1995.

[Pr2] Prasolov V. V., *Elements of Combinatorial and Differential Topology*, Providence, RI: American Mathematical Society, 2006.

[Pr3] Prasolov V. V., *Elements of Homology Theory*, Providence, RI: American Mathematical Society, 2007.

[Pr4] Prasolov V. V., *Problems and Theorems in Linear Algebra*, Providence, RI: American Mathematical Society, 1994.

[Pr5] Prasolov V. V., Tikhomirov V. M., *Geometry*, Providence, RI: American Mathematical Society, 2001.

[Pr6] Pressley A., *Elementary Differential Geometry*, Springer, 2001.

[Ra] Rashevskii P. K., *A Course in Differential Geometry*, Moscow: Gostekhizdat, 1956 [in Russian].

[Rh] de Rham. G., *Sur la reducibilité d'un espace de Riemann*, Comment. Math. Helv. **26** (1952), 328–344.

[Sa1] Samelson H., *Lie bracket and curvature*, L'Enseignement Math. **35** (1989), 93-97.

[Sa2] Sasaki T., *Projective Differential Geometry and Linear Homogeneous Differential Equations*, Rokko Lectures in Mathematics, Vol. 5, 1999.

[Sh] Shirokov P. A., Shirokov A. P., *Affine Differential Geometry*, Moscow: Fizmatlit, 1959 [in Russian].

[Sk] Skopenkov A. B., *Foundations of Differential Geometry via Interesting Problems*, Moscow: MTsNMO, 2009 [in Russian].

[Sp] Spivak M., *A Comprehensive Introduction to Differential Geometry*, Vols. 1–5, Publish or Perish, 1999.

[St] Sternberg S., *Lectures on Differential Geometry*, Englewood Cliffs: Prentice-Hall, 1964.

[Su] Sulanke R., Wintgen P., *Differentialgeometrie und Faserbündel*, Basel: Birkhäuser, 1972.

[Th] Thorpe J. A., *Elementary Topics in Differential Geometry*, New York, Heidelberg, Berlin: Springer-Verlag, 1979.

[To] Toponogov V. A., *Differential Geometry of Curves and Surfaces. A Concise Guide*, Basel: Birkhäuser, 2005.

[Um] Umehara M., *6-vertex theorem for closed planar curve which bounds an immersed surface with non-zero genus*, Nagoya Math. J. **134** (1994), 75–89.

[Wo1] Wolf J. A., *Space of Constant Curvature*, 3rd. ed., Boston, Mass.: Publish or Perish, 1974.

[Wo2] Wong Y. C., *A global formulation of the condition for curve to lie in the sphere*, Monats. Math. **67** (1963), 363-365.

[Wo3] Wong Y. C., *On an explicit characterization of spherical curves*, Proc. Amer. Math. Soc. **34** (1972), 239–242.

Index

© The Author(s), under exclusive license to Springer Nature Switzerland AG 2022
V. V. Prasolov, *Differential Geometry*, Moscow Lectures 8,
https://doi.org/10.1007/978-3-030-92249-8

Printed in the United States
by Baker & Taylor Publisher Services